T0188138

Computational Optimization

This textbook offers a guided tutorial reviewing the theoretical fundamentals while going through the practical examples used for constructing the computational frame, applied to various real-life models.

Computational Optimization: Success in Practice will lead the readers through the entire process. They will start from the simple calculus examples of fitting data and the basics of optimal control methods and end with the construction of a multi-component framework for running PDE-constrained optimization. This framework will be assembled piece by piece; the readers may apply this process at the levels of complexity matching their current projects or research needs.

By connecting examples with the theory and discussing the proper "communication" between them, the readers will learn the process of creating a "big house." Moreover, they can use the framework exemplified in the book as the template for their research or course problems – they will know how to change the single "bricks" or add extra "floors" on top of that.

This book is authored for students, faculty, and researchers.

Features:

- The main optimization framework builds through the course exercises and centers on MATLAB®.
- All other scripts to implement computations for solving optimization problems with various models use only open-source software, e.g., FreeFEM.
- All computational steps are platform-independent; readers may freely use Windows, macOS, or Linux systems.
- All scripts illustrating every step in building the optimization framework will be available to the readers online.
- Each chapter contains problems based on the examples provided in the text and associated scripts. The readers will not need to create the scripts from scratch, but rather modify the codes provided as a supplement to the book.

This book will prove valuable for graduate students of math, computer science, engineering, and all who explore optimization techniques at different levels for educational or research purposes. It will benefit many professionals in academic and industry-related research: professors, researchers, postdoctoral fellows, and the personnel of R&D departments.

Textbooks in Mathematics

Series editors:

Al Boggess, Kenneth H. Rosen

Geometry and Its Applications, Third Edition
Walter J. Meyer

Transition to Advanced Mathematics
Danilo R. Diedrichs and Stephen Lovett

Modeling Change and Uncertainty
Machine Learning and Other Techniques
William P. Fox and Robert E. Burks

Abstract Algebra
A First Course, Second Edition
Stephen Lovett

Multiplicative Differential Calculus
Svetlin Georgiev, Khaled Zennir

Applied Differential Equations
The Primary Course
Vladimir A. Dobrushkin

Introduction to Computational Mathematics: An Outline
William C. Bauldry

Mathematical Modeling the Life Sciences
Numerical Recipes in Python and MATLAB™
N. G. Cogan

Classical Analysis
An Approach through Problems
Hongwei Chen

Classic Vector Algebra
Vladimir Leptic

Introduction to Number Theory
Mark Hunacek

Probability and Statistics for Engineering and the Sciences with Modeling using R
William P. Fox and Rodney X. Sturdivant

Computational Optimization: Success in Practice
Vladislav Bukshtynov

https://www.routledge.com/Textbooks-in-Mathematics/book-series/CANDHTEXBOOMTH

Computational Optimization

Success in Practice

Vladislav Bukshtynov

CRC Press
Taylor & Francis Group
Boca Raton London New York

CRC Press is an imprint of the
Taylor & Francis Group, an **informa** business

A CHAPMAN & HALL BOOK

First edition published 2023
by CRC Press
6000 Broken Sound Parkway NW, Suite 300, Boca Raton, FL 33487-2742

and by CRC Press
4 Park Square, Milton Park, Abingdon, Oxon, OX14 4RN

CRC Press is an imprint of Taylor & Francis Group, LLC

ISBN: 978-1-032-22947-8 (hbk)
ISBN: 978-1-032-23005-4 (pbk)
ISBN: 978-1-003-27516-9 (ebk)

DOI: 10.1201/9781003275169

Typeset in CMR10
by KnowledgeWorks Global Ltd.

Publisher's note: This book has been prepared from camera-ready copy provided by the authors.

To my wife Katya and daughter Ksenia.

Contents

Preface

An idea to write a book visited me many times since I started advising graduate students as a postdoc at Stanford and later as a professor at Florida Tech. There are plenty of methods in numerical optimization nicely presented in many beautiful books. To say more, many computational packages with beautifully written manuals are open sources featuring optimization toolboxes. Now, let us imagine graduate students (or young researchers, postdocs, whoever!) who completed several courses in optimization, say linear and nonlinear ones, and even have good programming skills, being able to choose proper software and write their own scripts for running model simulations and optimization algorithms. They construct their computational framework assembling all parts responsible for modeling, optimization, visualization, performance analysis, etc. Then they try to run it and, in up to 99% of cases, see nothing. What is the problem? Bugs in one or multiple scripts? Incorrect logic in the algorithm(s) implementation? Improper communication between different software as parts of a multicomponent framework? If the entire framework works correctly, how do we know that some fragments are losing performance as not being carefully tuned up? What if the solutions are ready, but we misunderstand their quality by a faulty visualization part? How to create a suitable benchmark model, consider uncommon cases that might arise in future modeling, choose proper parameters, enhance the performance, and eventually display the solutions enabling any analysis for its quality and performance? If not thoroughly guided through all these questions to create a big picture and navigate confidently through all components – the students will be lost and completely frustrated.

In 2018, being unable to deal with individual requests and help many students at my university, I designed a new course, Selected Topics in Numerical and Computational Mathematics: Numerical Optimization. I also ran a mini-tutorial Computational Aspects of Numerical Optimization during various summer schools at my university and outside. These courses had great success, and I collected enormous feedback. I also received multiple requests to offer this course again or share the course materials as it resonates with similar problems seen by researchers at various levels, from undergraduates to professors and experienced code developers.

Approach

I realize that I am incapable of helping everyone, but the book describing my vision and the systematic methodology could do it. This book will lead the readers through the entire process. They will start from the simple calculus examples of fitting data and the basics of optimal control methods and end up by constructing a multicomponent framework for solving optimization problems constrained by differential equations. This framework will be assembled piece by piece, and the readers could stop this process whenever they see the complexity matches their current project or research needs and, potentially some of the future goals. Of course, sophisticated, multi-purpose, and ready-to-use software is good. But very soon, it becomes clear that the novices cannot utilize all the benefits, especially if this software is added "blindly" into the custom-made research platform. By connecting examples with the theory and discussing the proper "communication" between them, the readers will learn the process of creating a "big house." Moreover, they could use the framework exemplified in the book as the template for their research or course problems – they will know how to change the single "bricks" or add extra "floors" on top of that.

This book targets the audience at various levels. In particular, it will be a valuable textbook for graduate students of math, computer science, all kinds of engineering, and other majors, who explore optimization techniques at different levels for educational or research purposes. There is no doubt that this book will be used as a helpful supplement by undergraduate students as well as high schoolers in courses with practical components for optimization incorporated in the syllabi. Even more, it will benefit many professionals in academic and industry-related research: professors, researchers, postdoctoral fellows, and the personnel of R&D departments who make the first steps to move optimization strategies into practice.

Suggestions for Instructors and Self-Learners

This book is written using a combination of two basic strategies. First, the entire text is an **active dialogue** with the reader. The fields of computational mathematics and optimization are growing constantly and dynamically. The arrival of new methods forces the users to reconsider their preferences for solution approaches, and some of these approaches become obsolete very soon. The idea of this book is not to promote any techniques, e.g., gradient-based ones, but rather to discuss their usability to draft a working structure with a free space that is large enough for the readers to experiment with their own ideas and preferences. The latter requires a small preamble at this point as we acknowledge different programming styles and methods to optimize the computations within various programming languages. The goal of this book is not to teach the programming skills or optimal use of the selected software but to enhance the ability to create computational frameworks, having the optimality of this process as the guiding point. It is made possible through multiple discussions and questions to the readers.

? *This question mark sign is used throughout the book to pose questions for the readers. If not addressed immediately or explicitly, the readers are encouraged to find the answers in the provided examples or the rest of the chapter.*

! *The exclamation mark sign signifies a hint for a posed question or a point for discussion.*

Another strategy was developed and proved its efficiency, while teaching the associated graduate course in computational optimization. This book moves away from a common approach of describing the already created and ready-to-use computational framework in detail (structure, included methods, etc.). Instead, the framework is created almost from scratch based on a simple example, and the discussion focuses on the changes applied to this framework when it moves to new problems to be solved, alters the structure, or adds any new functionality. In a nutshell, we learn more and with better chances for success if we study the **dynamics of the structural changes** rather than the steady state structure itself, even if the latter is very detailed. In this context, active discussions (similar to teaching the course in person) are invaluable, and both strategies interplay successfully to help achieve our goals.

A typical piece of advice would be to start by reviewing the information in the Appendix containing concise topics selected from calculus and linear algebra to help the reader refresh the knowledge of mathematical notations, definitions, and formulas mentioned in this book. The entire book's material fits a regular semester, however, if taught rather extensively. **Graduate students** with sufficient background in multivariable calculus, linear algebra, and optimization basics are usually fine to explore Chapters 1–12, even if it includes some basics from the calculus of variations or perturbation analysis in Chapters 9–12. There is also a general remark about mathematical notations, definitions, and derivations in this book. Easing the math part by removing some properties and simplifying the definition structures could make those definitions incomplete. However, it is assumed that such simplifications do not prevent us from applying the related analysis correctly and help the reader focus on the technicalities related to the optimization. Given that, **undergraduate courses** using this book for the main or supplementary reading may benefit from covering Chapters 1–10 and using the rest (PDE-based optimization) if inspired to take it, e.g., for course projects. Similarly, **high school students** may freely study Chapters 1–8 and beyond, depending on their background and the course structure.

Students at all levels, including self-learners, could find more benefits by answering the question and solving additional problems posted in the *Homework* sections after each chapter. These sections also contain more information with references to materials for reading and review (refer to *Where to Read More* lists of sources). To pursue the most convenient way of using additional reading, the references are provided to the books and also to their chapters in relation to the material currently discussed, e.g.,

Nocedal (2006), [25]
  Chapter 1 (Introduction), Chapter 2 (Fundamentals of Unconstrained Optimization).

In addition, the book provides 5 lab assignments (after every 2–3 chapters) and a sample midterm exam (after Chapter 7).

Computational Software and Supplemental Materials

The students and self-learners are expected to be familiar with iterative (gradient-based and derivative-free) algorithms for unconstrained and constrained optimization. Having some prior experience or interest in computer programming and software for mathematical (computational) modeling, solving (differential) equations, and using any optimization toolboxes or packages might also benefit this study. During this course, students will have a lot of practice, especially with MATLAB® and FreeFEM (an open source[1] C++-based scientific environment). However, proficiency with MATLAB, C++, and FreeFEM is not required. It is expected that the suggested examples and individual practice will naturally stimulate learning of their functionalities in parallel with the increased level of confident use and practical needs. While particular versions of MATLAB (R2018a) and FreeFEM (3.61-1) are used to prepare all figures in this book, most of the associated codes will run in many of the earlier versions with no changes required. Other versions, however, will likely generate results that might slightly deviate from those seen in the book.

Finally, the readers may freely download the computer codes for practical examples in all chapters from the author's website at

http://www.bukshtynov.xyz/book_CompOPT.html

The course instructors are welcome to contact the author at

VladislavBukshtynov@yahoo.com

to obtain a **complimentary copy of the lecture slides**, provided they adopt this book as the course textbook.

---

[1] FreeFEM is licensed under the GNU General Public License v3.0.

Acknowledgments

I owe sincere thanks to many people who shared their ideas and provided constructive input and invaluable support that inspired me to write this book. In particular, I would like to thank my PhD advisor Bartosz Protas who brought me to the world of optimization and shared his extensive knowledge in various scientific fields. I am grateful to Khalid Aziz and Lou Durlofsky, my postdoc advisors at Stanford, whose tremendous experience and highest level of expertise encouraged me in future steps, for providing me with enough freedom to experiment and grow as a scientist and a teacher. I must give special thanks to my friend and colleague Oleg Volkov for his vision of optimization, shared experience, and a lot of advice he gave on my long way to this achievement. In addition, special thanks go to all my students of various courses in optimization for their valuable feedback and comments on the course components and structure.

I am also very grateful to my wife Katya and daughter Ksenia who perpetually inspire me with their understanding, infinite patience, and continuous care.

Finally, I heartily appreciate all comments and suggestions sent to me after this book is published and also errors found which, of course, are the sole responsibility of the author.

Vladislav Bukshtynov
Melbourne, Florida
November 18, 2022

Author

Dr. Vladislav Bukshtynov earned a PhD in computational engineering and science from McMaster University. He is an assistant professor in the Dept. of Mathematical Sciences at the Florida Institute of Technology. He completed a three-year postdoctoral term in the Dept. of Energy Resources Engineering at Stanford University. He actively teaches and advises students from various fields: applied and computational math, operations research, and different engineering majors. His teaching experience includes multivariable calculus, honors ODE/PDE courses for undergrad students, applied discrete math, and linear/nonlinear optimization for senior undergrads and graduates. As a researcher, Dr. Bukshtynov leads his research group with several dynamic scientific directions and ongoing collaborations for various cross-institutional and interdisciplinary projects. His current interests lie in, but are not limited to, areas of applied and computational mathematics focusing on combining theoretical and numerical methods for various problems in computational/numerical optimization, control theory, and inverse problems.

Acronyms and Abbreviations

Description

AD	automatic differentiation		LMS	line minimization search
BB	bracketing-Brent (method)		LP	linear programming (problem)
BF	brute-force (method)			
BFGS	Broyden–Fletcher–Goldfarb–Shanno (method)		LSO	large-scale optimization
			LV	Lotka–Volterra (model)
BLS	backtracking line search		MC	Monte-Carlo (method)
BVP	boundary-value problem		mN	modified Newton's (method)
CG	conjugate gradient (method)			
dN	discretized Newton's (method)		NLP	nonlinear programming (problem)
DOF	degrees of freedom		ODE	ordinary differential equation
DS-SD	diagonally scaled SD (method)		PCA	principal component analysis
			PDE	partial differential equation
EA	evolutionary algorithm		POD	proper orthogonal decomposition
EIT	electrical impedance tomography			
			PSO	particle swarm optimization
EOS	equations of states		qN	quasi-Newton (method)
FD	finite difference		QP	quadratic problem
FE	finite element		RK	Runge–Kutta (method)
FEM	finite element method		ROM	reduced-order model
GA	genetic algorithm		SD	steepest descent
GN	Gauss–Newton (method)		STUN	stochastic tunneling
GS	golden section		TRDL	trust region and dogleg (methods)
IO	input–output			
IVP	initial-value problem		TSVD	truncated singular value decomposition
KKT	Karush–Kuhn–Tucker (conditions)			

List of Algorithms

1

Introduction to Optimization

This chapter helps the reader review briefly the main mathematical concepts and notations used in the general theory of optimization developed later to be adopted for practical use in computational frameworks to solve optimization problems. For that reason, we establish a formal separation between the notations of mathematical and optimization problems and discuss the structure and main components of the latter. A practical example (in `MATLAB`) of the simplistic 2D data fitting initiates the discussion on the computational aspects of algorithms employed in optimization. We also provide a basic review of some fundamentals: feasibility, optimality, convexity, and convergence. These essentials will be further discussed while adopted for use in practice.

1.1 Optimization Models

To help the reader immediately build a big picture, we start with a brief discussion on the models created in the context of mathematical notations and how these models may be seen from the optimization perspective. A very general concept of a *mathematical model* implies creating a complete, to some extent, description of certain phenomena in any scientific field such as engineering, physics, chemistry, biology, etc. To create a model, we usually follow several main steps.

(1) Identifying variables responsible for changing the state of the modeled system.

(2) Making a set of reasonable assumptions about this system.

(3) Putting these assumptions into some mathematical form, e.g., in terms of algebraic or differential equations.

(4) Solving the model.

(5) Comparing the obtained solutions with any known facts to correct assumptions made in (2).

(6) Repeating, if necessary, steps (4) and (5) until we are satisfied with the results provided by this model.

DOI: 10.1201/9781003275169-1

We may correct model assumptions mentioned in (5) at different levels. For example, one could change the complete structure of the model by making it more, or sometimes less, complicated. The model's complexity would mean its overall **size** and **simplicity** (or complexity) in finding an accurate mathematical description. If the model's framework looks correct, we could identify a set of parameters that require tuning for the model to agree with desired conditions. In general, these conditions help arrive at the solution produced by the model and decide if this solution is the best possible one. In other words, we look for the model **optimized over selected parameters** to satisfy our needs in creating this model and taking the most benefits from using it in practice.

Various models to describe any medical, biological, or chemical processes may serve as *optimization models (problems)* if they incorporate any techniques to make them **accurate** and **reliable**. In particular, these models may use any data obtained as observations of the phenomena to compare with the model output provided in the same format. For example, mathematical models may provide a better understanding of natural phenomena (in the atmosphere, oceans, underground media, etc.) and serve as foundations for the optimization models that will **minimize** losses from natural disasters or **maximize** profit when using the power of our nature. In this example, one could substitute the mathematical part now to describe, for instance, a business structure (money flows within a corporation, production cycles, investment portfolios, etc.). Despite this crucial change, the optimization framework will continue working (with some, probably minimal, adjustments) and solving the problem of minimizing losses and maximizing profit but now in the new context.

In the same way, we could move from one model to another by preserving the main structure of the optimization framework once it is (a) constructed, (b) validated to work with different models (of different complexities), and (c) matured enough to fit all user's requirements and preferences. Of course, we do not pretend that we can construct a universal optimization algorithm to fit any mathematical model and all possible minimization-maximization targets. Not at all! But the goal of this book is much more ambitious: to learn the process of creating an optimization framework to be used individually by readers to fit (optimally) all their practical goals in education, research, at the current workplace, and in the future. To conclude, we reiterate that the main focus in this study is not on the mathematical models but rather on creating and understanding the optimal structure of the practical optimization frameworks.

1.2   General Notations for Optimization Problem

To start with the main notations and mathematical objects used for designing the structures of most optimization problems, we refer to a commonly used and rather general form.

$$\min / \max_{\mathbf{u} \in \mathbb{R}^n} \quad f(\mathbf{u})$$
$$\text{subject to} \quad h_i(\mathbf{x}; \mathbf{u}) = 0, \qquad i = 1, \ldots, p \tag{1.1}$$
$$g_j(\mathbf{x}; \mathbf{u}) \leq 0, \qquad j = 1, \ldots, m$$

The problem formulation is complete once the following components are discussed and properly defined.

(a) Are we solving the minimization or maximization problem? As they convert easily into each other, choosing the optimization problem type is up to us, users, and the overall convenience of being consistent with the rest notations. For most examples here, we will use the **min-problem** type.

(b) Next, we define *optimization* (also *control, decision, or design) variables* and appropriate space; e.g., n-dimensional vector[1] space for control $\mathbf{u} \in \mathbb{R}^n$ as shown in (1.1). Here, we have to note a difference between the control $\mathbf{u}$ and the *state variable* $\mathbf{x}$ if the latter appears only as a solution function in mathematical models not included in the list of optimization targets.

(c) *Objective (cost) function(al)* $f(\mathbf{u}) : \mathbb{R}^n \to \mathbb{R}$ is a scalar function to measure success while solving the optimization problem. When both state and control variables are present, we consider a broader notation $f(\mathbf{x}; \mathbf{u})$.

(d) The optimization problem is *constrained* if at least one constraint exists in its description. Otherwise, we solve an *unconstrained problem*. The constraints may be bounds set for both optimization and state variables, algebraic equations h_i, and inequalities g_j as shown in (1.1), also ordinary (ODEs) and partial differential equations (PDEs). There might be other requirements, e.g., functional spaces for $\mathbf{x}$ and $\mathbf{u}$ to be considered as constraints.

(e) A *feasible region* (also, *feasible set, control,* or *solution space*) $\mathbb{S}$ is represented by all possible solutions $\mathbf{u}$ that satisfy all constraints. Our particular interest is in obtaining an optimal solution that is also *feasible*, $\mathbf{u} \in \mathbb{S}$. However, some problems may produce *infeasible solutions*, $\mathbf{u} \notin \mathbb{S}$, characterized by better values of objective function $f(\mathbf{u})$.

(f) Finally, an *optimal solution*

$$\mathbf{u}^* = \operatorname*{argmin}_{\mathbf{u} \in \mathbb{S}} f(\mathbf{u}) \tag{1.2}$$

could be obtained after solving an optimization problem and confirming its feasibility. We could also call $\mathbf{u}^*$ a *minimizer (maximizer)* of $f(\mathbf{u})$ in $\mathbb{S}$, and a scalar value $f^* = f(\mathbf{u}^*)$ appears to be a *minimum (maximum)* of $f(\mathbf{u})$ in $\mathbb{S}$. We also notice that (1.1) and (1.2) are equivalent mathematical forms used in this book interchangeably.

---

[1]See *Vectors* on p. 377 for review.

Based on the components introduced in (a)–(f) above, we may classify optimization problems and algorithms by certain criteria.

- *Type of constraints:* unconstrained or constrained (e.g., ODE- or PDE-based optimization).

- *Nature of equations involved:* linear programming (LP), nonlinear programming (NLP) problems, quadratic problems (QP), etc.

- *Permissible values of controls:* continuous or discrete (e.g., integer programming).

- *Deterministic nature of variables:* deterministic or stochastic (or probabilistic).

- *Separability of functions:* separable or nonseparable problems.

- *Number of objectives:* single-objective or multi-objective problems.

- *Other types:* optimal control, nonoptimal control, etc.

Although we discuss some cases in that classification later (whenever we use it in our examples), we refer a diligent reader to comprehensive descriptions provided in various books referenced in Section 1.7. We close this brief overview of components to describe an optimization problem with a simple example.

Example 1.1 Constrained Optimization with Nonlinear Objective
Find the point on the line $x_1 + x_2 = 10$ closest to the point $(10, 10)$.

It is a 2D problem where the optimization variable is a two-component vector $\mathbf{x} = [x_1 \ x_2]^T \in \mathbb{R}^2$. The feasible region $\mathbb{S} \subset \mathbb{R}^2$

$$\mathbb{S} = \left\{ \mathbf{x} \in \mathbb{R}^2 : x_1 + x_2 = 10 \right\}$$

consists of all points on the line $x_1 + x_2 = 10$ that represents one (linear) constraint. The objective function could be derived by using a simple calculus formula computing the distance between points $A = (x_A, y_A)$ and $B = (x_B, y_B)$

$$d = \sqrt{(x_B - x_A)^2 + (y_B - y_A)^2}.$$

The instruction to find "the closest point" suggests minimizing this distance by setting $A = (10, 10)$ and $B = (x_1, x_2)$. We may also omit the square root as this will lead to the equivalent problem with the same minimizer x^*

(but different minimum f^*). Therefore, we created a nonlinear (*quadratic*) optimization problem constrained by one linear equation.

$$\min_{\mathbf{x}\in\mathbb{R}^2} \quad f(\mathbf{x}) = (x_1 - 10)^2 + (x_2 - 10)^2$$
$$\text{s.t.} \quad x_1 + x_2 = 10 \tag{1.3}$$

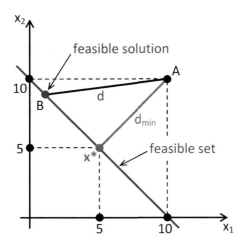

FIGURE 1.1
Geometry of optimization problem of Example 1.1.

The geometry of this problem is shown in Figure 1.1 leading to an obvious optimal solution $x^* = [5\ 5]^T$. We leave finding the analytic solution for problem (1.3) to the reader.

❗ *Despite various approaches available in 2D (e.g., via Lagrange multipliers or analytical geometry statements), eliminating x_1 or x_2 and reducing the problem to minimizing a one-variable function with no constraints may be seen as the simplest one.*

1.3   Data Fitting Examples

Let us now consider a simple example extended later in multiple directions to explore various options in constructing and solving associated optimization problems.

Example 1.2 Data Fitting
Given three points of 2D data $(2,1)$, $(4,9)$, and $(7,6)$, find coefficients a_1, a_2, and a_3 assuming quadratic fit via parabolic function $y(x) = a_1 + a_2 x + a_3 x^2$.

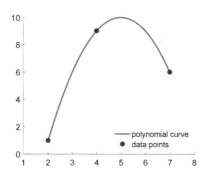

FIGURE 1.2
Solution for Example 1.2.

To some extent, this problem also qualifies to be an optimization problem as the solution requires the general form of the quadratic function $y(x) = a_1 + a_2 x + a_3 x^2$ to be optimized (or controlled) to find proper parameters, coefficients in $y(x)$. The optimal (exact) solution $\mathbf{a} = [-15\ 10\ -1]^T$ or $y(x) = -15 + 10x - x^2$ could be found after solving the following system of linear equations:

$$\begin{cases} a_1 + 2a_2 + 4a_3 = 1, \\ a_1 + 4a_2 + 16a_3 = 9, \\ a_1 + 7a_2 + 49a_3 = 6. \end{cases} \tag{1.4}$$

We could also use the matrix[2] notation, i.e.,

$$A\,\mathbf{a} = \mathbf{y} \quad \Rightarrow \quad \mathbf{a} = A^{-1}\mathbf{y},$$

where

$$A = \begin{bmatrix} 1 & x_1 & x_1^2 \\ 1 & x_2 & x_2^2 \\ 1 & x_3 & x_3^2 \end{bmatrix} = \begin{bmatrix} 1 & 2 & 4 \\ 1 & 4 & 16 \\ 1 & 7 & 49 \end{bmatrix},$$

---

[2]See *Matrices and Matrix Equations* on p. 378 for review.

$$\mathbf{a} = \begin{bmatrix} a_1 \\ a_2 \\ a_3 \end{bmatrix}, \quad \text{and } \mathbf{y} = \begin{bmatrix} y_1 \\ y_2 \\ y_3 \end{bmatrix} = \begin{bmatrix} 1 \\ 9 \\ 6 \end{bmatrix}.$$

Simple linear algebra analysis for (1.4) suggests that a unique solution (and, thus, optimal in the context of this problem) exists **conditionally** when the number of data pieces n_d matches the number of unknown parameters in $y(x)$, i.e., $n_d = 3$.

? *What are the* **conditions** *for data to guarantee the solution exists within the suggested framework?*

? *What if the problem appears* **overdetermined** *($n_d > 3$) or* **underdetermined** *($n_d < 3$)?*

To answer the first question, we suggest running the first MATLAB code `Chapter_1_data_fit.m` with multiple cases of data, see Figure 1.3 for some examples, and exploring the code structure to initiate the discussion on the general design of the computational framework further developed in Chapter 3.

MATLAB: `Chapter_1_data_fit.m`

Code structure:

- *input data:* `data = [2 1; 4 9; 7 6];`

- *solution #1:* by m-code in separate file `DataFitM.m`

- *solution #2:* by user-defined function `DataFitFn` in the same code

- *solution #3:* by MATLAB's built-in function `polyfit(x,y,n)`

- *solution conversion:* polynomial preparation for visualization

- *visualization:* data and results in various formats

While the reader may find many other computational approaches to solve this linear algebra problem in MATLAB, we limit our consideration to three of them for two reasons. First, using existing functions saves time on their implementation and debugging; it might be worth checking built-in functionalities before writing them manually. Second, considering different formats for repeated parts of the code will potentially help with the code structure. We will use our solutions #1 (keeping a piece of the code in a separate file, *m-code*) and #2 (using a user-defined MATLAB function, *m-function*) repeatedly in future coding depending on their suitability. We may think about the frequency of using and shared variables when making a final decision.

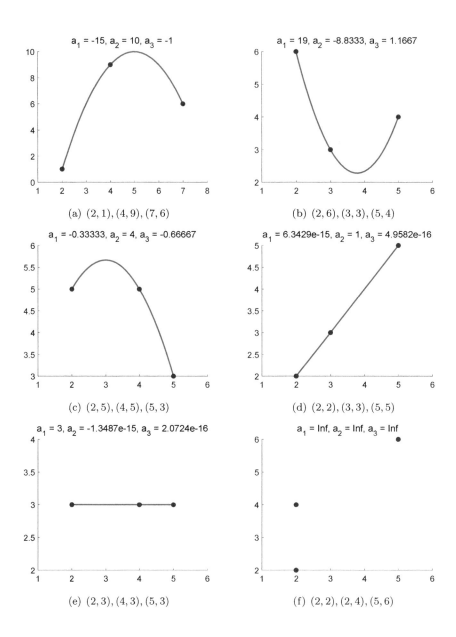

FIGURE 1.3
Test cases for Example 1.2 including "defective" one in (f).

- Solution #1 is good if some part of the code is used only a few times but shares too many variables (vectors and matrices) with the main code.

- Solution #2 is preferred when a function is used rather frequently and requires a limited number of input parameters.

In general, this consideration helps avoid computational overhead and memory misuse due to the necessity to replicate many code variables within the function environment. The reader may also find other reasons for their decisions while the code length in our future examples will grow continuously.

While playing with this first, and thus very simple, example in MATLAB, we would like the reader to pay attention to several more aspects usually considered as critical in successful computations.

- Improving code *readability* by utilizing block-wise structures, clear and comprehensive comments, self-explanatory variable naming, etc. It is vital for large projects where a computational team creates a complete set of programming concepts and policies to unify personal contributions and minimize time spent on code optimization. However, individual projects need the same for various reasons (code size, sharing with others, multipurpose use, etc.).

- User-defined interface to prevent unexpected and undesired *code breaking*, e.g., foreseeing MATLAB messaging in red whenever the computed results follow a wrong format or parameters use values not specified in the solution algorithm. Various messages to identify improper user input or incorrectly working parts of the code help localize the problem and minimize the debugging time.

- Proper *debugging* with multiple tests, including "special" cases. These cases may include any, even rare, occasions seen in the simulated results. If not adequately tested, the code part may become "a weak link in the chain" and produce errors that might be difficult to locate within the entire code.

- Investing in *visualization* for easy and comfortable analysis of the obtained results. For example, graphing intermediate solutions while the code is running helps in dynamic analysis of its correctness and locating a possible problem even in cases with no error messages displayed. Final plots for multiple solution characteristics contribute to our conclusion on the method's overall performance and computational efficiency.

After addressing some computational aspects, now we are ready to analyze the issues the reader may notice while running the code with different data to answer the first question posed on page 7. Figure 1.3 shows the results of some of these possible experiments. As seen in (d) and (e), provided data may cause the model to shift from quadratic form to the linear (or even constant) representation by setting respective components in the solution vector **a** to zeros (or machinery zeros of order 10^{-15} or even lower). Figure 1.3(f), in fact, shows the extreme case that the quadratic model obviously cannot accommodate. Thus, we could consider two data pieces $(2, 2)$ and $(2, 4)$ as "defective" as they share the same x-component. It will lead to redundancy in the system (1.4), inability to invert matrix A, and, as a result, inability to find a unique solution. This "defect" may be noted by the user-defined error message, or the data may be corrected. We leave deciding on the resolution of this problem to the reader.

Now it is easy to answer the second question on page 7. From our analysis, it appears that each equation in system (1.4) corresponds to one piece of data (one data point on the graph). Therefore, if the problem of fitting data with parabolic function is **underdetermined** ($n_d < 3$), we have only one or two equations. This situation is equivalent to the "defective" data case considered before. We may also find a practical analogy with data sets containing repeated elements.

Now, let us examine the case when we increase the data size to make the problem **overdetermined** ($n_d > 3$). We encourage the reader to run MATLAB code for Example 1.2 with added $(3, 6)$ point to explore the functionality of all methods. The new (4th) point fits the model $y(x) = a_1 + a_2 x + a_3 x^2$, and our solution method #3 (via MATLAB's polyfit function) can find it. Other methods use linear algebra based on updated matrix A (now 4–by–3) that cannot be inverted. Removing manually one of the equations in the new system (as linearly dependent on other equations) returns both approaches to fully functioning mode.

In general, if $y_4 \neq 6$ or other data points (considered as measurements for some quantity) suffer from errors, model equation $y(x)$ cannot be solved exactly in the context of Example 1.2 (quadratic data fitting). There is also a possibility to change the problem settings by allowing to fit data with nth degree *polynomial*[3], i.e., $y(x) = P_n(x) = a_1 + a_2 x + a_3 x^2 + \ldots + a_{n+1} x^n$, by taking n depending on the data size and structure. We also leave this exercise to a reader for revising linear algebra skills and extra practicing with MATLAB. Here, however, we intend to keep our main focus on the quadratic fit to preserve the dimensionality of vector $\mathbf{a} \in \mathbb{R}^3$. To discuss further the solution, we redesign our last example to project it onto more general settings.

Example 1.3 Least-Squares Data Fitting
Given $m \geq 3$ points of 2D data (x_1, y_1), (x_2, y_2), $\ldots$, (x_m, y_m), find coefficients a_1, a_2, and a_3 assuming quadratic fit via parabolic function $y(x) = a_1 + a_2 x + a_3 x^2$.

---

[3]See *Polynomials* on p. 377 for review.

A common approach used in different scientific fields consists of converting this problem into a new optimization problem of data fitting by minimizing the squared distances between the created model and data pieces, also known as *least-squares data fitting*. To evaluate these distances, we first construct the *residual vector* $\mathbf{r} \in \mathbb{R}^m$ for m data points

$$\mathbf{r} = \mathbf{y} - A\,\mathbf{a} = \begin{bmatrix} y_1 - (a_1 + a_2 x_1 + a_3 x_1^2) \\ y_2 - (a_1 + a_2 x_2 + a_3 x_2^2) \\ \cdots \\ y_m - (a_1 + a_2 x_m + a_3 x_m^2) \end{bmatrix} \tag{1.5}$$

given the same solution (optimization or control) vector $\mathbf{a} = [a_1 \ a_2 \ a_3]^T$. Ideally, we would like to minimize every component of vector $\mathbf{r}$. But practically, it is more convenient to use its scalarized version when defying the objective function $f(\mathbf{a})$ to represent accumulated (squared) mismatch between the model and data, i.e.,

$$f(\mathbf{a}) = r_1^2 + r_2^2 + \cdots + r_m^2 = \sum_{i=1}^{m} \left(y_i - (a_1 + a_2 x_i + a_3 x_i^2) \right)^2. \tag{1.6}$$

Finally, we arrive at the mathematical description of an unconstrained optimization problem in 3D:

$$\min_{\mathbf{a} \in \mathbb{R}^3} \ f(\mathbf{a}) \tag{1.7}$$

where objective $f(\mathbf{a})$ is given by (1.6). Here, we have to mention that if data from Example 1.2 is used (with possibly included extra $(3, 6)$ point), optimization problem (1.7) has optimal solution $\mathbf{a}^* = [-15 \ 10 \ -1]^T$ with "perfect match" as the residual vector $\mathbf{r} = 0$. Otherwise, data not matched "perfectly" will result in $\mathbf{r} \neq 0$ even with optimal solution found for minimized data mismatches, $r_i, \ i = 1, \ldots, m$.

1.4 Optimization Fundamentals

In this section, we review the main fundamentals of the optimization theory, such as feasibility, optimality, convexity, and convergence. To begin with, we use pretty straightforward definitions as they may appear in examples for simple calculus problems. Later, we will add more details once each notation is involved in the discussion for the practical implementation.

1.4.1 Feasibility

Although not considered necessary in many optimization problems, the *feasibility* issue is the main point to characterize the solution. Any constraints added to the problem's formulation automatically add complications to check if the obtained solution agrees with those constraints. Practical implementations vary depending on their complexity, the problem itself, and our preferences in choosing available techniques. For example, we may remove all restrictions and solve the problem by hoping the final solution will be feasible. However, a clever approach would consider adding constraints, and thus some knowledge of the feasibility, to the solution method. At this time, let us review the general concept and associated notations.

In Section 1.2, we referred to a notation of the feasible region regarding the optimization variable $\mathbf{u}$ only. In general, the term "feasibility" applies to state and control variables if they both appear in the description of the problem. In other words, the final (optimal) solution $\mathbf{u}^*$ proves to be feasible if it satisfies all its constraints, and the state variable $\mathbf{x}$ obtained with given $\mathbf{u}^*$ is also feasible. To make our initial review simpler, let us assume that in the following constrained optimization problem all states are also considered as optimization variables, i.e.,

$$\min_{\mathbf{x} \in \mathbb{R}^n} f(\mathbf{x}). \tag{1.8}$$

Similar to (1.1), we define the constraints as a set of p equations and m inequalities.

$$\begin{aligned} h_i(\mathbf{x}) &= 0, \quad i = 1, \ldots, p \\ g_j(\mathbf{x}) &\leq 0, \quad j = 1, \ldots, m \end{aligned} \tag{1.9}$$

Therefore, the *feasible region* (or *set*) $\mathbb{S}$ is the set of all solutions satisfying (1.9), and correspondingly the *feasible solution* is denoted $\mathbf{x} \in \mathbb{S}$.

We refer the reader to Figure 1.4 for an example of the feasible region and some solutions (feasible and infeasible) in 2D. In this figure, the boundary of closed region $\mathbb{S}$ is given by three segments associated with three different constraints (#1, #2, and #3). Therefore, solutions $\mathbf{x}_A$ and $\mathbf{x}_B$ are considered to be on the boundary, and solution $\mathbf{x}_C$ is in the interior of $\mathbb{S}$ (interior point). The current location of the solution with reference to all boundaries identifies all constraints as being *active* or *inactive*. For example, solution $\mathbf{x}_A$ makes constraints #1 and #3 active, solution $\mathbf{x}_B$ activates #2, and $\mathbf{x}_C$ makes all constraints inactive. In general, constraint $g_i(\mathbf{x}) \leq 0$ is active at $\mathbf{x}_0$ if $g_i(\mathbf{x}_0) = 0$, and all active constraints for solution $\mathbf{x}_0$ create an active set of constraints. This set is changing dynamically through the optimization process following the changes in the solution. And it becomes obvious that tracking these changes to maintain feasibility increases the problem's computational complexity with the number of constraints. We review selected methods for eliminating constraints by adding them to the objective function (e.g., by substitution or Lagrange multipliers) in Chapter 8.

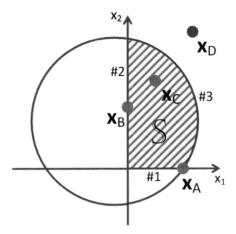

FIGURE 1.4
Example of feasible region $\mathbb{S}$ in 2D with feasible ($\mathbf{x}_A, \mathbf{x}_B$, and $\mathbf{x}_C$) and infeasible ($\mathbf{x}_D$) solutions (points).

1.4.2 Optimality

Now, after reviewing the notation of feasibility, we could move to the definition of *optimality* by considering the optimization problem (1.8)–(1.9) in a simpler form.

$$\min_{\mathbf{x} \in \mathbb{R}^n} \quad f(\mathbf{x})$$
$$\text{s.t.} \quad \mathbf{x} \in \mathbb{S} \tag{1.10}$$

We consider $\mathbf{x}^*$ as a *global minimizer* of f in $\mathbb{S}$ if

$$\forall \mathbf{x} \in \mathbb{S} \quad f(\mathbf{x}^*) \leq f(\mathbf{x}). \tag{1.11}$$

In case the global minimizers exist, solving an optimization problem aims at finding these minimizers. Ideally, we are interested in finding one global minimizer if its uniqueness is proved. In practice, most problems have multiple local minimizers considered to represent the solutions if global ones do not exist or may not be found or properly identified. We call $\mathbf{x}^*$ as a *local minimizer* of f in $\mathbb{S}$ if

$$\forall \mathbf{x} \in \mathbb{S} \quad \text{s.t.} \quad \|\mathbf{x} - \mathbf{x}^*\| < \epsilon \quad f(\mathbf{x}^*) \leq f(\mathbf{x}). \tag{1.12}$$

The analysis of local minimizers is relatively simple. It requires proving $\mathbf{x}^*$ to be the best possible solution only within its ϵ-neighborhood, where ϵ is a small positive number. Similar to solving problems analytically in calculus, computational (numerical) optimization algorithms may discover one or more solutions with the initial claim to be local minimizers. It requires further analysis to change the status of some of these solutions from local to global, which may not be straightforward. We will discuss the availability of such analysis for the selected models based on their known properties later.

Switching to $f(\mathbf{x}^*) < f(\mathbf{x})$ (with added extra statement $x \neq x^*$) in (1.11) and (1.12) gives us the definitions of, respectively, *strict local* and *global minimizers*. Figure 1.5(a) shows examples of local (infinitely many points from the interval denoted as x_3) and strict local (x_1 and x_2) minimizers. If there is no change in $f(\mathbf{x})$ behavior (when $x \to \pm\infty$), then solution x_1 may also serve as the global minimizer. Figure 1.5(b) illustrates the problem with a bounded feasible region, $\mathbb{S} = [a, b]$, by added constraints (bounds a and b). Here, solution x_1 is the local minimizer, while the bound $x = b$ is the global one. To confirm the latter, we use the following analysis: $f(x)$ is continuous over the interval $[a, b]$ that is a closed and connected (with no holes) region in 1D. We close this brief review of optimality by noting that similar notations (local, global, and strict) exist to define *maximizers* of any objective function $f(\mathbf{x})$.

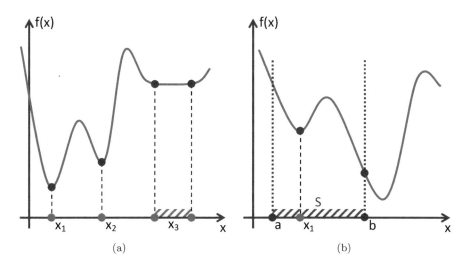

FIGURE 1.5
Examples of local, global, and strict minimizers for 1D functions with feasible regions (a) $\mathbb{S} \in \mathbb{R}$ and (b) $\mathbb{S} = [a, b]$.

1.4.3   Convexity

Reviewing the notation of *convexity* to be applied in the context of the solutions of optimization problems requires both definitions, namely a *convex set* and a *convex function*.

A set $\mathbb{S}$ is convex if

$$\forall \mathbf{x}, \mathbf{y} \in \mathbb{S} \quad \alpha\mathbf{x} + (1 - \alpha)\mathbf{y} \in \mathbb{S}, \quad \forall\, 0 \le \alpha \le 1. \tag{1.13}$$

Figure 1.6 provides examples of convex and nonconvex sets in 1D and 2D.

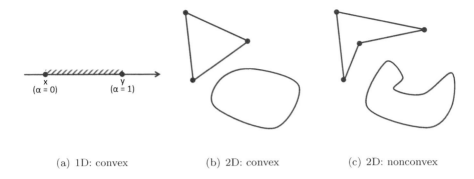

(a) 1D: convex   (b) 2D: convex   (c) 2D: nonconvex

FIGURE 1.6
Examples of (a,b) convex and (c) nonconvex sets in (a) 1D and (b,c) 2D.

A function $f(\mathbf{x})$ is convex on a convex set $\mathbb{S}$ if

$$\forall \mathbf{x}, \mathbf{y} \in \mathbb{S} \quad f(\alpha \mathbf{x} + (1 - \alpha)\mathbf{y}) \leq \alpha f(\mathbf{x}) + (1 - \alpha)f(\mathbf{y}) \qquad (1.14)$$

that should be valid for any α

$$0 \leq \alpha \leq 1. \qquad (1.15)$$

In case we change sign in (1.14) to ">," function $f(\mathbf{x})$ will be called *concave*. Use of signs "<" or ">" in (1.14) allows to define, respectively, *strictly convex* and *strictly concave* functions. Figure 1.7 illustrates the definition (1.14)–(1.15) of a convex function in 1D. We also have to mention that convexity established for both feasible region and objective function helps in the existence and uniqueness analysis of the solutions for various optimization problems to be discussed later in applications to our model examples.

1.5   General Optimization Algorithm

We postpone the discussion on one more notation, convergence, to discuss first the solution process for the optimization problem that requires iterations. Direct analytical solutions or solutions "in one iteration" exist for many problems. The reader may refer to our first examples (Examples 1.1, 1.2, and 1.3 in this chapter) solvable by known geometry or linear algebra formulas. However, many other problems will require the solution process executed in iterations starting from the initial (presumably feasible) solution and assuming that better solutions (enabled to decrease or increase the objective function) are possible.

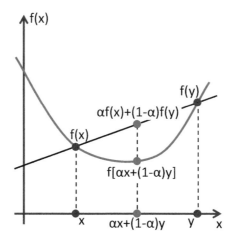

FIGURE 1.7
1D example of a convex function $f(x)$.

1.5.1   Solving Optimization Problems in Iterations

We could generalize the idea of solving the optimization (minimization) problem (1.10) *iteratively* by

$$
\begin{aligned}
\mathbf{x}^0 &= \mathbf{x}_0, \\
\mathbf{x}^{k+1} &= \mathbf{x}^k + \boldsymbol{\delta}^k, \quad k = 0, 1, 2, \ldots
\end{aligned}
\tag{1.16}
$$

where $\mathbf{x}_0$ is the initial guess chosen by any reasonable assumption, $\mathbf{x}^k$ represents the solution after kth iteration, and $\boldsymbol{\delta}^k$ is the *solution update* (*change* or *perturbation*) with proven potential for improvement, i.e.,

$$
f(\mathbf{x}^{k+1}) < f(\mathbf{x}^k).
\tag{1.17}
$$

Algorithm 1.1 provides a very general scheme for solving optimization problems iteratively.

Algorithm 1.1 (General Optimization Algorithm (Iterative))
To solve optimization problem:

$$
\begin{aligned}
\min_{\mathbf{x} \in \mathbb{R}^n} \quad & f(\mathbf{x}) \\
s.t. \quad & \mathbf{x} \in \mathbb{S}
\end{aligned}
$$

1. Choose initial guess $\mathbf{x}^0 = \mathbf{x}_0$ and other algorithm settings

2. *For $k = 1, 2, \ldots$, check (computational)* **optimality** *of $\mathbf{x}^k$, e.g.,*

   - $\mathbf{x}^k$ *reduces $f(\mathbf{x})$ up to a necessary level*
   - $\nabla f(\mathbf{x}^k) \cong \mathbf{0}$ *(local optimum condition)*
   - *consider other* **termination conditions**

3. *If solution $\mathbf{x}^k$ is optimal* $\rightarrow$ **STOP**

4. *Find solution* **update** $\boldsymbol{\delta}^k$,
   e.g., by means of **search direction** $\mathbf{d}^k$

5. *Update the solution*

$$\mathbf{x}^{k+1} = \mathbf{x}^k + \boldsymbol{\delta}^k = \mathbf{x}^k + \alpha^k \mathbf{d}^k \tag{1.18}$$

6. *Go to Step 2*

As exemplified in Algorithm 1.1, solution update $\boldsymbol{\delta}^k$ may be provided by the *search direction* $\mathbf{d}^k$ that improves the solution in some sense, e.g., $\mathbf{d}^k = -\nabla f(\mathbf{x}^k)$ in (1.18). Step size α^k in (1.18) is determined in general assumption for decreasing objective (1.17).

1.5.2 Termination Criteria

Even if we solve a problem with an existing unique global minimizer, no one guarantees to reach the exact solution. It is due to the inaccuracies accumulated through all computational steps within multiple iterations, truncation, or round-off errors. Thus, approaching the minimizer, local or global, by making smaller and smaller steps will be an endless process required to be terminated at some point. We consider the following common strategies to use as *termination conditions*:

1. Based on **sufficient changes in the solution:** (in some norm[4] $\| \cdot \|_N$)

   - absolute decrease

   $$\left\| \mathbf{x}^k - \mathbf{x}^{k-1} \right\|_N < \epsilon \tag{1.19}$$

   - relative decrease

   $$\frac{\left\| \mathbf{x}^k - \mathbf{x}^{k-1} \right\|_N}{\left\| \mathbf{x}^{k-1} \right\|_N} < \epsilon \tag{1.20}$$

---

[4]See *Vector and Matrix Norms* on p. 379 for review.

2. Based on **sufficient changes in the objective:**

- absolute decrease

$$\left| f(\mathbf{x}^k) - f(\mathbf{x}^{k-1}) \right| < \epsilon \tag{1.21}$$

- relative decrease

$$\left| \frac{f(\mathbf{x}^k) - f(\mathbf{x}^{k-1})}{f(\mathbf{x}^{k-1})} \right| < \epsilon \tag{1.22}$$

3. Based on allowable **computational efforts:**

- maximum number of optimization iterations $k_{\max}$

$$k > k_{\max} \tag{1.23}$$

- max number of objective evaluations
- limit on elapsed computational time T

$$t > T \tag{1.24}$$

The reader may establish other termination criteria or use their combinations as it is usually very dependent on the solved problem and individual preferences.

? *Among options (1.19)–(1.24), when dealing with practical examples within the rest of this book, the reader will be advised to use termination conditions (1.20), (1.22), and (1.23). Why?*

1.6   Convergence

In the last section of this chapter, we suggest the reader to review the notation of *convergence*. Following the general description of the iterative optimization algorithm (or framework) applied to solve a selected problem, we may question its computational complexity. In one of the simplest ways, with no analysis made on the performance, a measure for this complexity may be a number of basic arithmetic operations required to find a solution. By assuming that each evaluation of the objective function requires (approximately) the same number of arithmetic operations, we could also operate with the total number of objective evaluations required to find the solution. This number is a good indicator of the overall computational performance of the entire optimization algorithm. Subject to the computational hardware, the user may get the average time for one evaluation and, therefore, estimate the runtime for the same solution obtained on various computers. However, when solving a problem iteratively, evaluated objectives are often used multiple times to provide the solution updates. As such, these evaluations do not provide values decreased monotonically causing difficulty in proving the solution convergence.

? *While running the optimization, there are two obvious questions. Does it converge? If yes, how fast?*

Convergence analysis borrowed from calculus suggests studying the sequence of solutions obtained after each iteration

$$\{\mathbf{x}^k\} = \{\mathbf{x}^0, \mathbf{x}^1, \mathbf{x}^2, \ldots\}. \tag{1.25}$$

First, this sequence should converge to a limit value (in our case, optimal solution $\mathbf{x}^*$),

$$\{\mathbf{x}^k\} \to \mathbf{x}^*. \tag{1.26}$$

Figure 1.8 illustrates two optimization processes started from different initial

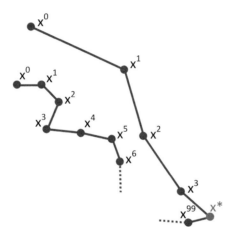

FIGURE 1.8
Example of two optimization solutions converged in four and 100 iterations.

solutions $\mathbf{x}_0$. They converge to the same optimal solution $\mathbf{x}^*$, each with its own speed. To evaluate this speed, we introduce the sequence of errors $\{\mathbf{e}^k\}$,

$$\mathbf{e}^k = \mathbf{x}^k - \mathbf{x}^*, \tag{1.27}$$

and assume that it converges, i.e.,

$$\lim_{k \to \infty} \mathbf{e}^k = \mathbf{0}. \tag{1.28}$$

As commonly used in calculus, we also state that sequence $\{\mathbf{x}^k\}$ converges to $\mathbf{x}^*$ with *rate r* and *constant C* if

$$\lim_{k \to \infty} \frac{\|\mathbf{e}^{k+1}\|}{\|\mathbf{e}^k\|^r} = \lim_{k \to \infty} \frac{\|\mathbf{x}^{k+1} - \mathbf{x}^*\|}{\|\mathbf{x}^k - \mathbf{x}^*\|^r} = C, \quad C < \infty. \tag{1.29}$$

Similarly, we define *linear convergence* when $r = 1$ or

$$\|\mathbf{e}^{k+1}\| = C\|\mathbf{e}^k\|. \tag{1.30}$$

Positive constant C plays a role technically confirming convergence ($0 < C < 1$) or divergence ($C > 1$). Having C close to 1 indicates slower (*sublinear*) convergence, while about 0 values characterize faster (*superlinear*) one. The latter may also suggest re-checking rate r that may be greater than 1. In this case, $r = 2$ determines *quadratic* convergence. In some (very rare) cases, $r = 3$ demonstrates *cubic* convergence.

Again, this calculus analysis is applied easily to any sequences (1.25) given analytically. However, in real computations limit value $\mathbf{x}^*$ is not always available, neither as prior information nor after terminating the optimization. Nevertheless, we will use a modified version of this analysis to evaluate computational convergence as discussed in the next chapter. We close this review by adopting two strategies for evaluating the performance of various computational methods to perform optimization:

(a) by the **rate of convergence** as discussed in this section, and

(b) by a total number of **objective function evaluations** to replace the elapsed time for computing the solution. In some cases, however, the total number of optimization iterations may still work if every iteration evaluates objectives more or less the same number of times.

1.7   Homework Problems

1. Solve the problem in Example 1.1 analytically.

2. Modify `MATLAB` code `Chapter_1_data_fit.m` for Example 1.2.

   (a) Apply new data (four or five points) and use all three solution approaches.
   (b) Repeat for data with m ($m > 5$) points.
   (c) Implement check-up to prevent *code breaking* in case the problem is under- or overdetermined.
   (d) Implement check-up to prevent `Inf/NaN` problem in case the data is "defective."

3. Show that a set is convex if and only if its intersection with any line is convex.

4. In general the product or ratio of two convex functions is not convex. However, there are some results that apply to functions on $\mathbb{R}$. Prove the following.

(a) If f and g are convex, both nondecreasing (or nonincreasing), and positive functions on an interval, then $f \cdot g$ is convex.

(b) If f is convex, nondecreasing, and positive and g is concave, nonincreasing, and positive, then f/g is convex.

5. Using second-order condition for convexity (Hessian[5] is positive definite[6] or positive semidefinite) for each of the following functions determine whether it is convex or concave:

   (a) $f(x) = e^x - 1$ on $\mathbb{R}$,

   (b) $f(x_1, x_2) = x_1 x_2$ on $\mathbb{R}_+ \times \mathbb{R}_+$,

   (c) $f(x_1, x_2) = \dfrac{1}{x_1 x_2}$ on $\mathbb{R}_+ \times \mathbb{R}_+$,

   (d) $f(x_1, x_2) = \dfrac{x_1^2}{x_2}$ on $\mathbb{R} \times \mathbb{R}_+$,

   (e) $f(x_1, x_2) = x_1^\alpha x_2^{1-\alpha}$, where $0 \le \alpha \le 1$, on $\mathbb{R}_+ \times \mathbb{R}_+$.

6. For each of the following sequences with given general term x^k, prove that the sequence converges, find its limit, and determine convergence parameters r and C:

   (a) $x^k = 2^{-k}$,

   (b) $x^k = 1 + 5 \cdot 10^{-2k}$,

   (c) $x^k = 3^{-k^2}$.

READ   **Where to Read More**

Bertsimas (1997), [5]
Chapter 1 (Introduction), Chapter 2 (The Geometry of Linear Programming)

Boyd (2004), [6]
Chapter 1 (Introduction), Chapter 2 (Convex Sets), Chapter 3 (Convex Functions)

Griva (2009), [15]
Chapter 1 (Optimization Models), Chapter 2 (Fundamentals of Optimization)

---

[5]See *Hessians* on p. 381 for review.
[6]See *Positive Definite Matrices* on p. 380 for review.

Nocedal (2006), [25]
Chapter 1 (Introduction), Chapter 2 (Fundamentals of Unconstrained Optimization)

| RUN | MATLAB **Codes for Chapter 1**

- Chapter_1_data_fit.m

- DataFitM.m

2

Minimization Approaches for Functions of One Variable

This chapter suggests the reader review various algorithms for minimizing the functions of one variable: namely, the bisection and golden section search approaches, the 2-order derivative search by Newton's method, the derivative-free brute-force, and a random search by the Monte Carlo method. The pros and cons of applying these methods to various functions (locally/globally convex/concave, monotonic, periodic, nonsmooth) are discussed based on the analysis of the solutions obtained in `MATLAB`. We also start discussing the methodology of practical assessment for the computational performance and the rate of convergence used in various applications for optimization.

2.1 Minimizing a Function in 1D

An obvious target of many optimization frameworks is their ability to solve problems defined in multiple dimensions ranging from a few optimization variables to millions of controls. The reader should not be disappointed with any negative experience when starting directly with a big problem that includes multiple components to support any computational optimization project. Testing separately functionality of each part and how properly the data are flowing between them is usually a very scrupulous process. Some components do not allow testing individually outside the entire framework, or test examples may not be simplified for checking functionality of a particular place in this framework. Many other reasons make the debugging process too complicated, even for experienced optimization professionals. Thus, it might be a good idea making a small step back and review some minimization approaches for one-variable functions possibly known to the reader from calculus. Here, we focus this review more on computational issues as many of the discussed algorithms will serve later as parts of the entire solution method enabled to solve "big and mature" problems.

Let us consider the following optimization problem to minimize a one-variable function $f(x)$, $x \in \mathbb{R}$,

$$
\begin{aligned}
\min_{x \in \mathbb{R}} \quad & f(x) \\
\text{s.t.} \quad & a \leq x \leq b
\end{aligned}
\tag{2.1}
$$

by assuming that $f(x)$ is (piecewise) continuous for all $x \in [a, b]$. We set the linear constraints (fixed bounds) $x \geq a$ and $x \leq b$ in (2.1) as required by algorithms discussed in this chapter. However, it may be reconsidered later subject to a particular problem or solution method. Two examples below may serve as particular examples of problem (2.1).

Example 2.1 Minimizing a Function Given Analytically
Minimize $f(x) = \sin^3 x + \cos^3 x$ on $[0, 2\pi]$.

In this example, $f(x)$ is (twice) continuously differentiable. Thus, the use of the "derivative tests" from calculus allows locating all extreme points (extremums) within given interval $[0, 2\pi]$

$$
x_1 = 0, \ x_2 = \frac{\pi}{4}, \ x_3 = \frac{\pi}{2}, \ x_4 = \pi, \ x_5 = \frac{5\pi}{4}, \ x_6 = \frac{3\pi}{2}, \ x_7 = 2\pi.
$$

We also identify two global minimizers

$$
x^* = x_4 = \pi, \ x^* = x_6 = \frac{3\pi}{2}, \ f^* = f_{\min} = f(x_4) = f(x_6) = -1.
$$

Example 2.2 Example 1.1 Revisited
Convert Example 1.1 into 1D optimization problem with extra linear (bound) constraints $-2 \leq x_1 \leq 2$.

We leave finding the analytic solution for problem in Example 2.2 to the reader.

Both problems in Examples 2.1 and 2.2 may be solved analytically as their objective functions and associated derivatives exist and are easily computable. For some other cases, when it may not be true, we could apply methods to locate (approximate) the solutions using the following numerical approaches.

- *Bisection (binary search)* method, see Section 2.2 for details.

- *Golden Section Search* method (Section 2.3).

- *Newton's (2-order)* method (Section 2.4).

- *Exhaustive (Brute-Force) Search* method (Section 2.5).

- *Stochastic* approaches, e.g., *Monte Carlo* (*Random Search*) method (Section 2.6).

- Other, may be *heuristic* to some extent, approaches.

Next, we are going to review these numerical approaches and apply them for solving minimization problems to discuss some computational issues.

? *While discussing these applications here for 1D ($x \in \mathbb{R}$) minimization only, in general, why do we need them for $\mathbf{x} \in \mathbb{R}^n$ cases?*

2.2   Bisection Method

By applying the *bisection* or *binary search method* to minimize function $f(x)$ over interval $[a, b]$ we assume that $f(x)$ is *unimodal* on that interval. Unimodality means $f(x)$ has exactly one global minimizer on $[a, b]$. To avoid any confusion, we have to mention the difference when bisecting intervals (dividing them into two equal parts) for finding roots of an algebraic equation $f(x) = 0$ and solving optimization problem (2.1). When used in the former, we evaluate $f(x)$ to find the interval where the function changes its sign. However, application for minimizing functions is less straightforward as the exact interval midpoint is not used to evaluate $f(x)$. Instead, we use two points in the close neighborhood of the midpoint to get an idea if $f(x)$ increases or decreases within this small region. We may also compare this technique with the finite difference approximation of the derivative at the midpoint. However, the bisection method does not require the slope, just the sign to choose between the left and right halves of the bisected interval. Figure 2.1 provides a schematic illustration of the method's idea.

Before applying the bisection method computationally we have to set two parameters:

- **small constant** δ, $0 < \delta < b - a$, to evaluate the sign of $f'(x)$ at midpoint $(a + b)/2$, and

- **tolerance** $\epsilon > 0$ to terminate iterations. Note that $\epsilon > \delta$.

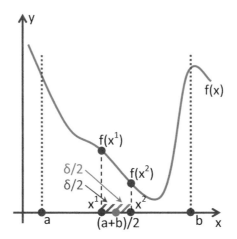

FIGURE 2.1
Schematic illustrating the bisection method used in minimizing unimodal function $f(x)$.

All computational steps to implement this method are provided in Algorithm 2.1.

Algorithm 2.1 (Bisection or Binary Search Method)

1. *For initial interval $I^0 = [a, b]$ compute*

$$x^1 = \frac{a + b - \delta}{2}, \quad x^2 = \frac{a + b + \delta}{2}$$

2. *Compare $f(x^1)$ and $f(x^2)$:*
   - *if $f(x^1) \leq f(x^2)$ then $a^1 = a, \ b^1 = x^2$*
   - *if $f(x^1) > f(x^2)$ then $a^1 = x^1, \ b^1 = b$*

3. *Get a new interval $I^1 = [a^1, b^1]$ of length*

$$d^1 = b^1 - a^1 = \frac{b - a - \delta}{2} + \delta$$

4. *Perform the same process iteratively. For kth step compute $f(x^{2k-1})$ and $f(x^{2k})$ for $x^{2k-1}, x^{2k} \in I^{k-1}$ where*

$$x^{2k-1} = \frac{a^{k-1} + b^{k-1} - \delta}{2}, \quad x^{2k} = \frac{a^{k-1} + b^{k-1} + \delta}{2}$$

5. *Compare* $f(x^{2k-1})$ *and* $f(x^{2k})$:

- *if* $f(x^{2k-1}) \leq f(x^{2k})$ *then* $a^k = a^{k-1}, \ b^k = x^{2k}$
- *if* $f(x^{2k-1}) > f(x^{2k})$ *then* $a^k = x^{2k-1}, \ b^k = b^{k-1}$

6. *Get a new interval* $I^k = [a^k, b^k]$ *of length*

$$d^k = b^k - a^k = \frac{b - a - \delta}{2^k} + \delta$$

7. *Terminate search if* $d^k = b^k - a^k < \epsilon$, *otherwise go to Step 4*

8. *Approximate* x^*, *e.g.*,

$$x^* = \frac{a^k + b^k}{2}$$

(extra $f(x^*)$ *evaluation required) with error*

$$e_B \sim \frac{b - a}{2^k} \tag{2.2}$$

In case the computational cost of $f(x)$ evaluation is high, the final (optimal) solution may be approximated as $x^* = a^k$ or $x^* = b^k$. By examining the termination condition $d^k = b^k - a^k < \epsilon$ used in the bisection method, it is pretty straightforward to make a prior estimation for the total **number of iterations**

$$k > \log_2 \frac{b - a - \delta}{\epsilon - \delta} \tag{2.3}$$

and the **number of objective function evaluations**

$$n = 2k > 2 \log_2 \frac{b - a - \delta}{\epsilon - \delta} \tag{2.4}$$

required to complete the search. The reader may find more details on the bisection method as well as other numerical methods discussed here in the *Where to Read More* section placed at the end of this chapter.

2.3   Golden Section Search

?   *Bisection method requires* $n = 2k$ *function* $f(x)$ *evaluations, where* k *is given by* (2.3). *Could we reduce* n *as it affects the computational costs?*

Let us answer this question right away by considering another approach to compete with the bisection search. We refer to Figure 2.2 to review the notation of the *golden section* and its property.

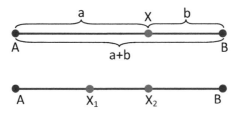

FIGURE 2.2
Illustrating the notation of golden section (ratio) and its property.

Any interval AB of finite length could be divided into two parts AX and XB with respective lengths a and b such that

$$\frac{a+b}{a} = \frac{a}{b} = \frac{1+\sqrt{5}}{2} \approx 1.6180\ldots$$

This ratio is called a *golden section* (GS) or a *golden ratio*. We could also introduce the following symbolic notation: $X = GS(AB)$, refer to Figure 2.2. Due to the symmetry, the following property is in place for golden section:

$$
\begin{aligned}
&\text{if} \quad && X_1 = GS(AB), \quad && X_2 = GS(AB), \ X_1 \neq X_2, \\
&\text{then} \quad && X_1 = GS(AX_2), \quad && X_2 = GS(X_1B).
\end{aligned}
\tag{2.5}
$$

Finally, by introducing a golden section constant $\alpha = \dfrac{\sqrt{5}-1}{2}$ we could express positions of both golden sections X_1 and X_2 in terms of the interval endpoints X_A and X_B:

$$x_1 = \alpha x_A + (1-\alpha)x_B, \quad x_2 = (1-\alpha)x_A + \alpha x_B.$$

Briefly, the *golden section search method* applied for solving the minimization problem (2.1) has the same structure as the bisection method discussed in the previous section. The only difference is that the truncated interval will be $[a^1, b^1] = [a, x^2]$ or $[a^1, b^1] = [x^1, b]$, where, respectively, x^1 or x^2 will be one of the golden sections in the new interval. As such, we have to compute the second golden section point and evaluate function $f(x)$ only one time at this point. Figure 2.3 provides a schematic illustration of the method's idea with two possibilities, (a) and (b), for each iteration.

Before applying the GS search method computationally we have to set only one parameter: **tolerance** $\epsilon > 0$ to terminate iterations. All computational steps to implement this method are provided in Algorithm 2.2.

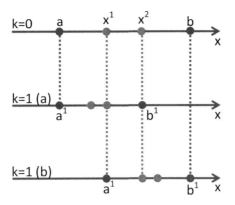

FIGURE 2.3
Schematic illustrating the golden section points propagated from initial interval $[a, b]$ to truncated interval $[a^1, b^1]$.

Algorithm 2.2 (Golden Section Search Method)

1. *For initial interval $I^0 = [a, b]$ compute:*

$$x^1 = \alpha a + (1 - \alpha)b, \quad x^2 = (1 - \alpha)a + \alpha b$$

2. *Compare $f(x^1)$ and $f(x^2)$:*

  (a) *if $f(x^1) \leq f(x^2)$ then $a^1 = a$, $b^1 = x^2$, $\bar{x}^1 = x^1$*
  (b) *if $f(x^1) > f(x^2)$ then $a^1 = x^1$, $b^1 = b$, $\bar{x}^1 = x^2$*

3. *Get a new interval $I^1 = [a^1, b^1]$ of length $d^1 = b^1 - a^1 = \alpha(b - a)$*

4. *For $k = 2$ use $\bar{x}^1$ as a new value x^1 or x^2*

5. *Perform the same process iteratively. For kth step*

  • *if $f(x^{2k-1}) \leq f(x^{2k})$ then $f(x^2) = f(\bar{x}^{k-1})$ is known, just compute $f(x^1)$*
  • *if $f(x^{2k-1}) > f(x^{2k})$ then $f(x^1) = f(\bar{x}^{k-1})$ is known, just compute $f(x^2)$*

6. *Get a new interval $I^k = [a^k, b^k]$ of length $d^k = b^k - a^k = \alpha^k(b - a)$*

7. *Terminate search if $d^k = b^k - a^k < \epsilon$, otherwise go to Step 5*

8. *Approximate x^*, e.g.,*

$$x^* = \bar{x}^k$$

with error

$$e_{GS} \sim \alpha^k(b-a) = \left(\frac{\sqrt{5}-1}{2}\right)^k (b-a) \qquad (2.6)$$

Establishing the termination condition $d^k = b^k - a^k < \epsilon$ (the same as in the bisection method) provides an estimation for the total **number of iterations** k, which is about the same as the **number of objective function evaluations** n (or more precisely, $n = k + 1$ due to Step 2 in Algorithm 2.2)

$$k > \log_{\frac{1}{\alpha}} \frac{b-a}{\epsilon}, \quad n = k+1, \quad \alpha = \frac{\sqrt{5}-1}{2} \approx 0.62\ldots \qquad (2.7)$$

Compared with the bisection method, the golden section search decreases the interval length d^k slower if referenced to the iteration count k. However, taking into account the total number of objective evaluations n, it is much faster – we may check it by comparing errors for both methods given by (2.2) and (2.6) for the same n (we assume $n = k$ for GS)

$$\frac{e_\mathrm{B}}{e_\mathrm{GS}} = \frac{2^{-\frac{n}{2}}(b-a)}{\left(\frac{\sqrt{5}-1}{2}\right)^n (b-a)} = \left(\frac{\sqrt{2}}{\sqrt{5}-1}\right)^n \approx (1.144\ldots)^n. \qquad (2.8)$$

For example, limiting n to 10 allows the golden section search to arrive at about four times better accuracy

$$\left(\frac{\sqrt{2}}{\sqrt{5}-1}\right)^{10} \approx 3.84. \qquad (2.9)$$

? *Is the unimodality assumption previously set for the bisection method is still in place for the golden section search?*

The reader may answer this question and check it computationally when working with practical examples in Section 2.7.

2.4 Newton's Method

? *In the previous section, we checked that the GS runs the search faster than the bisection method. Now, how to get even faster convergence assuming $f(x)$ is (twice) differentiable?*

We know that *Newton's method* attempts to construct sequence $\{x^k\}$ from initial guess x^0 that converges toward *stationary point* x^* of $f(x)$ satisfying condition $f'(x^*) = 0$. We may also think about Newton's method as about root-finding algorithm for solving $g(x) = 0$ where $g(x) = f'(x)$. Before applying this method computationally we have to set again only one parameter: **tolerance** $\epsilon > 0$ to terminate iterations. All computational steps to implement the Newton's method are provided in Algorithm 2.3.

Algorithm 2.3 (Newton's 2-order Method)

1. *Choose initial guess x^0*

2. *For kth step: provided that $f''(x^k) \neq 0$ update the solution by*

$$x^{k+1} = x^k - \frac{f'(x^k)}{f''(x^k)} \tag{2.10}$$

3. *Terminate if*

$$\left| \frac{f(x^{k+1}) - f(x^k)}{f(x^k)} \right| < \epsilon \quad or \quad \left| \frac{x^{k+1} - x^k}{x^k} \right| < \epsilon \quad or \quad |f'(x^k)| < \epsilon \tag{2.11}$$

4. *Approximate $x^* = x^k$ with error*

$$e_N \sim M(x^* - x_k)^2, \ where \ M = \frac{f'''(x^k)}{2f''(x^k)} \tag{2.12}$$

The geometrical interpretation of Newton's method could be described by the following statement. At each (kth) iteration function $f(x)$ is approximated by a quadratic function around x^k, and a step is taken toward the maximum or minimum of that quadratic function as shown by (2.10).

The review of calculus provides the main **properties** of Newton's method that make this approach very attractive for using while solving various optimization problems.

- It converges with the **quadratic rate** of convergence if initial guess x^0 is **close enough** to the solution, stationary point x^*. We refer the reader back to Section 1.6 for reviewing the notation of convergence.

- It allows minimizing (or maximizing) quadratic functions in just **one iteration**.

- Newton's method is easily extendable for solving **multidimensional** problems (shifting from $x^k \in \mathbb{R}$ to $\mathbf{x}^k \in \mathbb{R}^n$) by adjusting the update part in (2.10):

$$\frac{f'(x^k)}{f''(x^k)} \quad \rightarrow \quad \left[\nabla^2 f(\mathbf{x}^k)\right]^{-1} \nabla f(\mathbf{x}^k). \tag{2.13}$$

At the same time, we have to keep in mind several known **drawbacks**.

- Newton's method exhibits much **slower convergence** (linear or even slower) in case the stationary point x^* is the inflection point (or $f''(x^*) \rightarrow 0$).

- It does **not guarantee** convergence if not started reasonably close to x^*.

- Instead of the desired minimum, it may also converge to the maximum or inflection points.

Therefore, when applying Newton's method, especially in multidimensional optimization problems, computational efficiency should be questioned and properly checked. It is a method with the potential for fast convergence if its implementation is carefully aligned with the solved problem and properly governed. However, the performance may worsen abruptly with increased dimensionality and complexity to evaluate gradients[1] $\nabla f(\mathbf{x}^k)$ and Hessians[2] $\nabla^2 f(\mathbf{x}^k)$ as seen in (2.13).

2.5   Brute-Force Search

In Sections 2.2–2.4, we discussed approaches that may be grouped following the way they construct the sequence $\{x^k\}$. The bisection and golden section methods use this sequence to update the search intervals, while Newton's method treats each x^k value as an approximate solution. However, all three methods update $\{x^k\}$ dynamically as values x^k cannot be known before their search starts. Such methods are often called *active search* methods. On the other hand, *passive search* methods define $x^1, x^2, \ldots, x^n \in [a, b]$ before the computation phase, which allows very easy parallelization of the underlying algorithms.

---

[1] See *Gradient* on p. 381 for review.
[2] See *Hessians* on p. 381 for review.

Here, we review one of the simplest passive search methods, namely *exhaustive* or *brute-force* (BF) *search* algorithm. To apply it for minimizing $f(x)$ over interval $[a, b]$, one parameter should be set: **discretization step** $h < b - a$. Algorithm 2.4 provides the complete details on all computational steps.

Algorithm 2.4 (Exhaustive or Brute-Force Search Method)

1. For given $I = [a, b]$ compute $x^1, x^2, \ldots, x^n \in [a, b]$

$$x^1 = a + h, \quad x^2 = a + 2h, \quad \ldots, \quad x^n = \min\{a + nh, b\}$$

2. Compute sequentially

$$
\begin{aligned}
&f(x^1) \quad then \qquad\qquad\qquad\qquad\quad store: \quad \bar{x} = x^1 \quad and \quad f(\bar{x}) = f(x^1) \\
&f(x^2) \quad then \quad if\ f(x^2) < f(\bar{x}) \quad store: \quad \bar{x} = x^2 \quad and \quad f(\bar{x}) = f(x^2) \\
&\qquad\qquad\qquad\qquad\quad \ldots \\
&f(x^n) \quad then \quad if\ f(x^n) < f(\bar{x}) \quad store: \quad \bar{x} = x^n \quad and \quad f(\bar{x}) = f(x^n)
\end{aligned}
$$

3. Approximate $x^* = \bar{x}$ with error

$$e_{BF} \sim 2h \tag{2.14}$$

The brute-force search method has a simple geometrical interpretation. As we may see from Figure 2.4, this method provides the uniform approximation of $f(x)$ by linear splines. A simple modification of Algorithm 2.4 ensures that all intervals have the same size:

(a) choose the number of inner points n first, then

(b) compute parameter h.

Despite its apparent simplicity, the brute-force method still possesses several **properties**.

- It enables **global search** over the discretized region $[a, b]$.

- Different **speeding-up** strategies exist in the forms of combined passive-active search.

- A brute-force search is easily expandable for use with **multi-modal functions**. However, the error provided by (2.14) is not guaranteed and is hardly assessable.

Finally, we have to mention two evident **drawbacks**.

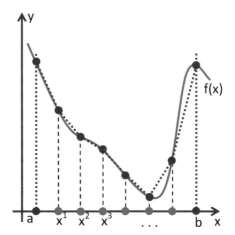

FIGURE 2.4
Schematic illustrating the uniform approximation of $f(x)$ by linear splines performed by the brute-force search method.

- The computational part of this method is **very expensive**, especially for multidimensional problems.

- We may easily miss the optimal solution if the applied discretization is **not fine enough**.

Despite these drawbacks, the brute-force search algorithms are very popular due to their simplicity in implementation and ability to perform a global search, which may be quite effective for certain optimization problems of low dimensionality ($\mathbf{x} \in \mathbb{R}^n$, $n < 10$).

2.6   Monte Carlo Method

The search methods discussed in Sections 2.2–2.5 appear to be *deterministic* in their nature. It means once repeated, they provide exactly the same results. It might also be a good idea to consider methods from the area of *stochastic approaches*, a broad class of computational algorithms that rely on repeated random sampling to obtain numerical results. The essential idea is to use randomness to solve problems that might be deterministic in principle.

    We will focus on a simple method based on so-called *Monte Carlo*(MC) experiments or *random search*. To apply it for minimizing $f(x)$ over interval $[a, b]$, one parameter should be set: **number of samples** n (number of objective function evaluations). Algorithm 2.5 provides the complete description and all computational steps involved.

Algorithm 2.5 (Monte Carlo or Random Search Method)

1. *For given $I = [a, b]$ generate a sequence $\{x^k\}, k = 1, \ldots, n$, of random values, e.g., **uniformly** or **normally** distributed*

$$x^1 \in I, \quad x^2 \in I, \quad \ldots, \quad x^n \in I$$

2. *Compute sequentially (similar to brute-force search)*

$$
\begin{aligned}
f(x^1) \quad &then & &store: \ \bar{x} = x^1 \ \ and \ \ f(\bar{x}) = f(x^1) \\
f(x^2) \quad &then \ \ if \ f(x^2) < f(\bar{x}) \ &store: \ &\bar{x} = x^2 \ \ and \ \ f(\bar{x}) = f(x^2) \\
&\ldots & & \\
f(x^n) \quad &then \ \ if \ f(x^n) < f(\bar{x}) \ &store: \ &\bar{x} = x^n \ \ and \ \ f(\bar{x}) = f(x^n)
\end{aligned}
$$

3. *Approximate $x^* = \bar{x}$ with error*

$$e_{MC} \sim (\bar{x}_+ - \bar{x}_-), \tag{2.15}$$

where $\bar{x}_+, \bar{x}_-$ are right and left "neighbors" of $\bar{x}$

The probabilistic interpretation of the random search (Monte Carlo) method is as follows. By the *Law of Large Numbers*, the expected value of some random variable can be approximated by taking the empirical mean of independent samples of the variable (refer to works of great mathematicians Gerolama Cardano, Jakob Bernoulli, and Pafnuty Chebyshev). When applied to optimization problems, we expect that by taking more experiments (objective evaluations) we have more chances to get closer to the optimal solution. A simple illustration of this fact is provided in Figure 2.5.

There exist many schemes and modifications to perform a random search to solve optimization problems. In particular, to decrease the computational load in a unimodal case, we could store $\bar{x}$, $\bar{x}_{\text{old}}$, and respective k's to control the random process, e.g., how often $\bar{x}$ is changed.

? *Bearing in mind the stochastic nature of the random search method, may its application be practical for solving deterministic problems?*

The immediate answer might be – yes. It is useful when it is difficult or impossible to use other approaches. We could also think of several properties

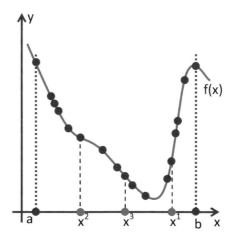

FIGURE 2.5
Schematic illustrating the random search (Monte Carlo) method.

that make the application of the Monte Carlo approach suitable for various optimization problems.

- In the same way, as for the brute-force method, the random search is also considered to be a **global search** method.

- Also, different **speeding-up** strategies exist in the forms of combined passive-active search.

- The Monte Carlo method could be easily applied to **large-scale** (multidimensional) optimization problems in case the solution space is efficiently explored or parameterized.

The **drawbacks** here also echo the brute-force search method: the computations are **very expensive**, especially for multidimensional problems, and the optimal solution may be missed if number of samples n is **not sufficiently large**.

2.7 Practical Examples

As we mentioned before, the list of methods discussed in Sections 2.2–2.6 is far from being comprehensive. We will use this small collection of algorithms to solve problems in 1D and analyze the first computational results with effortlessly processed visualization. Attempting to reveal as many computational issues as possible, we pick six different functions (locally/globally convex or concave, monotonic, periodic, nonsmooth, etc.). Figure 2.6 depicts the images obtained by running MATLAB script Chapter_2_min_1D_PlotFunctions.m (used along with the MATLAB's function fn_all_functions.m). The reader is encouraged to add other functions to this collection (by adding extra case statements in the switch block inside fn_all_functions.m) to get practical experience with analysis of methods' errors and convergence.

Now, we apply the *bisection* method to all six functions.

MATLAB: Chapter_2_min_1D_Bisection.m

Parameters used:

(f_1) fnNum = 1; a = -1; b = 3;

(f_2) fnNum = 2; a = 0; b = 5;

(f_3) fnNum = 3; a = -2.5; b = 3.5;

(f_4) fnNum = 4; a = -pi; b = 3*pi;

(f_5) fnNum = 5; a = -2; b = 4;

(f_6) fnNum = 6; a = -2; b = 2;

for all cases: delta = 0.0001; epsilon = 0.0002;

Figure 2.7 shows the results of this search: points on curves denote the optimal solutions after termination. For functions f_1, f_2, f_3, and f_6, the bisection method can find the unique solutions located either inside $[a, b]$ or on its boundary. The **unimodality** here plays the central role as nondifferentiability or nonsmoothness (function f_2) and nonconvexity (f_3 and f_6) obviously do not affect the result. However, functions f_4 (periodic) and f_5 (unimodal only in terms of its maximum) have two global minimizers, and only one is discovered for each function. It is easy to check the ability to capture other minimizers once the search intervals are modified.

? *Could we predict the presence of multiple global minimums and navigate the search interval modifications for any function?*

We will answer this question after analyzing the results of applying other search methods. We now try the *golden section search* to minimize the same functions over the same intervals using MATLAB code Chapter_2_min_1D_GoldenSection.m and keeping termination parameter $\epsilon = 0.0002$. After arriving at the same results (in terms of capturing one global minimizer), the reader has the answer for the unimodality question posed

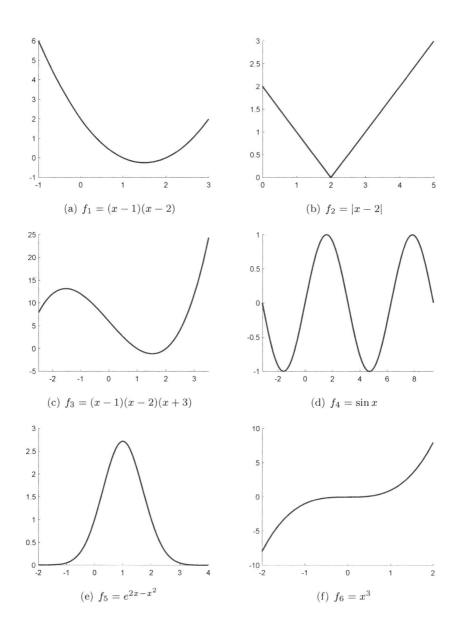

(a) $f_1 = (x-1)(x-2)$

(b) $f_2 = |x-2|$

(c) $f_3 = (x-1)(x-2)(x+3)$

(d) $f_4 = \sin x$

(e) $f_5 = e^{2x-x^2}$

(f) $f_6 = x^3$

FIGURE 2.6
Practical examples of function $f(x)$.

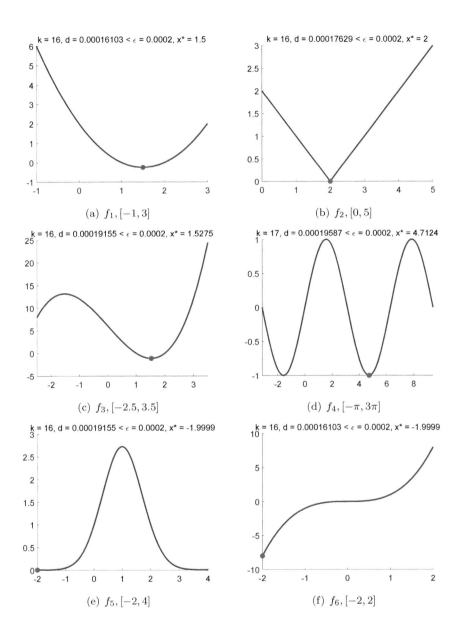

(a) $f_1, [-1, 3]$

(b) $f_2, [0, 5]$

(c) $f_3, [-2.5, 3.5]$

(d) $f_4, [-\pi, 3\pi]$

(e) $f_5, [-2, 4]$

(f) $f_6, [-2, 2]$

FIGURE 2.7
Computational results of applying bisection method to minimize $f(x)$ over $[a, b]$.

on p. 30. We also have to discuss the computational costs for both methods questioned on p. 27. Although all examples lead to the same conclusion, we choose f_3 to compare the number of objective evaluations as shown in Figure 2.8. Having $k_{GS} > k_B$, we conclude on the **superior computational performance** of the golden section method as $n_{GS} < n_B$. Obviously, the ratio n_B/n_{GS} will become larger once we allow more iterations k for search with both methods.

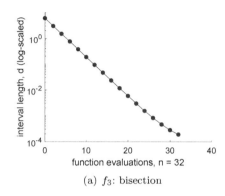

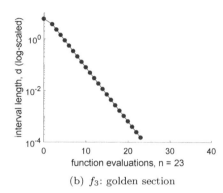

(a) f_3: bisection   (b) f_3: golden section

FIGURE 2.8
Comparison of computational performance, number of objective evaluations n, for (a) bisection and (b) golden section search methods.

Our next step is exploring the various peculiarities of *Newton's method* discussed in Section 2.4 by applying MATLAB code `Chapter_2_min_1D_Newton.m` to our six example functions and using different initial guesses x^0. First, we explore the inability to find a solution for **nondifferentiable** (at $x = 2$) f_2 with $x^0 = x^* = 2$. Even if we start the search with different guess, say $x^0 = 5$ (where f_2 is differentiable), the search cannot be completed due to $f''(x^0) = 0$. We may observe the same result in case of function f_3 and $x^0 = 0$.

MATLAB: `Chapter_2_min_1D_Newton.m`

```
• fnNum = 2; xOpt = 2;
  Attention! 1st derivative does not exist: terminated!

• fnNum = 2; xOpt = 5;
  Attention! 2nd derivative is 0 or does not exist:
  terminated!
```

- `fnNum = 3; xOpt = 0;`
  `Attention! 2nd derivative is 0 or does not exist:`
  `terminated!`

We continue with function f_3 by setting initial guess first to $x^0 = -1$ and then $x^0 = 1$. As Figure 2.9 shows, Newton's method cannot guarantee convergence if not started reasonably close to the desired minimum and may converge to the maximum point instead.

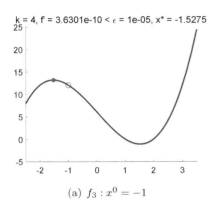

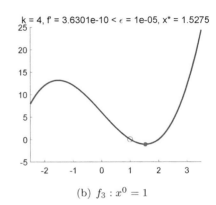

(a) $f_3 : x^0 = -1$ (b) $f_3 : x^0 = 1$

FIGURE 2.9
Convergence of Newton's method to (a) local maximum and (b) local minimum. Initial guess x^0 and optimal solution x^* are shown, respectively, by empty and filled circles.

At the same time, with the initial guess "properly placed" to enable capturing a local minimum (discussed with more theory for convergence in Chapter 5) Newton's method is really fast. Figure 2.10(a) shows its application to the quadratic function f_1 terminated after the first iteration. Re-applying the method to function f_3, locally convex on $[0, \infty]$, with $x^0 = 8$, which is really far (specifically in y-axis direction) from the optimal solution x^*, gives only 6 iterations to converge, compare Figures 2.9(b) and 2.10(b).

For our numerical experiments with Newton's method applied, we use termination criterion $|f'(x^k)| < \epsilon = 0.00001$ in MATLAB code `Chapter_2_min_1D_Newton.m`. The reader is encouraged to apply further this method to the rest functions and use various initial guesses and termination criteria from (2.11) to get more insights into its benefits, drawbacks, and general usability for different problems.

Let us move now to the passive search algorithms and discuss the *brute-force method*. We apply it, with MATLAB file `Chapter_2_min_1D_BruteForce.m`, to function f_1 with a very coarse discretization of interval $[-1, 3]$ by setting

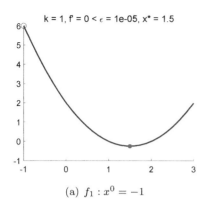

(a) $f_1 : x^0 = -1$

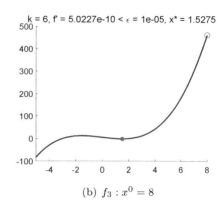

(b) $f_3 : x^0 = 8$

FIGURE 2.10
Fast convergence of Newton's method in case of (a) quadratic f_1 and (b) locally convex f_3 functions. Initial guess x^0 and optimal solution x^* are shown, respectively, by empty and filled circles.

the number of inner points $n = 3$. Figure 2.11(a) shows the geometrical interpretation of the method to provide the uniform approximation of $f(x)$ by linear splines (dashed line). This figure also illustrates the idea of global search or global exploration of the interval $[-1, 3]$ completed with a very rough approximation of the unique solution x^*. Figure 2.11(b) represents the results of minimizing function f_4 (with $n = 50$) to locate one of two global minimums. This solution is fairly accurate when compared with the results in Figure 2.7(d) obtained with $k = 17$ (34 function evaluations). Another observation is that even pretty modest discretization of the search interval enables the brute force method not only to solve the minimization problem but also to provide an approximated shape of the function over the entire interval. The analysis of this shape may offer different perspectives: e.g., to conclude if a function has several modes, and how to modify the search intervals to locate multiple minimizers. The main concern is still the computational costs discussed earlier.

Finally, we apply the random search by the *Monte Carlo* approach to minimize example functions with the number of samples $n = 50$. Due to the stochastic nature of this method, the returned result after each attempt approximate the solution x^* with different accuracy. For example, running MATLAB code `Chapter_2_min_1D_MonteCarlo.m` 10 times to minimize unimodal function f_1 gives 10 different values:

$$1.4802 \quad 1.4696 \quad 1.6237 \quad 1.4345 \quad 1.5010$$
$$1.4810 \quad 1.5571 \quad 1.6005 \quad 1.4613 \quad 1.5573$$

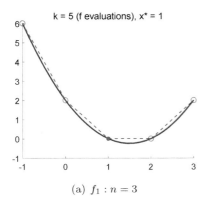
(a) $f_1 : n = 3$

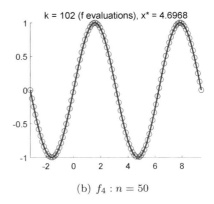
(b) $f_4 : n = 50$

FIGURE 2.11
Brute force search method applied to functions f_1 and f_4.

Applying the Monte Carlo algorithm to a function with multiple modes, in fact, proves it as a global search method enabled to discover multiple global minimizers. Two different attempts are shown in Figure 2.12. Similar to the brute force approach, the random search also suffers from being inaccurate if the number of samples n (function evaluations) is not sufficiently large.

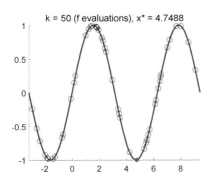

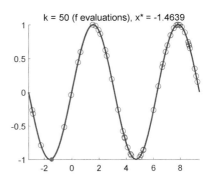

FIGURE 2.12
Random search (Monte Carlo) method applied to function f_4.

We close this section by referring back to question on p. 37. The last two methods (brute force and random search) prove, at least computationally, that investing in the global exploration of large search intervals has the potential to navigate modifications of these intervals. As such, a combination of different methods may be used successfully to optimize multimodal functions with several global solutions. For example, we may consider any passive search method for global (even rough) exploration, and then fast active search methods applied to all intervals where potential solutions are located.

2.8   Computational Analysis for Convergence

In the previous section, we discussed at length the performance of various computational algorithms in terms of computational time measured through the required number of objective function evaluations. Such assessment for performance is fair when compared algorithms are very different by their very nature. For instance, some methods may converge in a few iterations with numerous function calls per iteration and vice versa for some other approaches. In addition, we may want to check if the method's implementation is correct and all parameters are tuned up to expect the maximum performance generally claimed for this class of computational approaches.

To apply such analysis, we refer to our discussion on the convergence rate started in Section 1.6. At this time, we consider this notation from the computational point of view and advise the reader to review formula (1.29). We omit the norms and will use the following formulation when applied to any results obtained in the previous section, in 1D:

$$|e^{k+1}| = C|e^k|^r, \tag{2.16}$$

where r is the *rate of convergence* and C is the *constant*. After applying $\log_{10}$ to both parts of equation (2.16)

$$\log_{10} |e^{k+1}| = \log_{10} C + r \cdot \log_{10} |e^k|$$

and denoting

$$x = \log_{10} |e^k|, \quad y = \log_{10} |e^{k+1}|, \quad b = \log_{10} C,$$

we arrive at a linear function

$$y = b + rx. \tag{2.17}$$

Coefficients b and r in (2.17) may be easily approximated by MATLAB's polyfit function (using linear regression analysis). While in this analysis of *computational convergence* the numerical value for C does not play a significant role, approximated rate r is meaningful. If r is close to 1, it corresponds to the *linear convergence*. Rate r close to 2 identifies the method with *quadratic convergence*, etc. Again, as r values are approximated from the real computations that provide a sequence of solutions $\{x^k\}$, these rates are not expected to be ideally equal to 1, 2, 3, etc. Moreover, even small changes in rate r may give a reader some food for thought: correctness of the implementation, the right choice of parameters, suitability for the current problem, and possible space for improvement. Let us illustrate the application of this analysis with some examples.

Example 2.3 Analysis of the Computational Convergence
Apply computational convergence analysis to function $f_3 = (x-1)(x-2)(x+3)$ minimized by the bisection, golden section, and Newton's methods. Use termination tolerance $\epsilon = 0.0001$ for all methods.

Function $f_3 = (x-1)(x-2)(x+3)$ has the local minimizer $x^* = \sqrt{7/3} \approx 1.5275$ (the same as the global minimizer over interval $[-2.5, 3.5]$). MATLAB code `fn_convergence.m` provides the following function.

MATLAB: fn_convergence(xEx,xOpt)

Input parameters:

- xEx: x^*, if not provided (xEx = []), the last value in vector xOpt is used instead

- xOpt: $\{x^k\}$

Output:

- plot of available points $(\log_{10}|e^k|, \log_{10}|e^{k+1}|)$ with the line of linear regression (2.17)

- rate r and constant C (in the figure's title)

Function `fn_convergence(xEx,xOpt)` is called at the end of execution in three MATLAB codes

- Chapter_2_min_1D_Bisection.m

- Chapter_2_min_1D_GoldenSection.m

- Chapter_2_min_1D_Newton.m

Figures 2.13(a,b) provide the computational analysis for the convergence in applications of the bisection and the golden section search methods to function f_3. Both methods clearly show near-linear convergence. The bisection method converges a bit faster than the golden section search, $r_B = 1.0406$ vs. $r_{GS} =$

0.94021. We may explain this by the presence of one outlier point in the left bottom corner. In addition to this, we use the exact solution $x^* = \sqrt{7/3}$ "not visible" by all members of the solution sequence $\{x^k\}$, which converges (computationally) to its last value, x^{last}.

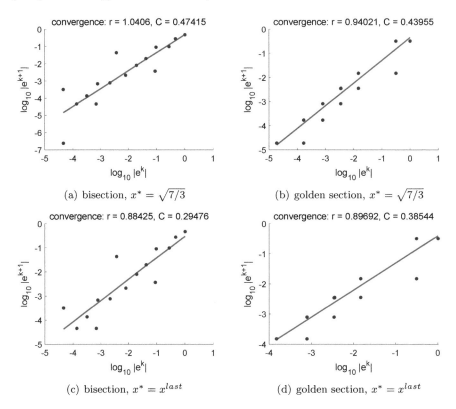

FIGURE 2.13
Computational analysis for the convergence in applications of (a,c) bisection and (b,d) golden section search methods when (a,b) $x^* = \sqrt{7/3}$ and $x^* = x^{last}$.

Application of this analysis in cases when the true solution is not known requires removing x^{last} from $\{x^k\}$ to be used in place of x^*. Results obtained in such a way may be considered even more trustful as it will reflect the natural convergence to a solution found by the method before its termination. Figures 2.13(c-d) provide the results when $x^* = x^{last}$. Here, we notice about the same rate of convergence, $r \approx 0.9$, expected for both methods. For all future numerical experiments, we will use the same concept to maintain consistency.

Now let us apply the computational analysis for the convergence in the case of using Newton's method. Figure 2.14 presents the results when we start the search from two different initial guesses: $x^0 = 1$ and $x^0 = 8$, refer to

Figures 2.9(b) and 2.10(b). Computational rates of convergence are, respectively, 2.0728 and 1.7821. It is reasonable taking into account the ability of Newton's method to show its claimed 2-order (quadratic) convergence only when started near the optimal solution.

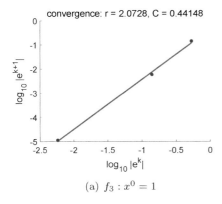

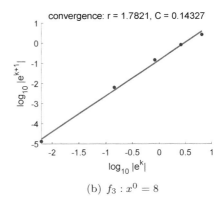

(a) $f_3 : x^0 = 1$ (b) $f_3 : x^0 = 8$

FIGURE 2.14
Computational analysis for the convergence in applications of Newton's method to function f_3.

Finally, we extend Example 2.3 by adding a numerical experiment to check the performance of Newton's method when it converges to an inflection point, e.g., $x = 0$ for f_6. As shown in Figure 2.15, we start with $x^0 = -1$ (pretty close to the inflection point) and exhibit convergence in 10 iterations with rate $r = 1.0527$ (linear convergence, refer to the discussion in Section 2.4). We close this section by advising the reader to apply the computational analysis discussed here to the rest functions and also by adding two final notes.

(a) The rate of convergence analysis is applicable for cases with three or more iterations (at least two points on the regression analysis graph). Otherwise, MATLAB function fn_convergence.m will generate the message: Convergence analysis cannot be completed due to the data size!

(b) Any obvious outliers on the regression analysis graph may be removed manually to make the analysis results more accurate.

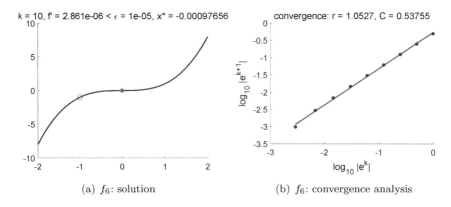

(a) f_6: solution   (b) f_6: convergence analysis

FIGURE 2.15
Convergence to the inflection point of function f_6 while applying Newton's method.

2.9 Homework Problems

1. Prove formulas (2.3) and (2.7) for estimating the number of iterations/function evaluations to terminate, respectively, the bisection and golden section search methods.

2. Explain the error estimations for all methods in (2.2), (2.6), (2.12), (2.14), and (2.15).

3. Apply all 1D search methods by running MATLAB codes for all six practical examples in Section 2.7, see Figure 2.6. Explain the obtained solutions and results of computational convergence analysis (whenever applicable).

4. Add functions from Examples 2.1 and 2.2 and your own ones to the collection practical examples by modifying MATLAB functions fn_all_functions.m and fn_all_derivatives.m. Apply 1D search methods, explain the obtained solutions and results of computational convergence analysis (whenever applicable).

5. Modify MATLAB code Chapter_2_min_1D_BruteForce.m to enable capturing multiple global minimums.

6. Modify MATLAB code Chapter_2_min_1D_MonteCarlo.m to generate a sequence $\{x^k\}$ of **normally** distributed values x^k. Explain the difference in the solutions obtained with normal and uniform distributions of random variable.

7. Modify `MATLAB` code `fn_convergence.m` to identify and remove outliers from the linear regression analysis. Compare the results before and after this removal, and conclude.

| READ | **Where to Read More**

Press (2007), [26]
    Chapter 9 (Root Finding and Nonlinear Sets of Equations), Chapter 10 (Minimization or Maximization of Functions)

| RUN | **MATLAB Codes for Chapter 2**

- `Chapter_2_min_1D_PlotFunctions.m`

- `Chapter_2_min_1D_Bisection.m`

- `Chapter_2_min_1D_GoldenSection.m`

- `Chapter_2_min_1D_Newton.m`

- `Chapter_2_min_1D_BruteForce.m`

- `Chapter_2_min_1D_MonteCarlo.m`

- `fn_all_functions.m`

- `fn_all_derivatives.m`

- `fn_convergence.m`

2.10   Lab Assignment #1: Review Chapters 1–2

Problem 1: Modify MATLAB code `Chapter_1_data_fit.m` for Example 1.2 on p. 6 to be able to fit **exactly** $m > 1$ data points by $P_{m-1}(x)$ polynomial. New code requirements:

- Input data should be downloaded from file `data.dat` (use command `data = load('data.dat');`).

- Modify all three solution approaches: (a) using plain m-code, (b) user-defined m-function, and (c) `MATLAB`'s built-in function `polyfit`.

- Check `format long` MATLAB's functionality to see if different methods (a)–(c) return different numbers and explain the results.

- Implement check-up to prevent `Inf/NaN` problem in case the data is "defective" by displaying the error message.

Show the results for the following test cases: (a) $m = 2$, (b) $m = 5$, (c) $m = 10$, (d) $m = 5$ with any portion of "defective" data.

Problem 2: Explore MATLAB code `Chapter_2_min_1D_Newton.m` that implements **Newton's (2nd-order)** method to solve 1D unconstrained optimization problem to minimize/maximize twice differentiable function $f(x)$ and perform computational analysis for rate of convergence using `MATLAB`'s `polyfit` function (see Sections 2.4 and 2.8).

1. Apply this code to find the local maximum of function $f(x) = (x-1)(x-2)(x+3)(x-4)$ by utilizing $f'(x)$ and $f''(x)$ obtained analytically with initial guess (a) $x_0 = -1$, (b) $x_0 = 1$, (c) $x_0 = 3$, and (d) $x_0 = 5$ to check convergence and to approximate convergence parameters r and C.

2. Find the initial guess(es) x_0 for which Newton's method will not converge. Show the result and explain.

3. Modify the code to obtain $f'(x)$ and $f''(x)$ using any finite difference (FD) approximation. Check the convergence and approximate convergence parameters r and C for $x_0 = 5$ when FD uses (a) $\Delta x = 10^{-4}$, (b) $\Delta x = 10^{-2}$, (c) $\Delta x = 10^{-1}$. Compare the results and make a conclusion.

For all computations, use the following termination parameter $\epsilon = 10^{-6}$.

3

Generalized Optimization Framework

This chapter initiates a discussion about the general structure and main computational elements of the generalized optimization framework. The concept of building an efficient system supported by various algorithms is illustrated by the problem of parameter identification for least-squares data fitting based on the example from Chapter 1. A solution for this 3D optimization problem is derived using a gradient-based iterative approach applied computationally. We compare the performance of practical computations using 1-order steepest descent and 2-order Newton's method supplied with the step of constant size and provided by the golden section search reviewed in Chapter 2. We also discuss the importance of visualization to analyze the quality of obtained solutions and the performance of the entire optimization framework. Another focus of this chapter is on testing implemented algorithms and dealing with various problems during the debugging phase. For example, we introduce the kappa-test for checking the correct computations for gradients. Finally, we review the structure and communication between multiple parts of our enhanced computational framework to allow further development most easily and optimally.

3.1 Parameter Identification for Least-Square Data Fitting

In this chapter, we discuss fundamental concepts commonly used while creating optimization frameworks consisting of multiple computational elements. Therefore, we advise the reader to revisit Example 1.3 on p. 10 as our first computational structure will be built around this simple problem. Later, we will continuously enhance this framework (in a component-wise manner) to allow solutions of increased complexity.

In short, we assume that we have m data points

$$(x_i, y_i), \quad i = 1, \ldots, m$$

to construct a model equation

$$y_f(x) = a_1 + a_2 x + a_3 x^2, \tag{3.1}$$

DOI: 10.1201/9781003275169-3

presumably describing a process associated with this data. Constant numbers a_1, a_2, and a_3 are the parameters in model $y_f(x)$ to be identified, while pursuing the best data fit in the "least-squares" sense. Here, we deal with the problem commonly known as *parameter identification* (or *parameter reconstruction*) for *least-squares data fitting*.

While we start with a piece of theory, our general approach is to consider the following *constrained optimization* (minimization) problem

$$\min_{\mathbf{a}} \sum_{i=1}^{m} (y_i - y_{f,i})^2$$
$$\text{s.t. } y_{f,i} = a_1 + a_2 x_i + a_3 x_i^2, \quad i = 1, \ldots, m \tag{3.2}$$

where we compare given data (y_i) with the data provided by model $y_f(x)$ in (3.1). Figure 3.1 schematically illustrates the least-squares data fitting concept.

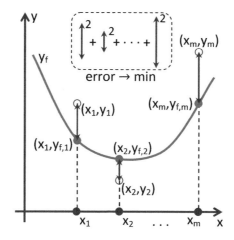

FIGURE 3.1
Schematic illustrating the least-squares data fitting concept.

As briefly discussed in Section 1.3, a computational form of the solution uses the residual vector $\mathbf{r} \in \mathbb{R}^m$ for m "pieces" of data

$$\mathbf{r} = \mathbf{y} - A\mathbf{a} = \begin{bmatrix} y_1 - (a_1 + a_2 x_1 + a_3 x_1^2) \\ y_2 - (a_1 + a_2 x_2 + a_3 x_2^2) \\ \cdots \\ y_m - (a_1 + a_2 x_m + a_3 x_m^2) \end{bmatrix}, \tag{3.3}$$

where

$$
A = \begin{bmatrix} 1 & x_1 & x_1^2 \\ 1 & x_2 & x_2^2 \\ \cdots & \cdots & \cdots \\ 1 & x_m & x_m^2 \end{bmatrix}, \quad \mathbf{y} = \begin{bmatrix} y_1 \\ y_2 \\ \cdots \\ y_m \end{bmatrix}.
$$

It allows us to convert problem (3.2) into its equivalent form of the *unconstrained optimization* problem

$$
\min_{\mathbf{a} \in \mathbb{R}^3} f(\mathbf{a}) \tag{3.4}
$$

with objective function

$$
f(\mathbf{a}) = \|\mathbf{r}\|^2 = r_1^2 + r_2^2 + \cdots + r_m^2 = \sum_{i=1}^m \left(y_i - (a_1 + a_2 x_i + a_3 x_i^2) \right)^2, \tag{3.5}
$$

where 3D vector $\mathbf{a} = [a_1 \, a_2 \, a_3]^T$ plays a role of the *control* (or *optimization*) *variable*. Using the discussion results for the problem in Example 1.2 on p. 6, we summarize possible outcomes in three cases.

- *Case 1: m = 3*, $y_1 \neq y_2 \neq y_3$. A **unique** solution could be found exactly.

- *Case 2: m = 1, 2*. The problem exhibits **infinitely many** solutions.

- *Case 3: m > 3*. The uniqueness of the solution **depends on data**.

Our particular interest is to solve optimization problem (3.4)–(3.5) iteratively starting from some initial guess $\mathbf{a}^0$ for control vector $\mathbf{a}$. Then, all obtained solutions could be easily compared with the exact solution $\mathbf{a}_{ex} = [-15 \ 10 \ -1]^T$ known back to Example 1.2.

By using the basics of multivariable calculus we compute the *gradient* of objective function $f(\mathbf{a})$ with respect to control vector $\mathbf{a}$

$$
\frac{\partial f}{\partial \mathbf{a}} = \nabla_{\mathbf{a}} f = \begin{bmatrix} \frac{\partial f}{\partial a_1} \\ \frac{\partial f}{\partial a_2} \\ \frac{\partial f}{\partial a_3} \end{bmatrix} = \begin{bmatrix} \sum_{i=1}^m 2 \left(y_i - (a_1 + a_2 x_i + a_3 x_i^2) \right) \cdot (-1) \\ \sum_{i=1}^m 2 \left(y_i - (a_1 + a_2 x_i + a_3 x_i^2) \right) \cdot (-x_i) \\ \sum_{i=1}^m 2 \left(y_i - (a_1 + a_2 x_i + a_3 x_i^2) \right) \cdot (-x_i^2) \end{bmatrix}. \tag{3.6}
$$

Then we could arrive at optimal solution $\mathbf{a}^*$ by using any *gradient-based* (e.g., *steepest descent*) iterative approach

$$
\mathbf{a}^{k+1} = \mathbf{a}^k + \alpha^k \cdot \mathbf{d}^k \tag{3.7}
$$

with search direction obtained by gradient (3.6)

$$
\mathbf{d}^k = -\nabla_{\mathbf{a}} f(\mathbf{a}^k) \tag{3.8}
$$

and optimal step size α^k computed by using one of the 1D minimization methods discussed in Chapter 2. For terminating the iterative search, we may use one of the criteria discussed in Section 1.5.2, e.g., a relative decrease of the objective function

$$
\left| \frac{f(\mathbf{a}^{k+1}) - f(\mathbf{a}^k)}{f(\mathbf{a}^k)} \right| < \epsilon. \tag{3.9}
$$

3.2   Generalized Optimization Framework

We will solve the problem discussed in the previous section computationally after a brief discussion on the structure of the optimization framework, its components, and the choice of proper software used to perform computations.

3.2.1   Computational Components

In Figure 3.2, we summarize the main structure of the *generalized optimization framework* commonly used for solving many problems. Before starting our discussion on its components (computational elements), we have to notice to the reader that this structure is conditional subject to used (general) methodology, description of the solved problem, etc. Due to the same reason, some components and connections may not be present. However, we will use this structure as a template leaving the reader the freedom to modify it to add more flexibility, complexity and to explore possibilities for enhanced optimality. Anyway, an optimization framework should show some structure; let us discuss it.

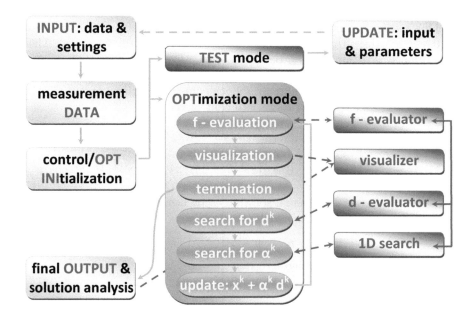

FIGURE 3.2
Computational elements of the generalized optimization framework.

Initialization. A few steps represent the initial phase of almost all algorithms to solve optimization problems. First, we need to supply all the involved methods with input data and settings to govern their work (see **INPUT** block on the diagram in Figure 3.2). The best practice is to organize this input in one place: in the separate file(s) (preferable) or at the very beginning of the solution code. The specific location for this information helps various users, including the code developer, check all input parameters without accessing, or even understanding, the rest of the entire framework for finding and changing individual settings. To be consistent with this rule of thumb, we will keep all parameters in the `MATLAB` files with the same name `params.m` in all our future examples.

It may be reasonable to keep some data, e.g., obtained as measurements, etc., in the separate file(s) (**DATA** block, Figure 3.2), especially when the size is substantial. It may also come from outside of the framework, e.g., as output from external software. The last step is to initialize the controls and the solution process itself (**OPT INI** block on the diagram). Simply, it includes all other necessary operations to start optimization: allocating memory for all variables (states, controls, parameters), setting controls to their initial guess values, initializing all other variables, creating function handlers, etc. The reader may check `MATLAB` files `params.m`, `data_main.m`, and `initialize.m` used in Chapter 3 for getting a better understanding of the *initialization procedure*.

TEST mode. It is a specific facility within the optimization framework to check the correctness of implementing various algorithms and the accuracy of associated computations. Adding this facility requires extra effort and time. Many users usually underestimate the importance of this tool, and, as a result, overlook an opportunity to catch multiple errors before allowing them to penetrate massive portions of computations. In Section 3.6, we provide an example of the TEST mode functionality for various gradient-based frameworks to check the correctness and accuracy of computed gradients. The results of running the algorithms in the TEST mode may also require updates on the input data and selected parameters (**UPDATE** block, Figure 3.2). Sometimes, these updates run in several loops until the result satisfies the requirements for accuracy and reliability.

OPT(imization) mode. Obviously, it is the central part of the constructed optimization framework. The reader may refer to Algorithm 1.1 in general and the gradient-based solution mentioned in Section 3.1 in particular for understanding the structure of the OPT mode (see also **OPT** block on the diagram of Figure 3.2). Again, the procedures and their order may change due to the optimization concept in use. We consider first the objective function evaluation (f-evaluation). For the initial iteration ($k = 0$), it gives a very first statement on the objective, while later ($k > 0$), it enables comparison between f-values and confirms the progress. f-evaluation usually completes updating all control and state variables. Consequently, the visualization and the termination check (for $k > 0$) procedures are natural after this. The core part of the optimization algorithm is changing optimization (or control) variables by finding the solution update. This part, again, depends on the chosen concept. We refer to the gradient-based methodology, so the next step is to find the search direction $\mathbf{d}^k$ and the size of the step α^k made in that direction (in the order depending on the algorithm, see Chapters 5 and 6). As seen in Figure 3.2, the OPT mode requires several computational facilities, such as f-evaluator, visualizer, d-evaluator (gradient-based evaluator), 1D search for optimal step size, etc. We will discuss the proper choice of software to fulfill these requirements in the next section.

Finally, if the termination criteria stop the search, the final output and the analysis of the obtained solutions are initiated (see **OUTPUT** block on the diagram in Figure 3.2). Depending on the problem, this stage may require formal messages on completed computations, saving parts of the solution in files, additional visualization for solution analysis (e.g., convergence), etc. Proper coding for this part is essential, as re-running the entire process for missing details in the final output may be very time-consuming.

3.2.2   Choice of Proper Software

Here, we shift the main focus on the OPT module of the generalized optimization framework to enable discussion on its structure and the proper choice of all components, see Figure 3.3. In pursuit of practicality, we usually start by choosing a particular type of software to create the central (core) part of this module. Ideally, the selected computer package or program is **self-contained** – it might be specialized software to solve a particular problem. However, many ready-to-use programs face difficulties when applied to specific needs or modified to fit the peculiarities of the optimization problem(s) the users try to solve. We also have to keep in mind that it may become critical if the selected software is a black box with limited ability to expand its functionalities whenever the problem solution may require it. Thinking proactively and acting in advance of a future situation, the best idea would be to decide on the core software to provide efficient **communication** and **data processing** functionality.

Besides communication and data processing, the core software may provide options for other functionalities (e.g., f- and d-evaluators, visualizer, etc.). Exploring these options in terms of working characteristics and expandability will lead to the conclusion of their suitability. The reader could also think about multiple options for the same functionality included in the same framework as it adds more flexibility for future experiments. For instance, the primary purpose of the f-evaluator in the framework is to evaluate objective function(s) $f(\mathbf{x})$. At the same time, it may require to solve (systems of) (non)linear equation(s), ODE(s), PDE(s), etc. The choice of this solver is crucial as it is very problem-dependent and requires arrival at some **trade-off** point between being fast and accurate. The reader may consider, e.g., switching between solving a single ODE and a system of PDEs to check the ability

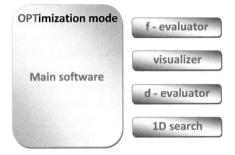

FIGURE 3.3
Generalized optimization framework: main (core) software part.

of the chosen software to support this change with minimal effort and time to incorporate.

We could continue with the d-evaluator in the same context. Its major purpose is to support iterative search by computing the search direction(s) $\mathbf{d}$. It also may require solving (systems of) (non)linear equation(s), ODE(s), PDE(s). In addition to this, we have to mention its necessity to **communicate effectively** with the f-evaluator to re-utilize its solutions.

The proper choice of facilities for 1D search is even more critical. It is used extensively in multidimensional optimization problems to find **optimal** step size α. As shown by multiple examples in Chapter 2, choosing the suitable method depends on the problem and its known properties: differentiability, convexity, constraints, etc. Also, as discussed above, it may require effective **communication** with f- and d-evaluators. As such, including a collection of 1D search methods (not just one) into the framework may benefit its flexibility, while finding the best fit for the solved problem.

The choice of visualizer is also very important as we may choose between the one provided by the same software used in optimization (not necessarily of good quality) and adding as an external facility from other packages. Generally speaking, it should allow a priori analysis of input data and a posteriori one for the obtained solutions. In addition to this, visualized solutions during the iterations help to control the progress of the entire optimization algorithm. Ideally, it should **not slow down** or **interrupt** the optimization process by having *fast and easy access* to stored intermediate data.

We close this discussion by providing several examples of commonly used core software platforms.

- `MATLAB` with access to parallel computing and various add-on toolboxes for mathematical, statistical, machine learning, image processing, and optimization functionalities.

- C++-based scientific environments with added libraries for linear algebra, solving PDEs, optimization, etc., e.g., FreeFEM [16].

- Separate solvers available in common formats for MATLAB, C++, PYTHON®, FORTRAN, etc.

3.3   Choosing and Adjusting Optimization Algorithms

In this section, we start creating a computational framework to solve the problem of Example 1.3 using the iterative gradient-based approach discussed earlier in Section 3.1. Here and for all examples later, we will use MATLAB as a core software platform (main software, refer to Figure 3.3). We start with the simplest version of the framework by templating all its computational elements with their simplest forms. The reader may explore this template by comparing MATLAB file Chapter_3_data_fit_by_gradient.m with the suggested structure of the generalized optimization framework given in Figure 3.2. Table 3.1 summarizes the results of this comparison.

MATLAB: Chapter_3_data_fit_by_gradient.m

```
close all; clc; clear;                                              1
                                                                   2
% setting INPUT parameters                                          3
params;                                                            4
                                                                   5
% loading DATA                                                      6
data = load(dataFile);                                             7
                                                                   8
% INITialization                                                    9
initialize;                                                        10
                                                                   11
% main OPTimization loop                                            12
while(k < kMax+1) % termination condition #2                       13
                                                                   14
  % f-evaluation                                                    15
  obj = [obj f(a, data)];                                          16
                                                                   17
  % visualization                                                   18
  visualize;                                                       19
                                                                   20
  % termination condition #1: checking optimality (by tolerance)    21
  if k > 0                                                          22
    err = abs(obj(end-1)-obj(end))/abs(obj(end-1));               23
    if (err < epsilon)                                             24
```

```
      break;                                          25
    end                                               26
  end                                                 27
                                                      28
  % search for d: computing gradient                  29
  d = -grad(a,data);                                  30
                                                      31
  % search for alpha                                  32
  alpha = alphaConst;                                 33
                                                      34
  % update for controls                               35
  a = a + alpha*d;                                    36
                                                      37
  % iteration counter increment                       38
  k = k + 1;                                          39
                                                      40
end                                                   41
```

element	implementation	software
main OPT	written manually *lines 13–41*	MATLAB
f-evaluator	m-function, analytically defined function $f(\mathbf{a})$ *fn_eval_f.m *called in line 16 via* *function handler defined in* initialize.m	MATLAB
d-evaluator	m-function, analytically defined gradient $\nabla_{\mathbf{a}} f(\mathbf{a})$ *fn_eval_grad.m *called in line 30 via* *function handler defined in* initialize.m	MATLAB
1D α search	constant value, $\alpha = $ const *line 33*	—
visualizer	plain m-code *visualize.m *called in line 19*	MATLAB

Refer to the code printout above.

TABLE 3.1
Computational elements of Chapter_3_data_fit_by_gradient.m.

Now, let us play with this framework to check its functionality and discuss ways for assessing and adjusting its performance. To keep the main focus on it, here we use the search direction $\mathbf{d}$ obtained through the analytically defined gradient $\nabla_{\mathbf{a}} f(\mathbf{a})$, as shown in (3.8), and the constant value for the step size α in (3.7). However, we explore more options for both $\mathbf{d}$ and α with associated discussions on the performance in Chapters 5 and 6. First, we suggest running MATLAB code Chapter_3_data_fit_by_gradient.m with the following settings in the `params.m` file.

MATLAB: params.m

Initial settings:

- *input data:* `dataFile = 'data_main.dat';`

- *initial guess:* `aini = [1 1 1];`

- *step size:* `alphaConst = 1e-3;`

- *termination #1:* `epsilon = 1e-6;` refer to (1.22)

- *termination #2:* `kMax = 5;` refer to (1.23)

We provide data in data-file `data_main.dat` exactly as given in Example 1.3. Solution analysis made in Example 1.2 confirms that vector $\mathbf{a}_{ex} = [-15\ 10\ -1]^T$ is a unique solution for this problem. We start the search from initial guess $\mathbf{a}^0 = [1\ 1\ 1]^T$, and make steps of sizes $\alpha^k = \alpha = 10^{-3}$ in the direction opposite to gradient $\nabla_{\mathbf{a}} f(\mathbf{a})$. We expect termination of this search when the relative change in the objective function $f(\mathbf{a})$ is less than $\epsilon = 10^{-6}$ or after completing 5 iterations, whatever comes first.

? *Why does it diverge (refer to Figure 3.4)?*

Let us assume that our f- and d-evaluators are properly implemented and work correctly (we will review these issues later in Sections 3.5–3.6 of this chapter). As the current solution approach has the simplest form, there are only two more options to associate the possible problem: initial guess and step size. We advise the reader to change the numbers a_1, a_2 and a_3 in the initial guess for solution vector $\mathbf{a} = [a_1\ a_2\ a_3]^T$ and conclude that this does not affect the result. Next, we experiment with various options for step size α by setting them consequently to $10^{-4}, 10^{-5}, 10^{-6}$, see Figure 3.5 for results (we also change $k_{\max}$ to 50 to observe convergence for all three processes).

? *Now it converges, what about performance?*

Results from Figures 3.4 and 3.5 explain the issue. Constant step size $\alpha = 10^{-3}$ is too big to allow approaching the optimal solution using the gradient-based direction. Making it smaller, 10^{-4}, shows a much better result, especially for the first 4-5 iterations. Then the iterative process enters the so-called *plateau* region with $f \approx 24$, where it continues to converge to the optimal solution but obviously at a much lower rate. Taking steps of even smaller sizes, 10^{-5} and 10^{-6}, also provides some progress seen in the solution

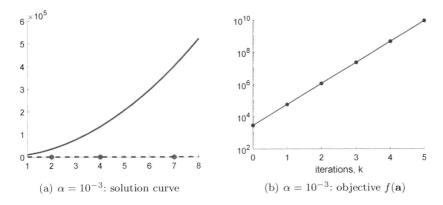

(a) $\alpha = 10^{-3}$: solution curve

(b) $\alpha = 10^{-3}$: objective $f(\mathbf{a})$

FIGURE 3.4
Divergence while running `Chapter_3_data_fit_by_gradient.m` with step size $\alpha = 10^{-3}$. Dots represent the data points; dashed and solid lines show solution curves obtained, respectively, with initial solution $\mathbf{a}^0$ and solution $\mathbf{a}^k$ after termination.

curve "traveling" from its initial shape (dashed line) "toward" the data points. Here, we observe a situation when making smaller steps, meaning providing the solution with more accurate updates, dissonates with an idea to get this solution faster. An apparent conclusion would be to find an optimal step size to guarantee the trade-off between being reasonably large and preventing divergence. As we will see from future examples, this optimality requires varying α for different iterations. Usually, it becomes smaller when we get closer to the optimal solution. And it is natural, as large steps could not allow "landing" on the exact values or get into the ϵ-neighborhood of that solution.

Now, when we establish the convergent iterative process, we could "help" the solution to converge "faster" by moving the initial guess closer to $\mathbf{a}_{ex} = [-15 \ 10 \ -1]^T$. Let us make it $\mathbf{a}^0 = [-14 \ 11 \ -2]^T$ and allow 5,000 iterations to check the ability of our algorithm to discover the optimal solution. As we see in Figure 3.6, now it does. The obtained solution $\mathbf{a}^* = [-14.6684 \ 9.8378 \ -0.98338]^T$ is not accurate to the extent that we probably expected. The objective function has not reached 0, meaning the data is not matched perfectly. However, we have confirmed that the computational framework, its first and very rough implementation, eventually works! For the rest of this chapter, we will try to answer, at least partially, the following question.

? *What could be done to check and increase further the performance?*

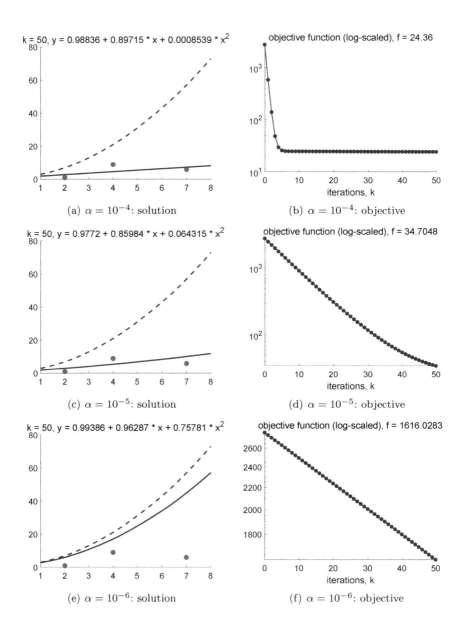

FIGURE 3.5

Convergence with step sizes $\alpha = 10^{-4}, 10^{-5}, 10^{-6}$ (refer to Figure 3.4 for notations).

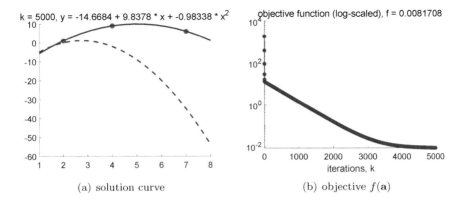

(a) solution curve

(b) objective $f(\mathbf{a})$

FIGURE 3.6
Convergence with $\alpha = 10^{-4}$ and $\mathbf{a}^0 = [-14\ 11\ -2]^T$ (refer to Figure 3.4 for notations).

3.4   Visualization and Analysis of Obtained Solutions

In Section 3.3, we obtained the first computational results for solving the problem of Example 1.3 using a simplified iterative gradient-based optimization framework. This framework has already proved its ability to converge to the known (unique) solution by using heuristic approaches for choosing the step size and initial guess. Although we already used some plotting results provided by MATLAB m-code `visualize.m` included in the framework, more findings are available by investing in visualization facilities and the analysis of the obtained solutions. Before getting back to practical computations, we have to discuss the data, depending on problems, that could and, potentially, should be visualized.

(a) Data to analyze optimization progress via **objective function**:

   (1) measurement data, e.g., if compared with the modeled data,

   (2) separate parts of the objective (how closely data is fitted),

   (3) change of the entire objective with iteration number k, e.g., to check monotonicity.

(b) Data to analyze optimization progress via **optimization (control) variables**:

   (1) "true" solution used, if available, to generate (synthetic) measurements, then forgotten,

   (2) current solution obtained after kth iteration,

(3) some measures for assessing how close they are: here, we note that monotonicity may not be expected.

(c) Other **optimization attributes**, e.g.,

(1) gradients,

(2) state variables (if different from control variables),

(3) dynamic parameters (optimal step size, weighting coefficients, etc.), and

(4) other parameters to control the performance of added optimization techniques (regularization, preconditioning, etc.).

This list, of course, is far from being comprehensive. Depending on the nature and complexity of the problem, the reader may freely add more options for visualization both dynamically when optimization is still in process and after it is complete. We also add a few more remarks as food for thought before the big project starts.

- It is very useful to save intermediate and final data in any **data format** instead of graphical images. Re-running optimization from scratch may take some time for updating these images or generating new ones.

- As now many graphing software packages are available, it is convenient to keep all data in **easily convertible formats**, e.g., in DAT or TXT files.

- It might be helpful to consider converting various data into high-resolution images employing external software instead of investing time into developing visualization facilities within the computational framework used for the optimization process.

To illustrate the concept of analyzing obtained solutions and optimization performance via visualization, we refer the reader to this functionality in MATLAB code `Chapter_3_data_fit_by_gradient.m` we used in Section 3.3. Figure 3.7 depicts four plots updated dynamically during optimization: (a) solution curves with data points, (b) objective $f(\mathbf{a})$ as a function of iteration count k, (c) structure of the current gradient-based search direction $\mathbf{d} = -\nabla_{\mathbf{a}} f(\mathbf{a})$, and (d) the history of step size α changes with k. This figure reflects the results of solving modified Example 1.3: the number of data points increased from 3 to 6, and the initial guess is set to the exact solution of original Example 1.3. The updated `params.m` file is shown here.

MATLAB: params.m

Updated settings:

- *input data:* `dataFile = 'data_6pt.dat';`

- *initial guess:* `aini = [-15 10 -1];`

- *step size:* `alphaConst = 1e-4;`

- *termination #1:* `epsilon = 1e-6;`

- *termination #2:* `kMax = 100;`

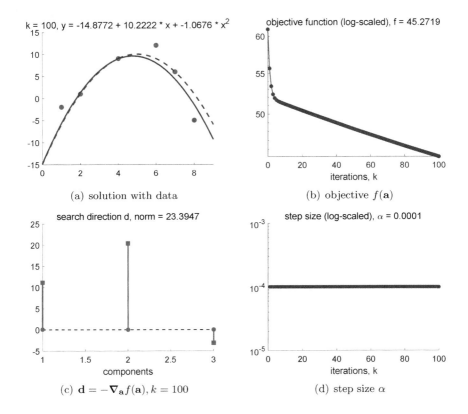

(a) solution with data (b) objective $f(\mathbf{a})$

(c) $\mathbf{d} = -\nabla_{\mathbf{a}} f(\mathbf{a}), k = 100$ (d) step size α

FIGURE 3.7
`MATLAB` window with four plots updated dynamically while solving modified
Example 1.3 with $m = 6$ data points.

In the same vein, as we discussed changes in the solution curve and numerical values of the objective in Section 3.3, the last example also shows multiple signatures of convergence. The same discussion applies to Figures 3.7(a,b). The gradient (search direction) structure is also subject to some useful analysis; thus, any fancy-style plotting of it may be included, as shown in Figure 3.7(c). First of all, visualized components of vector $\mathbf{d}$ demonstrate the

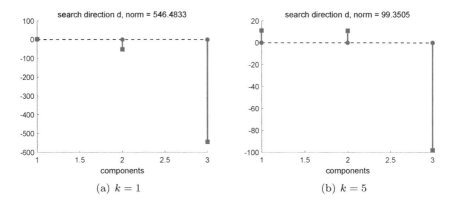

(a) $k = 1$ (b) $k = 5$

FIGURE 3.8
Modified Example 1.3 with $m = 6$ data points: **d**-components structure.

pattern currently used by the optimization framework to update all controls.
For instance, as seen in Figure 3.7(c), before termination ($k = 100$), optimization variables a_1 and a_2 were tried to be increased and a_3 – decreased. We
also note that the associated numbers provide the scheme of how this update
will be "distributed" among a_1, a_2, and a_3. Actual updates for the individual
controls are obtained from the components of search direction **d** after scaling
them with the chosen step size α.

Another interesting observation could be made while observing the gradient (search direction) change due to iterations: compare **d** structure obtained when $k = 1, 5$, Figure 3.9, and $k = 100$, Figure 3.7(c). The range of
d-component amplitudes changes from $[-600, 100]$ to $[-100, 20]$ and then finally to $[-5, 25]$. It means the optimization process eventually converges to
the local (in our problem, global) optimal solution characterized by the opti-
mality condition $\nabla_\mathbf{a} f(\mathbf{a}) = \mathbf{0}$. The easier way to control the convergence from
this point is to compute the norm $\|\mathbf{d}^k\| = \|\nabla_\mathbf{a} f^k(\mathbf{a})\|$, diminishing through
iterations (hopefully down to 0, but not necessarily uniformly). This analysis
(in the form of the 1-order optimality condition, see Chapter 5 for more details) also may be used to define a new termination condition in addition to
those discussed previously in Section 1.5.2.

As the last, but not the least, point of discussion on visualization here, we
would like to add more on the computational analysis for the convergence. In
the same manner, as done for multiple examples in Chapter 2, this facility
is included in MATLAB code `Chapter_3_data_fit_by_gradient.m` as the post-
processing tool. We advise the reader to review the concept of this analysis
applied to 1D optimization problems in Section 2.8 and given by (2.16)–(2.17).
Here, we perform optimization in 3D ($\mathbf{a} \in \mathbb{R}^3$) and have to be back to the
equivalent of the generalized form given by (1.29) also using $\|\cdot\|_2$ (Euclidean

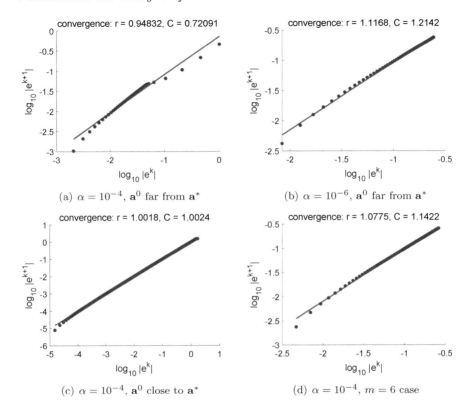

(a) $\alpha = 10^{-4}$, $\mathbf{a}^0$ far from $\mathbf{a}^*$

(b) $\alpha = 10^{-6}$, $\mathbf{a}^0$ far from $\mathbf{a}^*$

(c) $\alpha = 10^{-4}$, $\mathbf{a}^0$ close to $\mathbf{a}^*$

(d) $\alpha = 10^{-4}$, $m = 6$ case

FIGURE 3.9
Convergence analysis for cases shown in (a) Figures 3.5(a,d), (b) Figures 3.5(c,f), (c) Figure 3.6, and (d) Figure 3.7.

distance in $\mathbb{R}^n$) norm.

$$\lim_{k \to \infty} \frac{\|\mathbf{e}^{k+1}\|}{\|\mathbf{e}^k\|^r} = \lim_{k \to \infty} \frac{\|\mathbf{a}^{k+1} - \mathbf{a}^*\|_2}{\|\mathbf{a}^k - \mathbf{a}^*\|_2^r} = C, \quad C < \infty. \tag{3.10}$$

It requires minimal changes in the m-function used before: formally, use of MATLAB's `norm` function (in the new m-function `fn_convergence_sol_norm.m` to be used throughout entire Chapter 3) instead of `abs` (used before in `fn_convergence.m` for Chapter 2). Also, while exploring the convergence computationally we apply the idea of unknown optimal solution $\mathbf{a}^*$ by following the same, as in Section 2.8, concept, i.e.,

$$\mathbf{a}^* = \mathbf{a}^{last}.$$

Finally, Figure 3.9 collects the results of the computational convergence analysis applied to multiple cases we played within Sections 3.3 and 3.4 (refer to Figures 3.5 through 3.7). Numerical values for rate r vary slightly about 1, confirming 1-order or *linear convergence* for all cases. It justifies computationally the known fact: changes in the step size or initial guess cannot move the order of convergence (e.g., from *linear* to *quadratic* one) beyond the limits established by the computational method used to update controls. Here, we use the gradient-based 1-order *steepest descent* (SD) approach, which is consistent with the results in Figure 3.9. To make the convergence faster, we may consider two options.

(1) Investing further in the optimal step size search, e.g., by applying various 1D search methods discussed in Chapter 2.

(2) Changing the method itself, e.g., considering 2-order Newton's approach, see Section 2.4 for review.

? *What would be the best option for our current Example 1.3? For other problems?*

We review both options in application to Example 1.3 in Section 3.7.

3.5 Testing and Dealing with Problems (Debugging)

In Section 3.6, we will discuss a testing facility (TEST mode) that is suitable for various optimization frameworks employing the gradient-based methodology. Here, we briefly outline some procedures, commonly referred to as *debugging*, used for testing the general functionality, accuracy of the performed computations and for dealing with possible problems (coding "bugs") while running the framework.

We will start with f-evaluators. Among other tests, three of them are considered useful for checking the correctness of implementation and getting valuable insights into possible improvements for the accuracy of the provided computations.

- **Test case #1.** Choose a model with know solution, $\mathbf{x}^*$ and $f^* = f(\mathbf{x}^*)$, and run new f-evaluator with $\mathbf{x} = \mathbf{x}^*$ to check if $f = f^*$.

- **Test case #2.** If $f \neq f^*$, check your ability to **control** $|f - f^*| \to 0$ (meaning to get **gradually** closer to 0) by tuning solver parameters (refining mesh, applying higher-order schemes, etc.). Here, we note that due to many possible computational issues (finite discretization, accumulated numerical errors, etc.), getting **arbitrarily** close to 0 is often impossible.

- **Test case #3.** Run other trustful and commonly used *benchmark models* and compare the outcomes with available results, e.g., published in scientific journals, posted on the research websites, etc.

We could run test case #1 for Example 1.3 using MATLAB code `Chapter_3_data_fit_by_gradient.m` with the following settings in `params.m` file.

MATLAB: params.m

Settings for test case #1:

- *input data:* `dataFile = 'data_main.dat';`

- *initial guess:* `aini = [-15 10 -1];`

- *step size:* `alphaConst = 1e-4;` may be any

- *termination #1:* `epsilon = 1e-6;` may be any

- *termination #2:* `kMax = 1;` to allow checking gradient/search direction

---

As expected, the solution curve fits the data ideally. Objective function $f(\mathbf{a})$ returns the zero value, and all components of the search direction vector $\mathbf{d}$ are zeros, as seen in Figure 3.10. Although passing test case #1 (in fact, all test cases) **cannot guarantee** the absence of bugs in the f-evaluator, it helps find coding mistakes, if any. For even more experiments, the reader is advised to "spoil" MATLAB file `fn_eval_f.m` by purposely adding one or more "mistakes" (changing $\pm$ signs, messing up coefficients, etc.) and look into the results of test case #1 again. This practice provides a good experience for dealing with various problems to benefit, while working with structures at much more complicated levels.

As Figure 3.10(b) shows, running test case #1 for the f-evaluator contributes partially toward checking the correct implementation of the d-evaluator. In Section 3.6 we describe a complete test case suitable for various gradient-based methods. It consists of running "kappa-test" to check the gradient is accurate and consistent with its FD approximation. In general, debugging procedures for checking and tuning d-evaluators for accurate work are very problem- and method-dependent.

Finally, we could share a few steps used for testing the OPT part of the framework.

- Test every component **separately**. Usually, the general concept "change one part at a time" is applied here.

- Test **communication** within the entire framework: variables, dimensions of vectors/matrices, names, solution files, etc.

- **Tuning test.** For the same problem, change one parameter or technique at a time to see the sensitivity of the performance to this particular change.

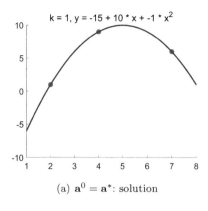
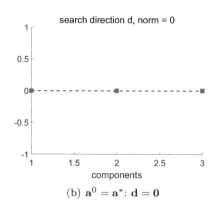

(a) $\mathbf{a}^0 = \mathbf{a}^*$: solution   (b) $\mathbf{a}^0 = \mathbf{a}^*$: $\mathbf{d} = \mathbf{0}$

FIGURE 3.10
Test-case #1 for f-evaluator.

- **Robustness test.** For a fixed set of parameters/techniques, run the framework for the same problem varying initial data; then explore the results and repeat tuning (if necessary).

- **Applicability test.** Apply a working framework to problems of complexity at different levels (low, moderate, high).

3.6   TEST Mode for the Gradient-based Framework

Figure 3.11 illustrates the general concept of the TEST mode in terms of parts of the generalized framework included in the testing facility.

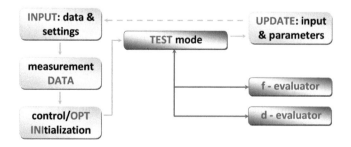

FIGURE 3.11
Generalized optimization framework: parts included in the TEST mode.

Before proceeding to multidimensional cases, we start with an example of applying a similar strategy to the 1D case and consider a 1-order FD approximation of derivative $f'(x)$

$$f'(x) \approx \frac{\Delta f}{\Delta x} = \frac{f(x + \Delta x) - f(x)}{\Delta x},$$

where Δx is a fixed (finite) step taken to re-evaluate $f(x)$ with the new argument $x + \Delta x$. To make Δf arbitrarily small, we scale step Δx with parameter $\epsilon \to 0$

$$f'(x) \approx \frac{f(x + \epsilon \Delta x) - f(x)}{\epsilon \Delta x}. \tag{3.11}$$

Then we introduce the ratio, *kappa-test* value, to compare the derivative obtained by its FD approximation with the one evaluated using an available analytical expression

$$\kappa = \frac{f(x + \epsilon \Delta x) - f(x)}{\epsilon \, \Delta x \, f'(x)} \to 1. \tag{3.12}$$

This value obviously should be close to 1 if

(1) *f*-evaluator works correctly and returns accurate values of function $f(x)$, **and**

(2) *d*-evaluator works correctly and returns accurate values of derivative $f'(x)$, **and**

(3) parameter ϵ is sufficiently small.

We could easily consider an extension of the kappa-test for use in multidimensional cases, $\mathbf{x} \in \mathbb{R}^n$,

$$\kappa(\epsilon) = \frac{f(\mathbf{x} + \epsilon \, \delta \mathbf{x}) - f(\mathbf{x})}{\epsilon \, \langle \nabla_{\mathbf{x}} f(\mathbf{x}), \delta \mathbf{x} \rangle}, \quad \epsilon \to 0, \tag{3.13}$$

where $\delta \mathbf{x} = [\Delta x_1 \; \Delta x_2 \; \ldots \; \Delta x_n]^T$ is the *perturbation* applied (componentwisely) to current solution $\mathbf{x}$. In (3.13), $\nabla_{\mathbf{x}} f(\mathbf{x})$ is the gradient of $f(\mathbf{x})$ defined analytically with respect to control $\mathbf{x}$ and evaluated for the current value of $\mathbf{x}$. We also use the notation $\langle \cdot, \cdot \rangle$ to define the inner product.

For practical reasons, we define two types of kappa-test.

(a) *Cheap test.* It requires **two** *f*-evaluations. For fixed $\delta \mathbf{x}$, e.g., $\delta \mathbf{x} = \mathbf{x}$, compute $\kappa(\epsilon)$ for a range of ϵ, e.g., $\epsilon = 10^{-12} \div 10^2$.

(b) *Expensive test.* It requires $n + 1$ *f*-evaluations. For fixed ϵ, e.g., $\epsilon = 10^{-6}$, perform kappa-test to check sensitivity for every component of $\mathbf{x}$ by changing $\delta \mathbf{x}$ in the following manner:
step #1: $\quad \delta \mathbf{x} = [x_1 \; 0 \; 0 \; \ldots \; 0]^T,$
step #2: $\quad \delta \mathbf{x} = [0 \; x_2 \; 0 \; \ldots \; 0]^T,$
$\ldots$
step #n: $\quad \delta \mathbf{x} = [0 \; 0 \; 0 \; \ldots \; x_n]^T.$

We also have to mention that choosing $\delta\mathbf{x} = \mathbf{x}$ is meaningful. The magnitudes of components in vector $\mathbf{x}$ may vary, sometimes by order. The suggested strategy helps apply the kappa-test to all components more a less uniformly.

Now we could check the correctness of implementing our algorithm for computing gradients using analytical formula in (3.6) (in MATLAB file fn_eval_grad.m) applied then for defining search direction $\mathbf{d}$. To explore the functionality of the TEST mode only, MATLAB code Chapter_3_data_fit_by_gradient.m has been modified by removing the OPT mode parts. The updated version is provided in the m-files Chapter_3_data_fit_by_gradient_test.m and kappa_test.m. The reader may also use the initial version of params.m file provided on p. 60 used to solve the problem in Example 1.3.

Figure 3.12 provides the results of running a cheap version of the kappa-test. It initially suggests that the gradient is computed **correctly**. By looking closely at the plateau region formed by the graph in Figure 3.12(a), we may identify the range of parameter ϵ (9-10 orders of magnitude) for those values where the κ value is very close to 1. Figure 3.12(b), in fact, provides more insights into this analysis: quantity $\log_{10}|\kappa(\epsilon) - 1|$ shows how many significant digits of accuracy are captured in gradient evaluation. We also comment on the regions where $\kappa(\epsilon)$ deviates from the unity (to the left and right sides of the plateau regions) as these effects are well-known. For very small values of ϵ subtractive cancelation (roundoff) errors start playing a significant role, while large ϵ values entail truncation errors.

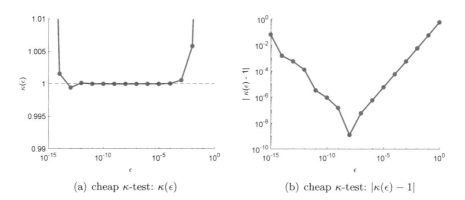

(a) cheap κ-test: $\kappa(\epsilon)$ (b) cheap κ-test: $|\kappa(\epsilon) - 1|$

FIGURE 3.12
Results of cheap kappa-test.

The results of the expensive version of the kappa-test are shown in Figure 3.13(a). It confirms the correctness of all three gradient components. To perform these computations, we use $\epsilon = 10^{-8}$ identified in Figure 3.12(b) as the best value to proceed with the component-wise sensitivity analysis (accuracy).

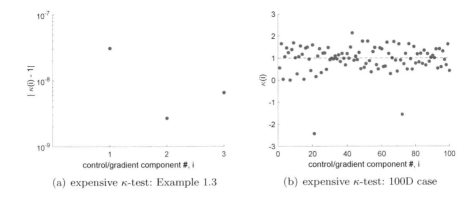

(a) expensive κ-test: Example 1.3   (b) expensive κ-test: 100D case

FIGURE 3.13
Results of expensive kappa-test.

? *What are the reasons to employ a computationally costly expensive version of the kappa-test?*

The answer is simple: it is for easy problem identification. The kappa-test run at its expensive mode allows evaluating the accuracy of computed gradients and associated sensitivities provided by single controls. In a multidimensional case, when n is large, sensitivity coming from controls with inaccurate gradient components may be overwhelmed by the sensitivity collected from other controls "presented" correctly in the gradient. The cheap version of the kappa-test in such a case will provide deceiving results. However, running the expensive version will help to look into the problem and locate the controls with incorrect or inaccurately computed gradient components.

Figure 3.13(b) shows how it typically appears, e.g., in a 100D case (Example 9.2 of Chapter 9) where some control/gradient components provide κ-values far outside the main "cloud" positioned naturally around 1. These components are not necessarily inaccurate, but their appearance requires some **reasonable explanation**. We close this discussion with the last comment that both tests (cheap and expensive) may be repeated periodically throughout the optimization process to control error or loss of sensitivity to avoid their propagation into the final solution.

3.7 Accuracy and Performance

After completing the tests discussed in Sections 3.5 and 3.6 for the correctness of implementing our f- and d-evaluators (for solving the problem of

Example 1.3), in the current section, we deal with the problem of improving the performance to answer the question posed on p. 68. First, review the computational elements of MATLAB code Chapter_3_data_fit_by_gradient.m provided in Table 3.1. The updated version Chapter_3_data_fit_by_gradient_ver_2.m allows using the *golden section* algorithm golden_section_search.m discussed in Section 2.3 and adapted from Chapter_2_min_1D_GoldenSection.m to perform the 1D search for optimal step size α; refer to Table 3.2 for a summary of implemented changes.

element	implementation	software
main OPT	written manually	MATLAB
f-evaluator	m-function, analytically defined function $f(\mathbf{a})$	MATLAB
d-evaluator	m-function, analytically defined gradient $\nabla_{\mathbf{a}} f(\mathbf{a})$	MATLAB
1D α search	**plain m-code for GS search**	MATLAB
visualizer	plain m-code	MATLAB

TABLE 3.2
Computational elements of Chapter_3_data_fit_by_gradient_ver_2.m.

In the context of our current optimization framework, implementing the GS method to run a 1D search for finding optimal α provides the mechanism commonly referred to as a *line minimization* search (LMS). Formally, at every optimization iteration k we have to find optimal step size α^k by solving a separate 1D constrained minimization problem

$$\alpha^k = \underset{\alpha > 0}{\text{argmin}} \; f\left(\mathbf{a}^k + \alpha \cdot \mathbf{d}^k\right). \tag{3.14}$$

In other words, solving problem (3.14) provides optimal scaling of vector $\mathbf{d}^k$ which provides the direction – "line." This scaling is **optimal** in terms of minimizing objective function $f(\mathbf{a}^k)$ at the next $(k+1)$ step. Here, we have chosen to experiment with the golden section search method to solve (3.14). However, other 1D methods, including those discussed in Chapter 2 could be easily adopted for the same issue, e.g., bisection, brute-force, Monte Carlo, etc.

To perform practical computations with the updated framework, we also have to modify two procedures:

(a) **initialization:** updated MATLAB code initialize_ver_2.m now contains the section for initializing GS search parameters, and

(b) **input settings:** updated MATLAB code params_ver_2.m now contains a new section for setting parameters for step size search via GS.

The added settings related to the golden section search include the search interval $[a, b]$ and termination parameter ϵ_α. We may choose to search within interval $[0, 0.01]$ and play with different termination options in the `params.m` file, see below.

MATLAB: params.m

Settings for golden section search:

- *input data:* `dataFile = 'data_main.dat';`

- *initial guess:* `aini = [-14 11 -2];`

- *search interval:* `alphaA = 0; alphaB = 0.01;`

- *tolerance for GS search termination:* `alphaEps = 1e-2;`

- *termination #1:* `epsilon = 1e-6;`

- *termination #2:* `kMax = 10;`

? *Running the optimization with these settings, including $\epsilon_\alpha = 10^{-2}$, will give a diverging result. Why?*

The condition set to terminate the golden section search is $b^k - a^k < \epsilon_\alpha$ (here, k is the iteration count in the GS algorithm), review the algorithm in Section 2.3. With the current settings, $b - a = \epsilon = 10^{-2}$ and the search is terminated immediately giving optimal $\alpha^k = b = 10^{-2}$ at every iteration. In that way, the golden section is equivalent to using constant α. Therefore, the proper initialization requires $\epsilon \ll b - a = 10^{-2}$. Figure 3.14 shows the results for running MATLAB code `Chapter_3_data_fit_by_gradient_ver_2.m` with the golden section search for optimal α terminated with $\epsilon_\alpha = 10^{-4}$ and $\epsilon_\alpha = 10^{-6}$.

The compared results in Figure 3.14 open some space for discussing parameters in the GS approach, e.g., if we have to change the search interval and termination parameter. We will see approximately the same pattern for sequence $\{\alpha^k\}$ if we allow optimization in both cases to run for $k > 10$. Also, $\alpha^k = 10^{-2}$ appears just once; it means having 10^{-2} as the upper bound for the search interval is appropriate. Evidently, with $\epsilon_\alpha = 10^{-6}$, the golden section method is better aligned with the gradient-based search by periodically providing α^k values close to 10^{-2}. It allows gradient-based optimization to move faster toward the optimal solution. However, this tolerance requires more objective function evaluations for every solution of problem (3.14). The number of f-evaluations is 21 opposing to 11 when $\epsilon_\alpha = 10^{-4}$ is in use; refer to (2.7) for proving these results.

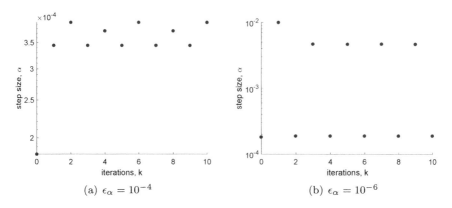

(a) $\epsilon_\alpha = 10^{-4}$ (b) $\epsilon_\alpha = 10^{-6}$

FIGURE 3.14
Results of applying golden section search for finding optimal step size α.

Now we fix ϵ_α in the GS method to 10^{-6} and compare convergence for cases when α^k is chosen from interval $[0, 10^{-2}]$ or preset to a constant value of 10^{-4}. The former is illustrated in Figure 3.15, and we refer the reader to Figures 3.6 and 3.9(c) for the latter. We stop optimization in the former case at $k = 200$. The result is approximately the same as that obtained at $k = 5,000$ when we used the constant value 10^{-4} for step size α. Even if the GS search takes 21 f-evaluations versus none for the case of constant α, the gradient-based algorithm converges 25 times faster when aligned with the golden section search. However, based on the computational convergence analysis, this algorithm is still in the group of 1-order methods.

? *How to improve further the performance of the GS method? Could we consider some "flexibility" in choosing a and b for different iterations?*

We will be back to the performance of the GS method and answer these questions in Chapter 6.

So far, answering the question on p. 68 about the best option for fast convergence, we considered optimal scaling of search direction vectors $\mathbf{d}^k$ obtained from computed gradients $\nabla_{\mathbf{a}} f(\mathbf{a}^k)$. Another option is to modify this direction itself, e.g., as in the case of *Newton's method*; refer to Section 2.4 for details. Here, we briefly review the implementation of this method applicable for multidimensional problems.

First, at every optimization iteration k, we evaluate gradient $\nabla_{\mathbf{a}} f(\mathbf{a}^k)$, e.g., following (3.6) while solving the problem in Example 1.3. Then we have to compute the *Hessian matrix*. For the same example, it takes the following

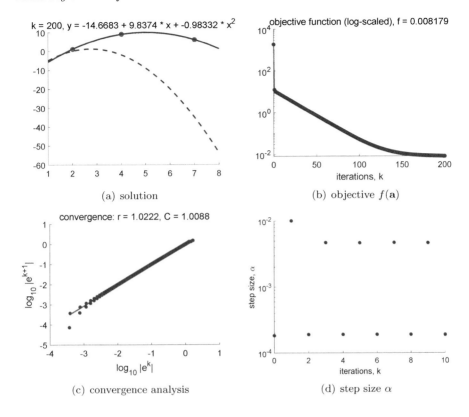

(a) solution

(b) objective $f(\mathbf{a})$

(c) convergence analysis

(d) step size α

FIGURE 3.15
Results of applying golden section search for finding optimal $\alpha^k \in [0, 10^{-2}]$ with $\epsilon_\alpha = 10^{-6}$.

form.

$$\nabla_{\mathbf{a}}^2 f(\mathbf{a}) = \begin{bmatrix} \frac{\partial^2 f}{\partial a_1 \partial a_1} & \frac{\partial^2 f}{\partial a_1 \partial a_2} & \frac{\partial^2 f}{\partial a_1 \partial a_3} \\ \frac{\partial^2 f}{\partial a_2 \partial a_1} & \frac{\partial^2 f}{\partial a_2 \partial a_2} & \frac{\partial^2 f}{\partial a_2 \partial a_3} \\ \frac{\partial^2 f}{\partial a_3 \partial a_1} & \frac{\partial^2 f}{\partial a_3 \partial a_2} & \frac{\partial^2 f}{\partial a_3 \partial a_3} \end{bmatrix} = 2 \cdot \begin{bmatrix} m & \sum_{i=1}^m x_i & \sum_{i=1}^m x_i^2 \\ \sum_{i=1}^m x_i & \sum_{i=1}^m x_i^2 & \sum_{i=1}^m x_i^3 \\ \sum_{i=1}^m x_i^2 & \sum_{i=1}^m x_i^3 & \sum_{i=1}^m x_i^4 \end{bmatrix} \tag{3.15}$$

The steepest descent search direction in (3.8) is then upgraded by using inverted Hessian from (3.15), i.e.,

$$\mathbf{d}^k = -\left[\nabla_{\mathbf{a}}^2 f(\mathbf{a}^k)\right]^{-1} \nabla f(\mathbf{a}^k). \tag{3.16}$$

In general, Newton's method works well with step size $\alpha^k = 1$. Let us check it together with its entire functionality in application to our problem in Example 1.3. The updated version of MATLAB code `Chapter_3_data_fit_by_gradient_ver_3.m` allows using Newton's method by adding m-function `fn_eval_hess.m` for Hessian computations as discussed above; refer to Table 3.3 for a summary of all implemented changes.

element	implementation	software
main OPT	written manually	MATLAB
f-evaluator	m-function, analytically defined function $f(\mathbf{a})$	MATLAB
d-evaluator	**m-function, analytically defined**	
	$\boldsymbol{\nabla}_{\mathbf{a}} f(\mathbf{a})$ **and** $\left[\boldsymbol{\nabla}_{\mathbf{a}}^2 f(\mathbf{a}^k)\right]^{-1}$	MATLAB
1D α search	**not required**	—
visualizer	plain m-code	MATLAB

TABLE 3.3
Computational elements of `Chapter_3_data_fit_by_gradient_ver_3.m`.

To start computations with the modified framework, we have to update three procedures:

(a) **initialization:** updated MATLAB code `initialize_ver_3.m` now contains a new handler for the function to compute Hessians and an option to choose between golden section search or constant value for the step size α (boolean variable `alphaUpdate` is `true` if GS is chosen),

(b) **termination:** a new condition, relative decrease in the solution norm by (1.20)

$$\frac{\|\mathbf{a}^{k+1} - \mathbf{a}^k\|_2}{\|\mathbf{a}^k\|_2} < \epsilon_2, \tag{3.17}$$

is added as relative decrease in the objective by (1.22)

$$\left|\frac{f(\mathbf{a}^{k+1}) - f(\mathbf{a}^k)}{f(\mathbf{a}^k)}\right| < \epsilon_1 \tag{3.18}$$

may fail if $f(\mathbf{a}^{k+1}) = f(\mathbf{a}^k) = 0$ (we choose $\epsilon_1 = \epsilon_2 = 10^{-6}$ but, in general, these tolerances may be set to different values), and

(c) **input settings:** updated MATLAB code `params_ver_3.m` now reflects these changes, see below.

MATLAB: `params.m`

Settings for Newton's method:

- *input data:* `dataFile = 'data_main.dat';`

- *initial guess:* `aini = [-14 11 -2];`

- *step size search:* `alphaUpdate = false;` declining update by GS and setting $\alpha^k = 1$

- *termination #1:* `epsilonF = 1e-6;` for ϵ_1, refer to (3.18)

- *termination #2:* `epsilonA = 1e-6;` for ϵ_2, refer to (3.17)

- *termination #3:* `kMax = 100;`

The results of running `MATLAB` code `Chapter_3_data_fit_by_gradient_ver_3.m` with initial guess $\mathbf{a}^0 = [-14\ 11\ -2]^T$ are shown in Figure 3.16.

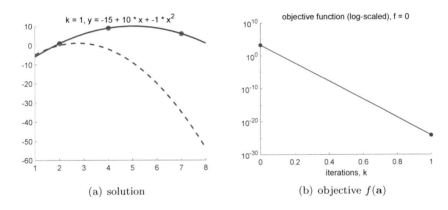

(a) solution   (b) objective $f(\mathbf{a})$

FIGURE 3.16
Results of applying Newton's method with $\mathbf{a}^0 = [-14\ 11\ -2]^T$.

? *Optimal solution* $\mathbf{a}^* = [-15\ 10\ -1]^T$ *is found in one iteration. Why?*

We already discussed this issue in Section 2.4. In addition, the reader will find more arguments and discussion points on the properties of various line search algorithms in Chapter 5.

Now let us explore the functionality of Newton's method a bit more. We encourage the reader to run `MATLAB` code `Chapter_3_data_fit_by_gradient_ver_3.m` one more time with an activated option for using golden section search to update step size α (by setting `alphaUpdate = true;` and, e.g., `alphaA = 0;` `alphaB = 10;` `alphaEps = 1e-6;` in `params_ver_3.m`). Check that α returns the optimal value in very close proximity to 1, e.g., 1.000000052835619, for the same settings.

> **!** *Use* MATLAB*'s* `format long` *functionality to access more digits after the decimal point.*

Finally, we could run the same Newton's method (with α preset to 1) to explore convergence in case the initial guess is far enough from the optimal solution, e.g., $\mathbf{a}^0 = [-100\ 1000\ 250]^T$, or with different data, e.g., $m = 6$ (`data_6pt.dat`), as used before for modified Example 1.3. Figure 3.17 shows the results of both experiments and allows us to conclude on the ability of Newton's method to converge in **one iteration** (computationally, in one or two iterations) if applied to **quadratic problems**.

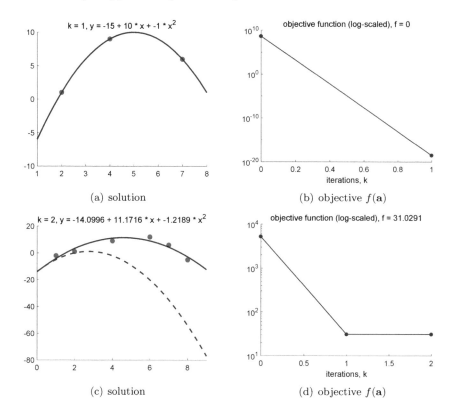

(a) solution   (b) objective $f(\mathbf{a})$

(c) solution   (d) objective $f(\mathbf{a})$

FIGURE 3.17
Results of applying Newton's method with (a,b) $\mathbf{a}^0 = [-100\ 1000\ 250]^T$, and (c,d) $m = 6$ data from `data_6pt.dat`.

We close this section with the following question answered by discussions initiated in Chapter 5.

? *How will convergence of Newton's method change if gradient $\nabla_{\mathbf{a}} f(\mathbf{a}^k)$ and Hessian $\nabla_{\mathbf{a}}^2 f(\mathbf{a}^k)$ are computed for quadratic/nonquadratic problems using any FD approximations?*

3.8 Communication within the Framework

We devote the last section of this chapter to a brief discussion on establishing effective communication between multiple parts of our optimization framework to allow its further development most easily and optimally. As the reader will see in the next chapters, this framework will continue its growth: we will add more optimizations methods, computational algorithms, associated parameters, initialization settings, etc. To implement some changes to our MATLAB code Chapter_3_data_fit_by_gradient.m, we created several new versions:

(a) **version 2** with added the golden section search for the step size α,

(b) **version 3** with implemented 2-order Newton's method for search direction **d**, and

(c) **test version** for performing the kappa-test to check the correctness of computed gradients.

The reader may probably notice that the approach we used for adding new functionalities by creating isolated versions is **not ergonomic**. If we keep going in the same direction, we will end up with numerous versions without a chance of using a combination of methods added separately by different code versions. Now we have to take another look at the structure of our framework enhanced quite a lot to the end of this chapter.

As discussed previously and also exemplified in Figure 3.18, an efficient structure of this framework is obtained by grouping multiple operations into blocks (computational parts), e.g., initialization, optimization, testing, visualization, etc. It helps in easy understanding this structure, making easier debugging and performance tunning processes. Our main goal is to allow our framework to grow inside itself by including new methods and functionalities within the structure of the same building concept and easiness.

Therefore, at this point, we stop creating new versions devoted to implementing single algorithms. Instead, we revise the general (block-wise) structure of our computational framework to enable its natural growth by establishing proper communication between its parts (blocks) and keeping it as easy as possible. This communication should allow newly added parts to integrate freely within the entire framework without any necessity to change the communication concept. Simply, this concept means "parameterizing" all

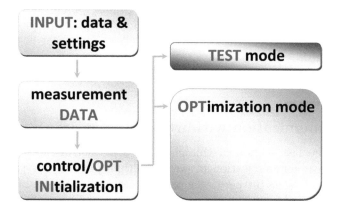

FIGURE 3.18
Generalized optimization framework: communication between parts.

existing modes, methods, functions by giving them unique and self-explanatory names, e.g.,

- **modes:** OPT, TEST,

- **methods:** SD, NEWTON, ..., future and your own methods,

- α-**search:** const, GS, ..., other algorithms,

- other parts (we will consider it later): main solver, regularization, etc.

These names, and thus associated modes, methods, functions with their settings described during the initialization phase, will be available within the whole framework to call them if the computation process needs it. Following this concept for proper communication, we create the final version of the existing MATLAB code naming it Chapter_3_data_fit_by_gradient_ver_final.m.

MATLAB: Chapter_3_data_fit_by_gradient_ver_final.m

```
close all; clc; clear;                                          1
                                                                2
params_ver_final;          % setting INPUT parameters           3
                                                                4
data = load(dataFile);     % loading DATA                       5
                                                                6
initialize_ver_final;      % INItialization                     7
```

```
                                                                        8
if strcmp(mode,'OPT')      % choosing mode OPT/TEST, based on           9
   mode_OPT;               % Chapter_3_data_fit_by_gradient_ver_3.m    10
elseif strcmp(mode,'TEST')                                             11
   mode_TEST;              % Chapter_3_data_fit_by_gradient_test.m     12
else                                                                   13
   disp(['error: Unknown mode ' mode ' is chosen!']); return;         14
end                                                                    15
```

We suggest the reader to explore the structure of the upgraded MATLAB code to familiarize with the general concept of splitting all computations into blocks and naming the parts of the entire framework and separate methods, functions, etc. We will use this structure for future experiments and further development.

3.9   Homework Problems

1. Run MATLAB code Chapter_3_data_fit_by_gradient.m to experiment with $m > 3$ (modified Example 1.3 using steepest descent and constant step size α) for different parameters α, k_{max}, and initial guess $\mathbf{a}^0$. Check the performance based on the analysis of the visualized solutions: solution curves, objective function, search direction (gradient structure), parameters for the computational convergence.

2. Modify MATLAB code Chapter_3_data_fit_by_gradient.m to use any FD approximations of $\nabla_{\mathbf{a}} f(\mathbf{a}^k)$ for the SD method. For constant step size α, check the convergence and approximate convergence parameters r and C for both cases: analytically defined and FD-approximated gradients $\nabla_{\mathbf{a}} f(\mathbf{a}^k)$. Compare the results and make a conclusion.

3. Modify MATLAB code Chapter_3_data_fit_by_gradient_ver_2.m and repeat the previous experiments (Problem 2) now with optimal step size α chosen by using the GS method.

4. Modify MATLAB code Chapter_3_data_fit_by_gradient_ver_3.m and apply Newton's method to check the convergence and approximate convergence parameters r and C for both cases: analytically defined and FD-approximated gradients $\nabla_{\mathbf{a}} f(\mathbf{a}^k)$ and Hessians $\nabla_{\mathbf{a}}^2 f(\mathbf{a}^k)$. Compare the results and conclude on the convergence when using 1-order, 2-order, mixed-order (e.g., 2-order for gradient and 1-order for Hessian) approximations.

5. Explore the structure of Chapter_3_data_fit_by_gradient_ver_final.m (the upgraded MATLAB code discussed in Section 3.8) to incorporate computations for FD-approximated gradients $\nabla_{\mathbf{a}} f(\mathbf{a}^k)$ and Hessians $\nabla_{\mathbf{a}}^2 f(\mathbf{a}^k)$.

Discuss the proper communication concept applied for using FD approximations throughout the entire framework.

6. In `Chapter_3_data_fit_by_gradient_ver_final.m`, upgrade the procedure for finding optimal step size α^k by solving 1D minimization problem (3.14) using the bisection, brute-force, and Monte Carlo methods.

READ **Where to Read More**

Press (2007), [26]
Chapter 9 (Root Finding and Nonlinear Sets of Equations), Chapter 10 (Minimization or Maximization of Functions), Chapter 15 (Modeling of Data)

RUN **MATLAB Codes for Chapter 3**

- `Chapter_3_data_fit_by_gradient.m`

- `Chapter_3_data_fit_by_gradient_test.m`

- `Chapter_3_data_fit_by_gradient_ver_2.m`

- `Chapter_3_data_fit_by_gradient_ver_3.m`

- `Chapter_3_data_fit_by_gradient_ver_final.m`

- `params.m`

- `initialize.m`

- `visualize.m`

- `fn_eval_f.m`

- `fn_eval_grad.m`

- `fn_convergence_sol_norm.m`

- `data_main.dat`

- `data_6pt.dat`

- `kappa_test.m`

- `golden_section_search.m`

- `params_ver_2.m`

- `initialize_ver_2.m`

- `params_ver_3.m`

- `initialize_ver_3.m`

- `fn_eval_hess.m`

- `params_ver_final.m`

- `initialize_ver_final.m`

- `mode_OPT.m`

- `mode_TEST.m`

4

Exploring Optimization Algorithms

This chapter continues discussing the structure of the optimization framework created previously in Chapter 3, with the main focus on future development. Here, we explore more the general concept of generating the sequence of suboptimal solutions in the context of the practical implementation of approaches with different solution mechanisms, namely gradient-based (line search and trust region) deterministic and (meta)heuristic (particle swarm optimization) methods.

4.1 Iterative Optimization Algorithms Revisited

In this chapter, we would like the reader to take a break from active computations started in Chapter 3 to solve the 3D problem of Example 1.3 by various techniques already included in the generalized optimization framework. Here, we shift the focus of our discussion into the theoretical plane to analyze more the structure of this framework in the context of future development. As advertised before, we will keep improving the techniques we have already used and add more (advanced) methods to enable our framework for dealing with various problems at various levels of complexity. In practice, as an experienced reader may notice, a computational structure designed for solving problems of a particular type does not have to feature too many solution methods with multiple options for each. Once a framework, or its rough draft, is started and checked for its ability to find some (may not be too accurate and fast) solution, it is time to decide on the core methods to include and develop. Every user makes this decision individually based on the problem itself, own experience, and preferences. As we have already reviewed the logic and associated practical patterns used in constructing computational frameworks, we could now mimic this decision-making process by exploring various concepts for choosing the "engine" to drive iterative algorithms while solving optimization problem (1.10).

Generally speaking, we have to arrive at a concept of generating a sequence of *suboptimal solutions* (iterates)

$$\mathbf{x}^0, \ \mathbf{x}^1, \ \mathbf{x}^2, \ \ldots, \ \mathbf{x}^k, \ \mathbf{x}^{k+1}, \ \ldots \tag{4.1}$$

DOI: 10.1201/9781003275169-4

constructed in assumption (1.17), facilitating monotonicity in decreased objective function $f(\mathbf{x})$, satisfied preferably (but not necessarily) for every k.

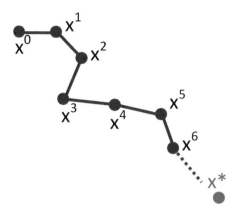

FIGURE 4.1
Suboptimal $\mathbf{x}^k$ and optimal $\mathbf{x}^*$ solutions.

We advise the reader to review Algorithm 1.1, focusing on solution update δ^k in (1.18) of Step 5. Mathematically, this update could be provided by the chosen method in any form consistent with the structure of solution $\mathbf{x}$. For instance, the user may apply a derivative-free approach with clear instructions for computing update δ^k, i.e.,

$$\mathbf{x}^{k+1} = \mathbf{x}^k + \delta^k.$$

In that case, there is no need in the 1D search part, and the d-evaluator has to be formally "converted" to the "δ-evaluator" (refer back to Figure 3.3 for details and established notations). Alternatively, for gradient-based methods,

$$\mathbf{x}^{k+1} = \mathbf{x}^k + \alpha^k \mathbf{d}^k, \tag{4.2}$$

and we could keep the structure of the core part unchanged, as depicted in Figures 3.2 and 3.3. By designing this structure before complicating the problem and implementing new algorithms, we invest in the adaptability of the entire framework.

! *Think of the structure prior to advancing the details, debugging, and check-ing performance. For example, how is it easy to adapt the whole framework to work with both gradient-based and derivative-free algorithms or their combinations? The selection of methods may not be evident or trivial, es-pecially at early stages. However, many users later will face the problem of expanding the framework to some extent. Rigid structures may not allow to upgrade them and attain the desired performance.*

The rest of this chapter does not provide a comprehensive overview of all existing optimization algorithms. Instead, we focus on just a few of them to encourage the reader to think broadly about implementing methods of similar formats or ones with completely different solution mechanisms to make a big picture of adding new functionalities to the existing framework.

4.2 Gradient-based Strategies: Line Search vs. Trust Region

Here, we review two fundamental strategies based on the information provided by evaluated gradients: line search and trust region. The purpose of this concise review is to focus on the conceptual difference between these two *gradient-based* approaches to understand the level of computational scheme modification to incorporate both. We will look into the details for the practical implementation in Chapters 5 and 7.

4.2.1 Line Search

First, let us review the basic concept for the *line search*. It consists of two steps made at every iteration k:

- *Step 1:* choose a search direction $\mathbf{d}^k$, then

- *Step 2:* search along this direction.

In practice, we use formula (4.2) where optimal step size α^k is found as a solution of the following 1D minimization problem

$$\alpha^k = \underset{\alpha > 0}{\operatorname{argmin}} \ f\left(\mathbf{x}^k + \alpha \cdot \mathbf{d}^k\right) \tag{4.3}$$

that provides optimal scaling of vector $\mathbf{d}^k$, the direction (or "line"), see Figure 4.2. The *exact line search* finds the **global** solution of problem (4.3), which is usually computationally expensive and, in most cases, is not necessary. A **practical** approach is to approximate the minimum by generating a limited number of trial step lengths α, as discussed before in Section 3.7. We will review this approach with more methods considered for practical computations in Chapter 6. The general performance of any line search algorithm depends on the quality of used gradient-based search directions. We will add more details on reaching a trade-off between accuracy and computational expenses in Chapter 5.

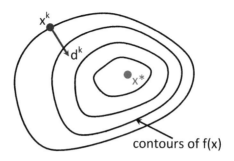

FIGURE 4.2
Schematic illustrating the concept of the line search.

4.2.2 Trust Region

The *trust region method* also uses information obtained from the computed gradients $\nabla f(\mathbf{x}^k)$, but its concept for updating the solution $\mathbf{x}^k$ is different. It also consists of two steps made at every iteration k:

- *Step 1:* construct a model function m^k to approximate function f near $\mathbf{x}^k$, then

- *Step 2:* minimize m^k within some (trust) region around $\mathbf{x}^k$.

The practical implementation comprises finding the candidate step $\mathbf{p}^k$ (used in place of solution update $\boldsymbol{\delta}^k$) such that $\mathbf{x}^{k+1} = \mathbf{x}^k + \mathbf{p}^k$ by solving

$$\mathbf{p}^k = \operatorname*{argmin}_{\mathbf{p}} m^k \left(\mathbf{x}^k + \mathbf{p} \right), \tag{4.4}$$

see Figure 4.3 for clarity. In fact, 1D minimization problem (4.4) comes with the following condition: $\mathbf{x}_k + \mathbf{p}$ should be inside the trust region which could be a ball, or ellipse, box, etc. For instance,

$$\|\mathbf{p}\|_2 \le \Delta^k, \tag{4.5}$$

where $\Delta^k > 0$ is a trust-region radius. The added condition (4.5) converts (4.4) to a constrained optimization problem.

Model m^k is often defined to be a quadratic function for easy minimization, i.e.,

$$m^k(\mathbf{x}^k + \mathbf{p}) = f(\mathbf{x}^k) + \mathbf{p}^T \nabla f(\mathbf{x}^k) + \frac{1}{2}\mathbf{p}^T B^k \mathbf{p}. \tag{4.6}$$

It is also required for trust-region radius Δ^k to be adjusted to ensure a sufficient decrease in objective $f(\mathbf{x})$; see Chapter 7 for more details.

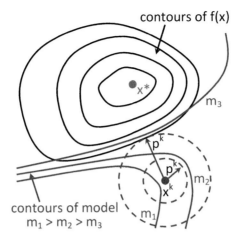

FIGURE 4.3
Schematic illustrating the concept of the trust region.

4.2.3 Comparing Strategies

Now, let us compare both gradient-based approaches to help the reader conclude on variations to have both methods implemented, possibly within the same computational framework. The fundamental difference is the order of processing two objects, namely the **direction** and the **distance**. In the case of line search, we first fix direction and then find the appropriate distance. For the trust region, we do the opposite: choosing maximum distance and then finding a direction with the step to update the solution, subject to distance constraint (4.5).

In addition to the fact that both methods are gradient-based, we have to mention the following property of the trust region approach relating it closely to the group of the line search methods. By setting $B^k = 0$ in (4.6) and defining the trust region by Euclidian norm (4.5), we end up with the problem

$$\min_{\|\mathbf{p}\|_2 \le \Delta^k} f(\mathbf{x}^k) + \mathbf{p}^T \boldsymbol{\nabla} f(\mathbf{x}^k) \tag{4.7}$$

with the solution for candidate step

$$\mathbf{p}^k = -\Delta^k \frac{\boldsymbol{\nabla} f(\mathbf{x}^k)}{\|\boldsymbol{\nabla} f(\mathbf{x}^k)\|} = -\frac{\Delta^k}{\|\boldsymbol{\nabla} f(\mathbf{x}^k)\|} \boldsymbol{\nabla} f(\mathbf{x}^k). \tag{4.8}$$

We leave it for the reader to check that (4.8) is a **steepest descent step** $\mathbf{p}^k = -\alpha^k \boldsymbol{\nabla} f(\mathbf{x}^k)$, where $\alpha^k = \dfrac{\Delta^k}{\|\boldsymbol{\nabla} f(\mathbf{x}^k)\|}$.

Another exciting property appears from choosing $B^k = \nabla^2 f(\mathbf{x}^k)$ (exact Hessian) in (4.6). In this case, problem

$$\min_{\|\mathbf{p}\|_2 \leq \Delta^k} f(\mathbf{x}^k) + \mathbf{p}^T \nabla f(\mathbf{x}^k) + \frac{1}{2}\mathbf{p}^T \nabla^2 f(\mathbf{x}^k)\mathbf{p} \qquad (4.9)$$

is guaranteed to have a solution $\mathbf{p}^k$ (unlike in the steepest descent case) because of the trust region restriction (4.5). Here, the trust region method has a bonus: to guarantee the convergence, Hessian $\nabla^2 f(\mathbf{x}^k)$ does **not** need to be positive definite[1]! We also leave it to the reader to prove these two properties and think on the following question.

? *How to adjust the structure of the optimization framework created in Chapter 3 to allow optimal placement of both line search and trust region methods assuming equal access to the facilities for computing gradients and Hessians?*

The reader will find all necessary details and examples to answer this question in Chapter 5. Chapter 7 will also add more clarity by considering some associated methods; e.g.,

- highly effective (in practice) **trust-region Newton,** and

- **trust-region quasi-Newton** method where matrix B^k arrives through any quasi-Newton approximation.

4.3   Heuristic Algorithms

Now, let us consider a different methodology to support the search for updating a solution. Adding *heuristic methods* to the existing framework could be a good option for many reasons: difficulty or inability to compute accurate gradients and Hessians, simplicity of implementation, or simple desire to compare the performance of algorithms with very distinct driving mechanisms. Among other pros and cons known about various heuristic approaches, we could mention a few common features:

- They **do not need** evaluating (approximating) gradients and Hessians.

- Heuristic methods usually **do not guarantee** (mathematically) an ability to find the optimal solution, either local or global.

- Many of them are claimed to be advantageous in applications only to **certain** practical situations (problem-dependent algorithms).

---

[1]See *Positive Definite Matrices* on p. 380 for review.

Some readers may also have prior experience with so-called *metaheuristic methods* characterized by the following.

- These methods make **few no assumptions** about the problem to be optimized.

- They can search within **very large spaces** of candidate solutions.

- Still, these approaches **cannot guarantee** an optimal solution to be ever found.

The reader may refer to some examples of the metaheuristic approaches below.

(a) *Evolutionary algorithms* (EA) uses mechanisms inspired by biological evolution.

(b) *Genetic algorithms* (GA), as a subclass of EAs, are commonly used to generate high-quality solutions to optimization by relying on bio-inspired operators such as mutation, crossover, and selection.

(c) *Particle swarm optimization* (PSO), as a subclass of GAs, solves a problem by having a population of candidate solutions (particles) and moving these particles around in the search space according to simple mathematical formulas over the particle's position and velocity.

(d) *Stochastic tunneling* (STUN) is a global optimization approach based on the Monte Carlo method.

To enable the analysis of how gradient-based algorithms (e.g., discussed in Section 4.2) and (meta)heuristic approaches could share the computational facilities within the structure of the same optimization framework, Section 4.4 provides a brief overview for the PSO concept. We will also note that various combinations of some deterministic (e.g., gradient-based) and heuristic approaches are possible by sharing useful functionalities for mutual improvement, also discussed briefly in Section 4.4. It gives rise to the so-called *hybrid methods* with continuously growing popularity nowadays. As the readers may find it practical, we encourage them to explore the heuristic approaches in more detail.

4.4 Particle Swarm Optimization

Particle swarm optimization is now a very popular method for solving various problems introduced firstly by Kennedy and Eberhart in 1995; see [19] for more details. Although multiple forms are available to enhance the performance of PSO since then, here, we consider a very basic version of it.

The basic concept behind the PSO method is as follows. It performs the *global stochastic* search in a way how it explores the solution space. The reader may recall our previous discussions to differentiate between the global search and the global solution. Here, we reiterate that performing the former does not necessarily imply the existence and ability to find the latter. The PSO compiles its solution from the individual solutions x_i, $i = 1, \dots, n$, provided by n particles p_i in a "swarm." The solution (position) of ith particle at kth iteration is updated by

$$\mathbf{x}_i^{k+1} = \mathbf{x}_i^k + \mathbf{v}_i^{k+1} \cdot \Delta t, \tag{4.10}$$

where particle's velocity is computed in three different parts

$$
\begin{aligned}
\mathbf{v}_i^{k+1} &= \omega \cdot \mathbf{v}_i^k & (\textit{inertial part}) \\
&+ c_1 \cdot D_1^k \cdot \left(\mathbf{x}_i^{k,pbest} - \mathbf{x}_i^k \right) & (\textit{cognitive part}) \\
&+ c_2 \cdot D_2^k \cdot \left(\mathbf{x}_i^{k,nbest} - \mathbf{x}_i^k \right). & (\textit{social part})
\end{aligned}
\tag{4.11}
$$

The users could tune up the performance of the PSO by varying multiple constant parameters, namely inertia weight $\omega > 0$ and acceleration coefficients $c_1, c_2 > 0$ in (4.11). An additional parameter $\Delta t > 0$ in (4.10) plays a role of a constant scaler for the direction, ith particle velocity $\mathbf{v}_i^{k+1}$, in which the solution $\mathbf{x}_i^k$ is updated. We refer the reader to formula (4.2) for comparison and the similarity analysis. As mentioned before, the PSO is a stochastic method: D_1^k and D_2^k are $n \times n$ diagonal matrices with diagonal entries taken as random numbers uniformly distributed in the interval $[0, 1)$, $D_1^k, D_2^k \sim U(0, 1)$.

In short, the inertial part in (4.11) forces the updated velocity $\mathbf{v}_i^{k+1}$ not to deviate too much from the one, $\mathbf{v}_i^k$, computed for using at the previous iteration. The cognitive part instructs the velocity $\mathbf{v}_i^{k+1}$ to move closer to the best position $\mathbf{x}_i^{k,pbest}$ previously found by particle p_i itself. And, finally, the social part provides an update based on the results of communication between all particles. It informs the velocity $\mathbf{v}_i^{k+1}$ to move closer to the best position $\mathbf{x}_i^{k,nbest}$ ever found by all particles in the neighborhood of particle p_i. Although (4.10) and (4.11) do not show the explicit necessity for evaluating objective function $f(\mathbf{x})$, the cognitive and social parts require such computations for all particles to collect information about their best positions. Figure 4.4 provides a schematic illustration for the basic PSO concept.

In the same vein, as discussed in Sections 2.5 and 2.6 for, respectively, the brute-force and Monte Carlo methods, the PSO's ability to perform the global search depends on the number of particles, n, in the swarm. However, we note again that even large n cannot guarantee the convergence when applying PSO to various problems.

? *Following the discussion on hybrid methods at the end of Section 4.3, how the performance of the basic PSO approach given in (4.10)–(4.11) might be improved?*

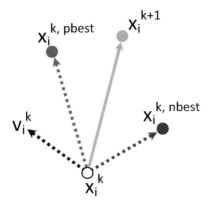

FIGURE 4.4
Schematic illustrating the concept of the PSO.

As a simple solution, one could suggest applying line minimization search for step size parameter Δt in (4.10), i.e.,

$$\Delta t = \underset{\Delta t > 0}{\operatorname{argmin}} \; f\left(\mathbf{x}_i^k + \mathbf{v}_i^{k+1} \cdot \Delta t\right),$$

e.g., by using any approaches discussed in Chapter 2 and applied practically in Chapter 3.

Another solution may be to represent constants ω, c_1, and c_2 in (4.11) as optimization variables enabled to control the performance of the PSO. The reader may notice that function $\mathbf{v}_i^{k+1}$ is easily differentiable with respect to all these new controls. As such, any gradient-based method could be easily employed to identify optimal values of these parameters at each iteration k for every particle p_i.

> **?** *In addition to the solutions provided above, many other ones may similarly address the performance issue in the last question. Does the structure of the optimization framework created in Chapter 3 allow the optimal placing of these solutions within the same framework along with the PSO method assuming equal access to all necessary facilities for computing all required components to support the new hybrid methods?*

We leave this question open. Despite being fairly challenging, students may find this problem a good topic, e.g., for their course projects.

4.5   Overview of Optimization Algorithms

In the final part of this chapter, we provide a brief overview of some methods for solving optimization problems using the structure of the generalized framework designed in Chapter 3. This overview provided in Table 4.1 is not comprehensive and may be easily broadened by adding new algorithms and factors considered for comparison.

TABLE 4.1
Overview of optimization algorithms.

Algorithms	Examples	Strengths	Limitations
Classical Methods	Gradient-based methods, Line Search, Pattern Searches, etc.	Optimality guaranteed (KKT condition)	Global optimum guaranteed only in convex cases
Evolutionary Algorithms	PSO, GA, Differential Evolution, etc.	Can be customized and easily adapted	Optimality guaranteed in limited cases (if any)
Global Search Optimizers	Branch & Bound, Cutting Plane, etc.	Can find global optimum of nonconvex problems	Can be computationally intractable
Hybrid Methods	PSO-MADS, rGA-SQP, etc.	Combine strengths of different algorithms	Heuristically done through trial and error

4.6   Homework Problems

1. Construct the general structure and describe the computational algorithm for the trust region method (to be discussed in detail in Chapter 7).

2. How to adjust the structure of the optimization framework created in Chapter 3 to allow optimal placement of both line search and trust region methods assuming equal access to the facilities for computing gradients and Hessians?

3. Derive analytically the solution for constrained optimization problem (4.7).

4. Modify MATLAB code Chapter_3_data_fit_by_gradient_ver_final.m for Example 1.3 by adding the brute-force and Monte Carlo methods extended from 1D to 3D search.

5. Derive analytically the solution for constrained optimization problem (4.9).

6. Show that Hessian $\nabla^2 f(\mathbf{x}^k)$ does **not** need to be positive definite to guarantee the convergence in Problem 5.

7. Conclude if the optimization framework created in Chapter 3, MATLAB code Chapter_3_data_fit_by_gradient_ver_final.m, allows adding the PSO approach by checking functionalities of all computational elements (input parameters, initialization, optimization, visualization) and providing any necessary upgrades.

8. How the performance of the basic PSO approach described in Section 4.4 might be improved? Does the structure of the optimization framework created in Chapter 3 allow these improvements within the same framework assuming equal access to all necessary facilities for computing all required components to support the new (hybrid) methods?

▢READ▢   Where to Read More

Griva (2009), [15]
Chapter 11 (Basics of Unconstrained Optimization)

Nocedal (2006), [25]
Chapter 2 (Fundamentals of Unconstrained Optimization), Chapter 4 (Trust-Region Methods)

5

Line Search Algorithms

This chapter focuses on the gradient-based line search algorithms such as the steepest descent, pure, discretized, and modified Newton's, diagonally scaled SD, quasi-Newton, Gauss-Newton, and conjugate gradient methods. We analyze their performance, computational efficiency, and applicability based on the practical example developed in Chapter 3. We build our discussion around two main concepts linked to each other: the principal generalization of line search methods and effective "communication" of these methods within the same computational framework to allow its optimal growth.

5.1 Local and Global Minimums: Theory Review

We remind the reader of our earlier references to the properties of the gradient-based line search algorithms, e.g.,

(a) **convergence** (review examples for Newton's method in Sections 2.7 and 3.7, also question on p. 79),

(b) **optimality** (recall the gradient structure discussed in Section 3.4), and

(c) **performance** while using approximated gradients and Hessians (revisit Section 4.2.1 and question on p. 81).

In this chapter, we return to active computations started in Chapters 2 and 3 to examine these properties by answering these questions and addressing other issues through discussions on different implementations of the line search algorithms. Before proceeding to the practical examples, we briefly review some theory facts known for local and global minimums to help us better understand the results obtained in practice.

5.1.1 Necessary Conditions

To simplify our review, let us consider the following n-dimensional *unconstrained optimization* problem

$$\min_{\mathbf{x} \in \mathbb{R}^n} f(\mathbf{x}), \tag{5.1}$$

DOI: 10.1201/9781003275169-5

similar to (1.8), with no distinction between the state and optimization variables being both represented by control $\mathbf{x}$. Assuming that minimized function $f(\mathbf{x})$ is differentiable in the neighborhood of point $\mathbf{x} = \mathbf{a}$, well-known Taylor series expansion provides the following results for

- **1D case, $f(x) : \mathbb{R} \to \mathbb{R}$**

$$f(x) = \sum_{k=0}^{\infty} \frac{1}{k!} f^{(k)}(a)(x-a)^k, \qquad (5.2)$$

- and **multidimensional cases, $f(\mathbf{x}) : \mathbb{R}^n \to \mathbb{R}, \ n > 1$**

$$f(\mathbf{x}) = f(\mathbf{a}) + \sum_{i=1}^{n} \frac{\partial f}{\partial x_i}(\mathbf{a})\,(x_i - a_i)$$

$$+ \frac{1}{2!} \sum_{i=1}^{n} \sum_{j=1}^{n} \frac{\partial^2 f}{\partial x_i\,\partial x_j}(\mathbf{a})\,(x_i - a_i)(x_j - a_j) + \ldots + R \qquad (5.3)$$

$$= f(\mathbf{a}) + \boldsymbol{\nabla} f(\mathbf{a})^T \Delta\mathbf{x} + \frac{1}{2}\Delta\mathbf{x}^T \boldsymbol{\nabla}^2 f(\mathbf{a})\,\Delta\mathbf{x} + \ldots + R,$$

where R is the remainder term, also called a residual or the tail of this expansion.

Based on (5.3), a 1-order approximation for the *variation* (or *perturbation by small* $\Delta\mathbf{x}$) of objective function $f(\mathbf{x})$ over its local minimum $\mathbf{x} = \mathbf{x}^*$ is given by

$$\Delta f = f(\mathbf{x}^* + \Delta\mathbf{x}) - f(\mathbf{x}^*) \approx \boldsymbol{\nabla} f(\mathbf{x}^*)^T \Delta\mathbf{x} \geq 0. \qquad (5.4)$$

We set $\mathbf{a} = \mathbf{x}^*$ and $\mathbf{x} = \mathbf{x}^* + \Delta\mathbf{x}$ in (5.3) to obtain (5.4). Well-known from calculus *1-order necessary condition* at a local minimum point $\mathbf{x}^*$ implies a "zero slope" there, i.e.,

$$\boldsymbol{\nabla} f(\mathbf{x}^*) = \mathbf{0}. \qquad (5.5)$$

We could move forward in the same manner by keeping more terms in (5.3) to get more accurate approximations of $f(\mathbf{x})$ variations Δf. For example, the 2-order approximation is derived as

$$f(\mathbf{x}^* + \Delta\mathbf{x}) - f(\mathbf{x}^*) \approx \boldsymbol{\nabla} f(\mathbf{x}^*)^T \Delta\mathbf{x} + \frac{1}{2}\Delta\mathbf{x}^T \boldsymbol{\nabla}^2 f(\mathbf{x}^*)\Delta\mathbf{x} \geq 0. \qquad (5.6)$$

Now, using (5.5), we arrive at a special property of the Hessian $\boldsymbol{\nabla}^2 f(\mathbf{x})$ evaluated at the local minimum point $\mathbf{x}^*$, namely

$$\Delta\mathbf{x}^T \boldsymbol{\nabla}^2 f(\mathbf{x}^*)\Delta\mathbf{x} \geq 0. \qquad (5.7)$$

This property defines the *2-order necessary condition*: the Hessian matrix H is *positive semidefinite* when evaluated at $\mathbf{x}^*$

$$H(\mathbf{x}^*) = \boldsymbol{\nabla}^2 f(\mathbf{x}^*) = \left[\frac{\partial^2 f}{\partial x_i\,\partial x_j}(\mathbf{x}^*) \right] \succeq 0. \qquad (5.8)$$

Practically, all *eigenvalues*[1] λ_i, $i = 1, \ldots, n$, of matrix $H(\mathbf{x}^*)$ should be non-negative. We refer the reader to our use of sign "$\succeq$" in (5.8) in the meaning of the positive semidefinite matrices, $\lambda_i \geq 0$.

Finally, back to calculus notations, we review that $\mathbf{x}^*$ is a *stationary point* if it satisfies (5.5). We also note that all conditions, 1- and 2-order, mentioned in this section are only **necessary** (**not sufficient!**). We elaborate more on this statement in the next section by presenting two theorems for both necessary and sufficient optimality conditions.

5.1.2   Necessary vs. Sufficient Conditions

Here, we invite the reader to review the mathematical formulation for the necessary and sufficient *optimality conditions* given by the following theorems.

Theorem 5.1 Necessary Optimality Conditions
Let $\mathbf{x}^*$ be an unconstrained local minimizer of $f(\mathbf{x}) : \mathbb{R}^n \to \mathbb{R}$ and $f(\mathbf{x})$ is continuously differentiable in an open set $\mathbb{S} \subseteq \mathbb{R}^n$, $\mathbf{x}^* \in \mathbb{S}$, then

- $\nabla f(\mathbf{x}^*) = \mathbf{0}$ *(**1-order necessary** condition).*

If $f(\mathbf{x})$ is twice continuously differentiable within $\mathbb{S}$, then

- $\nabla^2 f(\mathbf{x}^*) \succeq 0$ *(**2-order necessary** condition).*

Theorem 5.2 Sufficient Optimality Conditions
Let $f(\mathbf{x}) : \mathbb{R}^n \to \mathbb{R}$ be twice continuously differentiable in an open set $\mathbb{S} \subseteq \mathbb{R}^n$. Suppose a vector $\mathbf{x}^* \in \mathbb{S}$ satisfies both

- $\nabla f(\mathbf{x}^*) = \mathbf{0}$, *and*

- $\nabla^2 f(\mathbf{x}^*) \succ 0$ *(**2-order sufficient** condition).*

*Then $\mathbf{x}^*$ is a **strict** unconstrained local minimizer of $f(\mathbf{x})$. In particular, there exist $\gamma > 0$ and $\epsilon > 0$ to satisfy*

$$f(\mathbf{x}) \geq f(\mathbf{x}^*) + \frac{\gamma}{2}\|\mathbf{x} - \mathbf{x}^*\|^2 \tag{5.9}$$

for all $\mathbf{x}$, such that $\|\mathbf{x} - \mathbf{x}^*\| < \epsilon$.

In the same manner, as done in Section 5.1.1, we refer the reader to our use of sign "$\succ$" in Theorem 5.2 in the meaning of the Hessian $H(\mathbf{x}^*) = \nabla^2 f(\mathbf{x}^*)$ being a *positive definite matrix*, i.e., $\lambda_i > 0$, $i = 1, \ldots, n$. By comparing both theorems, it is noticeable that accurate computations for constructing Hessians and performing eigenvalue analysis on them allow justified conclusions on $\mathbf{x}^*$ as local minimizers of $f(\mathbf{x})$. We also note that when the dimension n of the solutions space is heavily increased, such computations are costly if ever

---

[1]See *Eigenvalues and Eigenvalue Decomposition* on p. 379 for review.

available. It adds extra logic in performing applicability tests mentioned in Section 3.5 in the manner to move slowly from low to high values of n. It allows extrapolating the ability of the obtained framework to find minimizers for very large problems and assuming (practically) their strict locality. The reader may easily find proofs for both theorem, e.g., in [4].

? *What about global minimizers and practical ways to identify them?*

5.1.3   Existence of Global Minimums

To answer the question posed at the end of the previous section, we ask the reader to review the mathematical definition of the global minimizer given by (1.11). The use of this definition is evidently not tractable in practice as it requires computing $f(\mathbf{x})$ for all $x \in \mathbb{S} \subseteq \mathbb{R}^n$. Still, there are some problems, even of large size, known to have *global minimizers*. The reader may revisit Section 1.4.3 to review the notation of *convexity* before proceeding to the following theorem.

Theorem 5.3 Convex Cost Function
Let $f(\mathbf{x}): \mathbb{R}^n \to \mathbb{R}$ be a convex function over the convex set $\mathbb{S} \subseteq \mathbb{R}^n$.

(a) *Then a local minimizer of $f(\mathbf{x})$ over $\mathbb{S}$ is also a* **global minimizer** *over* $\mathbb{S}$.

(b) *In case $\mathbb{S}$ is open, then (5.5) is a* **necessary and sufficient** *condition for* $\mathbf{x}^* \in \mathbb{S}$ *to be a* **global minimizer** *of $f(\mathbf{x})$ over $\mathbb{S}$.*

Back to calculus again, we remind the reader that if function $f(\mathbf{x})$ is twice differentiable, then for proving its **global** convexity over set $\mathbb{S}$ we have to show that its Hessian is a *positive definite matrix*, i.e., $H(\mathbf{x}) = \nabla^2 f(\mathbf{x}) \succ 0$ for all $\mathbf{x} \in \mathbb{S}$. An example of any size globally convex problem to satisfy Theorem 5.3 could be any unconstrained *quadratic problem*

$$\min_{\mathbf{x} \in \mathbb{R}^n} \frac{1}{2}\mathbf{x}^T Q \mathbf{x} + \mathbf{b}^T \mathbf{x}, \tag{5.10}$$

where $Q \in \mathbb{R}^{n \times n}$ is a symmetric matrix, and $\mathbf{b}$ is a vector in $\mathbb{R}^n$. To ensure global solution for (5.10) matrix Q, which is also a Hessian, should be positive definite.

Finally, we may refer to other mathematically known facts to guarantee that global minimums exist for certain problems. For example,

- global minimum exists if $f(\mathbf{x})$ is a continuous function on $\mathbb{S}$, which is compact (closed and bounded), and

- global minimum also exists if $\mathbb{S}$ is closed and $f(\mathbf{x})$ is continuous and coercive (in the meaning $f(\mathbf{x}) \to \infty$ if $\|\mathbf{x}\| \to \infty$).

These two statements are also known as parts of *Weierstrass' Theorem*, and they may serve as a good source for analysis to identify the problem's potential for having global solutions.

5.2   Selecting Search Direction

5.2.1   Principal Generalization

In the preceding chapters, we considered several algorithms with the concept of choosing search directions $\mathbf{d}^k$ to improve solutions $\mathbf{x}^k$ based on available gradients. The reader may refer, e.g., to Algorithm 1.1 and Section 4.1 for a general review of iterative methods to perform a gradient-based search. We also used the steepest descent in (3.7)–(3.8) and Newton's method (3.16) for solving the 3D optimization problem of Example 1.3. To move forward, we need to generalize the principal idea of using gradients as a "driving force" for updating solutions. This generalization will also help complete the structure of our computational framework by allowing all gradient-based methods to coexist for optimal use in our future examples.

In the context of such generalization, we reconsider the iterative approach for improving solution $\mathbf{x}$ starting from its initial guess $\mathbf{x}_0$

$$
\begin{aligned}
\mathbf{x}^{k+1} &= \mathbf{x}^k + \alpha^k \cdot \mathbf{d}^k, \quad k = 0, 1, \dots \\
\mathbf{x}^0 &= \mathbf{x}_0,
\end{aligned}
\tag{5.11}
$$

by substituting it with its principal generalization in the following form

$$
\mathbf{x}^{k+1} = \mathbf{x}^k - \alpha^k \cdot D^k \, \boldsymbol{\nabla} f(\mathbf{x}^k).
\tag{5.12}
$$

In (5.11)–(5.12), gradient-based search direction $\mathbf{d}^k = -D^k \, \boldsymbol{\nabla} f(\mathbf{x}^k)$ must satisfy the *descent condition*

$$
\boldsymbol{\nabla} f(\mathbf{x}^k)^T \mathbf{d}^k < 0,
\tag{5.13}
$$

as the angle between computed gradient $\boldsymbol{\nabla} f$ and chosen direction $\mathbf{d}$ should be obtuse, i.e.,

$$
\angle(\boldsymbol{\nabla} f, \mathbf{d}) > 90°.
\tag{5.14}
$$

Mandating descent condition (5.13) at each iteration guarantees selection of direction $\mathbf{d}$ enabled to perform the minimization of objective function $f(\mathbf{x})$. Figure 5.1 illustrates the concept of the descent condition.

In (5.12), matrix D^k modifies the search direction originally associated with gradient $\boldsymbol{\nabla} f(\mathbf{x}^k)$. To ensure descent condition (5.13), D^k should be a positive definite symmetric matrix. Applying $\mathbf{d}^k = -D^k \, \boldsymbol{\nabla} f(\mathbf{x}^k)$ to (5.13) gives

$$
\boldsymbol{\nabla} f(\mathbf{x}^k)^T D^k \, \boldsymbol{\nabla} f(\mathbf{x}^k) > 0,
$$

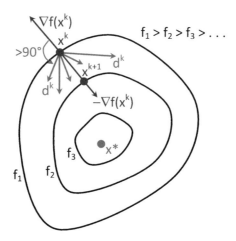

FIGURE 5.1
Schematic illustrating the concept of the descent condition (5.13)–(5.14).

which is, in fact, a definition of a positive definite matrix applied to D^k. The principal generalization concept (5.12) is closed by selecting optimal step size $\alpha > 0$. We refer the reader to our first practical experiments in Section 3.7. In Chapter 6, we will further develop the idea of choosing optimal α and consider other algorithms to perform such search. Finally, the general principle for the line search algorithms (5.12) assumes that the initial guess x_0 is provided feasible. Otherwise, the user should consider adding a methodology for obtaining a *feasible solution* before initiating the line search by any gradient-based method.

> **?** *Could we guarantee the convergence of the line search methods within their generalized concept (5.12) while choosing different "modifiers" D^k for gradients $\nabla f(\mathbf{x}^k)$?*

5.2.2   Some Theory for Convergence

Here, let us discuss briefly two classical theorems for the convergence of the iterative approach (5.12) and known conditions under which this convergence is guaranteed. For both theorems, we assume that for solving unconstrained minimization problem (5.1), the sequence of solutions $\{\mathbf{x}^k\} = \{\mathbf{x}^0, \mathbf{x}^1, \mathbf{x}^2, \ldots\}$ is generated by using line search general principle (5.11)–(5.12) starting with feasible $\mathbf{x}^0 = \mathbf{x}_0$.

Theorem 5.4 Stationarity of Limit Points for Gradient Methods
*Let $\{\mathbf{x}^k\}$ be generated by (5.11), where $\mathbf{d}^k$ is **gradient related** and α^k is*

chosen by **Armijo or (limited) line minimization rule**. *Then every limit point* $\mathbf{x}^*$ *of* $\{\mathbf{x}^k\}$ *(meaning* $\{\mathbf{x}^k\} \to \mathbf{x}^*$*) is stationary, i.e.,* $\boldsymbol{\nabla} f(\mathbf{x}^*) = \mathbf{0}$.

We have to emphasize the importance of the two conditions stated in Theorem 5.4 to guarantee the stationarity of limit points for sequence $\{\mathbf{x}^k\}$. First, the search direction $\mathbf{d}^k$ should be gradient-based as given in the generalized concept (5.12). Also, the role of optimal α^k is **not** of lesser importance. The convergence is proven only for selected methods to be considered later, with both theory and practical applications, in Chapter 6.

Theorem 5.5 Capture Theorem
Let $f(\mathbf{x})$ *be continuously differentiable. Also, let* $\{\mathbf{x}^k\}$ *satisfy nonincreasing condition*

$$f(\mathbf{x}^{k+1}) \leq f(\mathbf{x}^k),$$

and be generated by any gradient method (5.11)–(5.12). *Assume that there exist* $s > 0$ *and* $c > 0$ *such that for all* k

$$\alpha^k \leq s \quad and \quad \|\mathbf{d}^k\| \leq c\|\boldsymbol{\nabla} f(\mathbf{x}^k)\|.$$

Let $\mathbf{x}^*$ *be a local minimizer of* $f(\mathbf{x})$ *(meaning the only stationary point within some open set). Then there exists an open set* $\mathbb{S}$ *containing* $\mathbf{x}^*$ *such that if* $\mathbf{x}^{\bar{k}} \in \mathbb{S}$ *for some* $\bar{k} \geq 0$, *then*

$$\mathbf{x}^k \in \mathbb{S}$$

for all $k \geq \bar{k}$, *and*

$$\{\mathbf{x}^k\} \to \mathbf{x}^*.$$

Similar to the previous case, Theorem 5.5 requires the iterative solution process to be governed by any gradient-based method. In a nutshell, this theorem states that any "isolated" local minimums tend to attract gradient methods if any solution $\mathbf{x}^k$, in the sequence $\{\mathbf{x}^k\}$, is placed sufficiently close to these minimums. Although it does not require using any particular method for choosing step size, α^k should be able to get a value small enough, $\alpha^k \leq s$, to get into the neighborhood $\mathbb{S}$ of local minimizer $\mathbf{x}^*$ and keep the solution $\mathbf{x}^k$ there. We also note that the quality of search directions $\mathbf{d}^k$ is also important – they should be bounded, $\|\mathbf{d}^k\| \leq c\|\boldsymbol{\nabla} f(\mathbf{x}^k)\|$, to enable proper scaling by step size α^k.

5.2.3   Steepest Descent

To conclude on the principal generalization discussed in this section we would like to exemplify the *steepest descent method* to examine its fit to the generalization concept (5.12). The simplest version of the SD method (3.7)–(3.8) could be written in the following form.

$$\mathbf{x}^{k+1} = \mathbf{x}^k - \alpha^k \boldsymbol{\nabla} f(\mathbf{x}^k). \tag{5.15}$$

Comparing (5.15) with (5.12) provides a simple result

$$D_{SD}^k = \mathcal{I}^{n \times n}, \tag{5.16}$$

where $\mathcal{I}$ is an identity matrix of dimension n. This matrix is known to be symmetric and positive definite ($\lambda_i = 1$, $i = 1, 2, \ldots, n$). It confirms the steepest descent's ability to converge, refer to Theorems 5.4 and 5.5, assuming a proper choice of step size α^k.

We already experimented with the SD method, while solving the optimization problem of Example 1.3 in Sections 3.4 and 3.7 and examined its convergence, which is naturally **very slow**, of *linear* or *sublinear* type. At the same time, the SD does not require Hessian evaluations seen as a benefit allowing **easy implementation** and **light computations** per iteration.

We compare the SD with Newton's method by applying both to Example 1.3 in Section 3.7. The latter showed the opposite characteristics: fast convergence but nontrivial implementation due to the necessity to provide and evaluate Hessian structures. Both methods could be treated as extreme cases when a particular user decides to choose an approach to solve an optimization problem with gradients. In fact, there are many other methods derived from both the SD and Newton's one that may be placed in between them in terms of speed of convergence and implementation simplicity. Figure 5.2 provides a cartoon picture to illustrate this concept.

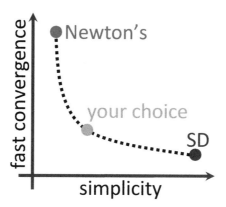

FIGURE 5.2
Schematic illustrating the concept of comparing gradient-based methods (SD and Newton's) by their speed of convergence and implementation simplicity.

In the next two sections, 5.3 and 5.4, we consider some of the options to choose between the SD and Newton's method:

- discretized and modified Newton's methods,

- diagonally scaled SD,

- quasi-Newton and

- Gauss-Newton methods, and

- conjugate direction approaches.

Moreover, in Section 5.5, we will evaluate all those methods in terms of their performance and the complexity of adding them into our existing computational framework.

5.3 Newton's Method and Newton-based Modifications

5.3.1 Pure Newton's Method

Here, we start with reviewing *Newton's method* for better comparison with all other derivatives of it to be considered later. Back to formula (3.16) used for our practical computations in Chapter 3, the gradient modifier matrix D^k for the pure Newton's method involves inverted Hessian, i.e.,

$$D_N^k = \left[\boldsymbol{\nabla}^2 f(\mathbf{x}^k)\right]^{-1}, \tag{5.17}$$

so that the iteration process in (5.12) is turned into the following form

$$\mathbf{x}^{k+1} = \mathbf{x}^k - \left[\boldsymbol{\nabla}^2 f(\mathbf{x}^k)\right]^{-1} \boldsymbol{\nabla} f(\mathbf{x}^k). \tag{5.18}$$

As noted before and also checked computationally in Section 3.7, setting $\alpha^k = 1$ is the best option for better convergence in case Hessian is computed exactly. Subject to accuracy in computing Hessians (or their approximations), the convergence is usually of *quadratic* or *superlinear* type. As we saw it in practice in Section 3.7, Newton's method is pretty fast due to the quadratic approximation of $f(\mathbf{x})$ at $\mathbf{x} = \mathbf{x}^k$.

On the other hand, pure Newton's method has several known drawbacks. First, the inverse $\left[\boldsymbol{\nabla}^2 f(\mathbf{x}^k)\right]^{-1}$ may be **hardly computable** and **fail to exist**. Second, the pure form (5.18) is not a (guaranteed) descent method. Therefore, the situation when $f(\mathbf{x}^{k+1}) > f(\mathbf{x}^k)$ may arise at any stage within the iterative process. This feature is known as "blindness" while converging toward the stationary solution. The pure form tends to be attracted by either local minimum or maximum as Newton's method just solves $\boldsymbol{\nabla} f(\mathbf{x}) = \mathbf{0}$.

Finally, the method in the pure form is "naughty" when it comes to the initial guess for the solution. It converges very fast if it starts from close proximity to the local minimum. Otherwise, it may converge very fast as expected or **may not converge** at all.

All mentioned above disadvantages prevent pure Newton's method from considering it for *global convergence* in the meaning of its *robustness* – ability to show its performance and converge starting from any feasible solution. There are many known modifications to improve the global convergence of Newton's method. Here, we mention just a few of them.

- **Switching** to SD method whenever Newton's direction $-\left[\nabla^2 f(\mathbf{x}^k)\right]^{-1} \nabla f(\mathbf{x}^k)$ is not a descent direction, refer to (5.13).

- Proper **scaling** by using

  (a) proper (optimal) step size α^k (different from $\alpha^k = 1$), or
  (b) gradient modifier D^k, e.g.,

  $$D^k = \left[\nabla^2 f(\mathbf{x}^k) + \Delta^k\right]^{-1}, \qquad (5.19)$$

  where Δ^k is a diagonal matrix to ensure that matrix $\nabla^2 f(\mathbf{x}^k) + \Delta^k$ is positive definite.

- Applying various **factorizations** to matrix $\nabla^2 f(\mathbf{x}^k)$ to ensure its easy invertibility, e.g., modified Cholesky factorization.

- Switching to **trust region methods** (will be discussed in Chapter 7).

Now let us proceed to other methods seen, to some extent, as derivatives of pure Newton's method.

5.3.2 Discretized Newton's Method

Discretized Newton's (dN) method consists of computing the gradient modifier matrix

$$D_{dN}^k = \left[\tilde{H}(\mathbf{x}^k)\right]^{-1}, \qquad (5.20)$$

where $\tilde{H}(\mathbf{x}^k)$ is an FD approximation of exact Hessian $\nabla^2 f(\mathbf{x}^k)$. As it obviously follows, the accuracy of the dN method depends entirely on the chosen approach to approximate all 2-order (mixed) partial derivatives

$$\left[\nabla^2 f(\mathbf{x}^k)\right]_{ij} = \frac{\partial^2 f(\mathbf{x}^k)}{\partial x_i \, \partial x_j}, \quad i = 1, 2, \ldots, n; \; j = 1, 2, \ldots, n.$$

This method may be **fast** enough to exhibit *superlinear* (sometimes, even faster) convergence, which is the behavior comparable to original pure Newton in case high-order FD schemes are applied. On the other hand, due to the number of objective function $f(\mathbf{x})$ evaluations required to fulfill the chosen FD scheme, this approach is inevitably **very exhausting** computationally. However, some readers may find quite efficient practical approaches to parallelize all computations for $\tilde{H}(\mathbf{x}^k)$. It is also possible to to make all partial derivatives available, e.g., by automatic differentiation (AD) techniques.

? *Suggest any ideas on optimal parallelization of discretized Newton's approach and evaluate the performance gain using units associated with the time necessary to evaluate objective function $f(\mathbf{x})$.*

5.3.3 Modified Newton's Method

For *modified Newton's (mN) method* Hessian $\nabla^2 f(\mathbf{x})$ is computed in full, either by pure or discretized Newton's methods, at $\mathbf{x} = \mathbf{x}^0$ and then used for some part or the rest of the entire optimization, i.e.,

$$D_{mN}^k = \left[\nabla^2 f(\mathbf{x}^0)\right]^{-1}. \tag{5.21}$$

Despite its evident simplicity, the mN method may be **very effective** computationally if appropriate analysis on Hessian updates is applied.

? *Assume that modified Newton's method is chosen to update Hessians periodically during the optimization process. Could you suggest any concepts and practical implementation ideas for analyzing the necessity of such updates?*

Depending on the problem to be solved, the behavior of modified Newton's method may be different. However, in case Hessian is not updated, it is **much slower** than pure and discretized Newton's methods, almost down to *linear convergence* or even slower.

5.3.4 Diagonally Scaled SD

The next modification of pure Newton's method uses the distinct role of the main diagonal elements of square matrices. Basic linear algebra properties provide that any scalar matrix (having only zeros outside its main/principal diagonal) with all main diagonal elements represented by nonzero numbers is easily invertible. Examination of large Hessians also provides that these matrices are usually sparse; their numerous nonzero elements are also very close to 0. As such, the *diagonally scaled SD (DS-SD) method* uses a diagonal approximation of inverted Hessian as its gradient modifier, i.e.,

$$D_{DSSD}^k = [\text{diag}(H)]^{-1}$$

$$= \begin{bmatrix} \dfrac{1}{\left[\nabla^2 f(\mathbf{x}^k)\right]_{11}} & 0 & \cdots & 0 \\ 0 & \dfrac{1}{\left[\nabla^2 f(\mathbf{x}^k)\right]_{22}} & \cdots & 0 \\ \cdots & \cdots & \cdots & \cdots \\ 0 & 0 & \cdots & \dfrac{1}{\left[\nabla^2 f(\mathbf{x}^k)\right]_{nn}} \end{bmatrix}. \tag{5.22}$$

By utilizing (partially) information provided by Hessian H, the DS-SD method is also a Newton-based method. It is generally known to be **more effective** computationally than discretized Newton's approach if updated at every iteration. The DS-SD may be used in combination with pure, discretized, and modified Newton's methods.

? *Suggest any scheme of using combinations of pure, discretized, modified Newton's, and diagonally scaled SD methods. Guess on expected benefits and performance.*

For general comparison, the DS-SD method is **less accurate** and **slower** than pure Newton's method and dN with accurately assembled Hessians. At the same time, depending on the problem, it may still show good convergence up to *superlinear* type, e.g., in case exact Hessian $\nabla^2 f(\mathbf{x}^k)$ is close to being a diagonally dominant matrix.

5.3.5   Quasi-Newton Methods

Before discussing another modification of Newton's approach, the *quasi-Newton (qN) method* with *superlinear* "Newton-like" convergence, we refer the reader to known from calculus 1-order Hessian approximation

$$\nabla^2 f(\mathbf{x}^k)(\mathbf{x}^k - \mathbf{x}^{k-1}) \approx \nabla f(\mathbf{x}^k) - \nabla f(\mathbf{x}^{k-1}). \tag{5.23}$$

We note that (5.23) could be easily derived from Taylor series expansion for multidimensional cases given in (5.3). The original version of Hessian $\nabla^2 f(\mathbf{x}^k)$ in (5.17) is replaced by its approximated version B^k computed using the so-called *secant equation*

$$B^k \mathbf{s}^k = \mathbf{y}^k, \tag{5.24}$$

where

$$\mathbf{s}^k = \mathbf{x}^k - \mathbf{x}^{k-1}, \quad \mathbf{y}^k = \nabla f(\mathbf{x}^k) - \nabla f(\mathbf{x}^{k-1}). \tag{5.25}$$

Then this approximation is used for obtaining the gradient modifier matrix D^k in (5.12)

$$D^k_{qN} = \left[B^k \right]^{-1}. \tag{5.26}$$

Formulas (5.23) through (5.26) describe the concept of quasi-Newton approaches very broadly. To make these approaches practical, one should consider multiple aspects, at least the following four.

(1) Similar to cases utilizing exact Hessian, additional **symmetry requirements** should be imposed on B^k; refer to Section 5.3.1 for details.

(2) If making computations **less costly** is desirable, update for D^k in (5.26) has to re-use previous version of D evaluated at iteration $k-1$, i.e.,

$$D^k = D^k \left(D^{k-1}, \mathbf{s}^k, \mathbf{y}^k \right). \tag{5.27}$$

(3) In the context of the previous aspect, for iterative updates of D^k from D^{k-1}, the user should choose the **initial approximation** for D^0 in (5.26), e.g., $D^0 = B^0 = \mathcal{I}$.

(4) Also, when running through multiple iterative updates, it is most likely that D^k will accumulate numerical **errors**. To improve the convergence, one should think about the *restart* strategy; for instance, setting $D^k = \mathcal{I}$ every n_r iterations or depending on the *condition number*[2] of D^k.

Many methods exist to provide updates for B^k based on B^{k-1} properly addressing aspects (1) and (2) above. Here, we will mention two of those commonly used in practice. *Symmetric-rank-one formula* (SR1)

$$B^k = B^{k-1} + \frac{(\mathbf{y}^k - B^{k-1}\mathbf{s}^k)(\mathbf{y}^k - B^{k-1}\mathbf{s}^k)^T}{(\mathbf{y}^k - B^{k-1}\mathbf{s}^k)^T\mathbf{s}^k} \tag{5.28}$$

is the first one. Another approach appears, in fact, much more popular nowadays. It is commonly known as *Broyden–Fletcher–Goldfarb–Shanno (BFGS) method*

$$B^k = B^{k-1} - \frac{B^{k-1}\mathbf{s}^k(\mathbf{s}^k)^T B^{k-1}}{(\mathbf{s}^k)^T B^{k-1}\mathbf{s}^k} + \frac{\mathbf{y}^k(\mathbf{y}^k)^T}{(\mathbf{y}^k)^T\mathbf{s}^k} \tag{5.29}$$

which allows updating D^k from D^{k-1} by

$$D^k = \left(\mathcal{I} - \rho^k\mathbf{s}^k(\mathbf{y}^k)^T\right) D^{k-1} \left(\mathcal{I} - \rho^k\mathbf{y}^k(\mathbf{s}^k)^T\right) + \rho^k\mathbf{s}^k(\mathbf{s}^k)^T, \tag{5.30}$$

where

$$\rho^k = \frac{1}{(\mathbf{y}^k)^T\mathbf{s}^k}. \tag{5.31}$$

The well-known fact about many quasi-Newton methods is that they are **not too sensitive** to the accuracy of computing step size α^k compared with 1-order methods. However, if investments into these computations are made, and this search for step size α is accurate, the approximations for D^k appear to be very close to $\left[\nabla^2 f(\mathbf{x}^k)\right]^{-1}$; it implies **faster convergence**.

Finally, we have to note that even if quasi-Newton methods provide some relief from exhausting computations (related to exact Hessians in case of pure Newton's scheme), they are still **computationally demanding**. The reader may think of $\mathcal{O}(n^2)$ operations to approximate Hessians compared to $\mathcal{O}(n^3)$ operations necessary to compute $\nabla^2 f(\mathbf{x}^k)$ exactly.

5.3.6 Gauss-Newton Approach

Last but not least method we consider here as a modification of Newton's approach is the *Gauss–Newton (GN) method*. It is commonly used to solve nonlinear least-squares optimization problems of the following form

$$\min_{\mathbf{x}\in\mathbb{R}^n} \ f(\mathbf{x}) = \frac{1}{2}\|\mathbf{g}(\mathbf{x})\|^2 = \frac{1}{2}\sum_{i=1}^{m} g_i^2(\mathbf{x}), \tag{5.32}$$

---

[2]See *Condition Number* on p. 380 for review.

where $\mathbf{g}(\mathbf{x}) : \mathbb{R}^n \to \mathbb{R}^m$ is an m-component vector function. In other words, the GN modification minimizes sums of squared function values $g_i(\mathbf{x})$ with no second derivatives required. To see this, we apply the solution approach provided by the pure Newton's method in (5.18) to solve problem (5.32):

$$\mathbf{x}^{k+1} = \mathbf{x}^k - \left[(\boldsymbol{\nabla}\mathbf{g}(\mathbf{x}^k))^T \boldsymbol{\nabla}\mathbf{g}(\mathbf{x}^k)\right]^{-1} (\boldsymbol{\nabla}\mathbf{g}(\mathbf{x}^k))^T \mathbf{g}(\mathbf{x}^k). \tag{5.33}$$

In (5.33), $(\boldsymbol{\nabla}\mathbf{g}(\mathbf{x}^k))^T \mathbf{g}(\mathbf{x}^k)$ is the gradient of $f(\mathbf{x}) = \frac{1}{2}\|\mathbf{g}(\mathbf{x})\|^2$ obtained using the chain rule (in multivariable calculus). Gradient $\boldsymbol{\nabla}\mathbf{g}(\mathbf{x}^k)$ is also called a *Jacobian matrix*[3] (gradient of a vector function)

$$\mathcal{J} = \boldsymbol{\nabla}\mathbf{g}(\mathbf{x}).$$

This notation helps describe the Gauss–Newton method through direct reference to Newton's approach. Here, Hessians are approximated via Jacobians $\mathcal{J}$, i.e.,

$$\boldsymbol{\nabla}^2 f(\mathbf{x}^k) \approx (\boldsymbol{\nabla}\mathbf{g}(\mathbf{x}^k))^T \boldsymbol{\nabla}\mathbf{g}(\mathbf{x}^k) = \mathcal{J}^T \mathcal{J}. \tag{5.34}$$

For simplicity, we illustrate the use of the GN method in application to the optimization problem of Example 1.3

$$\min_{\mathbf{a}\in\mathbb{R}^3} \; f(\mathbf{a}) = \|\mathbf{r}(\mathbf{a})\|^2 = r_1^2 + r_2^2 + \cdots + r_m^2$$

$$= \sum_{i=1}^m \left(\tilde{y}_i - (a_1 + a_2\tilde{x}_i + a_3\tilde{x}_i^2)\right)^2. \tag{5.35}$$

The reader may notice that (5.35), in fact, has the same format (except missing coefficient $\frac{1}{2}$) as the nonlinear least-squares optimization problem given in (5.32). Therefore, applying the same analysis to solve this problem by the Gauss-Newton approach gives iterations in the following form

$$\mathbf{a}^{k+1} = \mathbf{a}^k - \left[(\boldsymbol{\nabla}\mathbf{r}(\mathbf{a}^k))^T \boldsymbol{\nabla}\mathbf{r}(\mathbf{a}^k)\right]^{-1} (\boldsymbol{\nabla}\mathbf{r}(\mathbf{a}^k))^T \mathbf{r}(\mathbf{a}^k). \tag{5.36}$$

In Section 5.5, we will check the performance of the GN method (5.36) along with other gradient-based approaches discussed in Chapter 5. In some cases, it may show *quadratic convergence* under certain regularity conditions. However, in general (under weaker conditions or getting large condition numbers for $\mathcal{J}^T\mathcal{J}$), the GN method converges *linearly*.

Following the discussion in Section 5.3.1 for improved convergence of pure Newton's method, similar suggestions may apply to the Gauss-Newton schemes. For example, there is a modification of this approach to ensure descent direction (5.13) and prevent $(\boldsymbol{\nabla}\mathbf{g}(\mathbf{x}^k))^T \boldsymbol{\nabla}\mathbf{g}(\mathbf{x}^k)$ from being singular

$$\mathbf{x}^{k+1} = \mathbf{x}^k - \alpha^k \left[(\boldsymbol{\nabla}\mathbf{g}(\mathbf{x}^k))^T \boldsymbol{\nabla}\mathbf{g}(\mathbf{x}^k) + \Delta^k\right]^{-1} (\boldsymbol{\nabla}\mathbf{g}(\mathbf{x}^k))^T \mathbf{g}(\mathbf{x}^k). \tag{5.37}$$

This modification includes proper choice of step size α^k and diagonal matrix

---

[3]See *Jacobian* on p. 382 for review.

Δ^k to ensure matrix $(\boldsymbol{\nabla}\mathbf{g}(\mathbf{x}^k))^T \boldsymbol{\nabla}\mathbf{g}(\mathbf{x}^k) + \Delta^k$ is positive definite. For computing the latter, the reader might be interested in one of the simplest options, *Levenberg–Marquardt method*:

$$\Delta^k = m\mathcal{I}, \quad m > 0, \tag{5.38}$$

where $\mathcal{I}$ is the identity matrix.

5.4   Conjugate Gradient

The review of the Gauss–Newton method in Section 5.3.6 provides that the GN approach avoids the overhead typically associated with many Newton-like algorithms. It relates to exhausting computations to evaluate or approximate Hessians seen in Sections 5.3.1 through 5.3.5. At the same time, we would like to have a method with convergence better than linear (generally shown in applications with the SD and GN approaches). Being a good candidate, here, we discuss the *conjugate gradient (CG) method* that targets both computational easiness and faster convergence.

First, we have to establish the notation of conjugacy applied to vector spaces. Given matrix $Q^{n \times n} \succ 0$ (positive definite), a set of nonzero vectors $\mathbf{d}^1, \mathbf{d}^2, \ldots, \mathbf{d}^k$ are called Q-*conjugate* if

$$(\mathbf{d}^i)^T Q \, \mathbf{d}^j = 0 \tag{5.39}$$

for any i and j, but assuming $i \neq j$.

? *Could we prove that these vectors $\mathbf{d}^1, \mathbf{d}^2, \ldots, \mathbf{d}^k$ are linearly independent?*

Yes, the proof is pretty straightforward, and we leave this exercise to the reader. As such, every vector $\mathbf{d}$ in (5.39) could minimize $f(\mathbf{x})$ **independently in its own direction**. As an example, the $Q^{n \times n}$ quadratic problem

$$\min_{\mathbf{x} \in \mathbb{R}^n} f(\mathbf{x}) = \frac{1}{2}\mathbf{x}^T Q \, \mathbf{x} + \mathbf{b}^T \mathbf{x} \tag{5.40}$$

could be solved with **at most** n iterations by applying iterative search

$$\mathbf{x}^{k+1} = \mathbf{x}^k + \alpha^k \cdot \mathbf{d}^k, \tag{5.41}$$

where $\mathbf{d}^k = \mathbf{d}^0, \mathbf{d}^1, \ldots, \mathbf{d}^{n-1}$ are Q-conjugate vectors.

The practical implementation of the CG method may vary depending on the chosen approach to design the set of Q-conjugate vectors. The choice of the initial direction is simple. As discussed previously in Section 5.2.1, $\mathbf{d}^0$

should satisfy descent condition (5.13) and may be chosen, for example, using the plain steepest descent method, i.e.,

$$d^0 = -\nabla f(x^0). \tag{5.42}$$

At other steps ($k > 0$), we derive conjugate direction d^k from descent direction $-\nabla f(x^k)$, such that d^k is Q-conjugate to all vectors computed before, i.e., $d^0, d^1, \ldots, d^{k-1}$. Commonly used *Gram–Schmidt method* may ideally serve as a procedure to obtain the new direction:

$$d^k = -\nabla f(x^k) + \sum_{j=0}^{k-1} \frac{(\nabla f(x^k))^T Q\, d^j}{(d^j)^T Q\, d^j} d^j. \tag{5.43}$$

As mentioned before, quadratic problems require n iterations, or even less, to stop the iterative search. For all other problems, however, we could terminate the conjugate gradient method based, e.g., on the gradient structure, i.e., when

$$\nabla f(x^k) \approx 0.$$

As not using 2-order derivatives, the CG approach cannot compete with Newton's method. Usually, it shows *linear* or *superlinear convergence* with convergence rate

$$1 \le r < 2.$$

For many cases, however, these results are much better than those demonstrated by the steepest descent method. Finally, we provide the practical approach which allows generating the Q-conjugate directions for the CG method using the procedure based only on search directions and the gradients computed at the current step k and stored from the previous iteration $k - 1$:

$$\begin{aligned} d^0 &= -\nabla f(x^0), \\ d^k &= -\nabla f(x^k) + \beta^k d^{k-1}, \quad k = 1, \ldots, n - 1, \end{aligned} \tag{5.44}$$

where

$$\beta^k = \frac{(\nabla f(x^k))^T \nabla f(x^k)}{(\nabla f(x^{k-1}))^T \nabla f(x^{k-1})}. \tag{5.45}$$

5.5   Line Search Performance Comparison

In this section, it is logical to start comparing the performance of different gradient-based algorithms from the group of line search methods discussed at length in Chapters 3–5. As the reader noticed from these discussions, various components are to be computed to assist with the search: gradients (in all methods), exact or approximated Hessians (pure Newton, dN, mN, DS-SD, and BFGS), Jacobian matrices (GN), and conjugate vectors (CG).

First, we focus on the idea developed earlier in Section 3.8 to allow the existing code (created in Chapter 3 for solving Example 1.3) to grow up and include new methods by preserving the same structure as the building concept for our computational framework. To some extent, upgrading this framework is seen as an advanced-level "optimization problem" targeted to minimize the total changes in the code, while implementing any additions. We invite the reader to compare the final version of the MATLAB code used in Chapter 3 (`Chapter_3_data_fit_by_gradient_ver_final.m`) and the updated version named `Chapter_5_data_fit_by_line_search.m` containing all line search methods considered in the current chapter; refer to Table 5.1 for a summary for all implemented changes.

element	implementation	software
main OPT	written manually	MATLAB
f-evaluator	m-function, analytically defined function $f(\mathbf{a})$	MATLAB
d-evaluator	**m-code, various line search approaches**	MATLAB
1D α search	plain m-code for GS search	MATLAB
visualizer	plain m-code	MATLAB

TABLE 5.1
Computational elements of `Chapter_5_data_fit_by_line_search.m`.

In particular, we advise the reader to check the following MATLAB scripts to verify the **minimal changes** in the structure required for adding new functionalities:

- new functions `fn_eval_hess_approx.m`, `fn_eval_g.m`, and `fn_eval_g_grad.m` are added for computing, respectively, approximated Hessians (parameter `stepH` as an FD approximation step added to `params.m`), $\mathbf{g}(\mathbf{x})$ and $\nabla \mathbf{g}(\mathbf{x})$ in (5.33), with associated function handlers added to `initialize.m`

- computational facility to obtain search direction **d** has been moved from m-file `mode_OPT.m` to the new one, `direction_search.m`, and re-organized by using MATLAB's `switch` block to allow easy additions of any other **d**-evaluator methods (by adding extra `case` statements).

MATLAB: `direction_search.m`

```
switch method                                              1
                                                           2
    case 'SD'                                              3
```

```
      d = -grad_f(a,data);                                             4
                                                                       5
  case 'NEWTON'                                                        6
      d = -inv(hess(a,data))*grad_f(a,data);                           7
                                                                       8
  case 'DN'                                                            9
     d = -inv(hess_approx(a,data,stepH,f))*grad_f(a,data);            10
                                                                      11
  case 'MN'                                                           12
    if (k==0)                                                         13
       d = -inv(hess_approx(a,data,stepH,f))*grad_f(a,data);          14
    end                                                               15
                                                                      16
  case 'DSSD'                                                         17
    d = -(diag(1./diag(hess_approx(a,data,stepH,f))))...             18
        *grad_f(a,data);                                              19
                                                                      20
  case 'BFGS'                                                         21
    if (k==0)                                                         22
       D = eye(length(a),length(a));                                  23
       grad_old = grad_f(a,data);                                     24
    else                                                              25
       grad_curr = grad_f(a,data);                                    26
       s = a - a_old;                                                 27
       y = grad_curr - grad_old;                                      28
       grad_old = grad_curr;                                          29
       r = 1/(y'*s);                                                  30
       D = (eye(length(a),length(a))-r*s*y')*D...                     31
           *(eye(length(a),length(a))-r*y*s')+r*s*s';                 32
    end                                                               33
    d = -D*grad_f(a,data);                                            34
                                                                      35
  case 'GN'                                                           36
    d = -inv((grad_g(a,data))'*grad_g(a,data))...                     37
        *(grad_g(a,data))'*g(a,data);                                 38
                                                                      39
  case 'CG'                                                           40
    grad_curr = grad_f(a,data);                                       41
    if (k==0)                                                         42
       d = -grad_curr;                                                43
    else                                                              44
       d = -grad_curr + (grad_curr'*grad_curr)/...                    45
           (grad_old'*grad_old)*d;                                    46
    end                                                               47
    grad_old = grad_curr;                                             48
                                                                      49
  otherwise                                                           50
    disp(['Unknown search direction method ' method ' chosen!']);     51
    exCode = 0; % termination due to error                            52
    return;                                                           53
                                                                      54
end                                                                   55
```

Now, we could check all methods discussed in this chapter by running `Chapter_5_data_fit_by_line_search.m` with the following settings (to maintain consistency with the results obtained in Chapter 3 for the SD and Newton's methods).

MATLAB: params.m

Settings for all line search methods:

- *line search method:* `method = 'DN';` choosing from the list: DN, MN, DSSD, BFGS, GN, and CG

- *input data:* `dataFile = 'data_main.dat';`

- *initial guess:* `aini = [-14 11 -2];`

- *FD approximation step (for Hessian):* `stepH = 1e-4;`

- *step size search:* `alphaUpdate = true;` update by GS method

- *search interval:* `alphaA = 0; alphaB = 10;`

- *tolerance for GS search termination:* `alphaEps = 1e-6;`

- *termination #1:* `epsilonF = 1e-6;` for ϵ_1, refer to (3.18)

- *termination #2:* `epsilonA = 1e-6;` for ϵ_2, refer to (3.17)

- *termination #3:* `kMax = 200;`

First, we run the code with the discretized Newton's method discussed in Section 5.3.2, where the 1-order (forward) FD scheme applies to approximate exact Hessian $\nabla^2 f(\mathbf{a})$ with step size parameter Δa (`stepH`) set to 10^{-4}, i.e.,

$$
\begin{aligned}
\left[\nabla^2 f(\mathbf{a})\right]_{11} = \frac{\partial^2 f(\mathbf{a})}{\partial a_1^2} &\approx \frac{1}{(\Delta a)^2} \left[f(a_1 + 2\Delta a, a_2, a_3) \right. \\
&\left. -2f(a_1 + \Delta a, a_2, a_3) + f(a_1, a_2, a_3) \right],
\end{aligned}
$$

$$
\begin{aligned}
\left[\nabla^2 f(\mathbf{a})\right]_{12} = \frac{\partial^2 f(\mathbf{a})}{\partial a_1 \partial a_2} &\approx \frac{1}{(\Delta a)^2} \left[f(a_1 + \Delta a, a_2 + \Delta a, a_3) \right. \\
&\left. -f(a_1 + \Delta a, a_2, a_3) - f(a_1, a_2 + \Delta a, a_3) + f(a_1, a_2, a_3) \right],
\end{aligned}
$$

etc. The reader is advised to derive the rest of mixed partial derivatives or refer to m-code `fn_eval_hess_approx.m` to explore the symmetric structure of the approximated Hessian $\tilde{H}(\mathbf{a})$ in (5.20).

Figures 5.3(a,b) shows the results in terms of the obtained solution, a curve representing $\mathbf{a}^* = [-15 \ 10 \ -1]^T$, and the objective function $f(\mathbf{a})$ decreased to the level of machinery zero, 10^{-30}, after just three iterations. This result is similar to the one obtained by applying pure Newton's method; refer to Figure 3.16. As also seen in Figure 5.5(a), the computational analysis for the rate of convergence provides $r = 1.9853$, which is consistent with the superlinear (close to quadratic) type naturally expected from the dN method assuming that perturbation parameter Δa is sufficiently small. We make a final notice that applying the golden section search for step size α returns values very close to 1 seen in Figure 5.4(a), confirming the accuracy of the used approximation and ability to save the computational time by not investing into the GS search.

? *What would be the result if we change the FD scheme to be the 2-order accurate? What would happen to the computational complexity of the dN approach?*

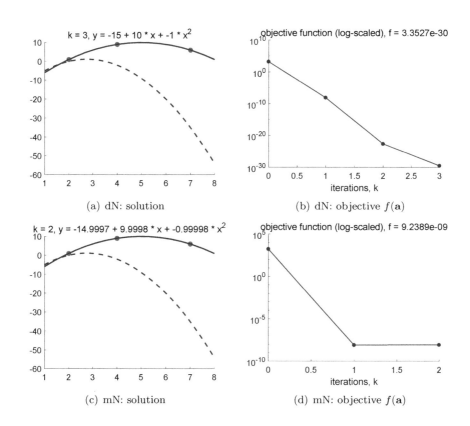

(a) dN: solution   (b) dN: objective $f(\mathbf{a})$

(c) mN: solution   (d) mN: objective $f(\mathbf{a})$

FIGURE 5.3
Results of applying the (a,b) dN and (c,d) mN methods.

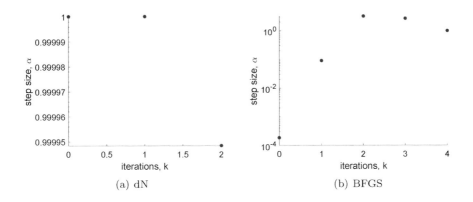

FIGURE 5.4
Results of applying the GS search for finding optimal step size α in applications of the dN and BFGS methods.

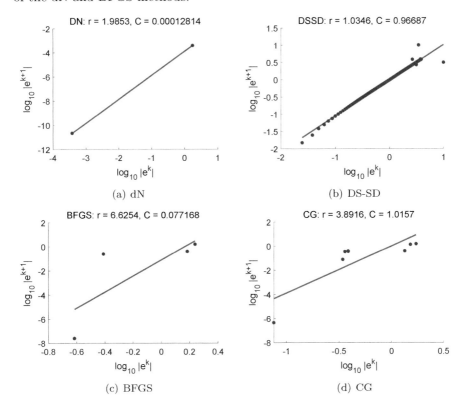

FIGURE 5.5
Convergence analysis for (a) dN, (b) DS-SD, (c) BFGS, and (d) CG methods.

The next turn is the modified Newton's method implemented as presented in Section 5.3.3 with the results provided in Figures 5.3(c,d). The discretized Hessian is used to perform the first iteration. However, the second one shows the inability to improve further this result, $f \sim 10^{-8}$, as the Hessian is no longer updated. Back to the question on p. 109, we may suggest updating the Hessian, e.g., every 3-5 iterations or whenever the solution does not show any (substantial) improvement.

Our next approach, the diagonally scaled SD method, also utilizes (partially) the approximated Hessian, namely its diagonal elements. As shown in Figure 5.6, this approach, as expected, is less accurate and much slower (refer to Figure 5.5(b), providing the computational rate of convergence $r = 1.0346$) than the dN and mN methods (both based on the same type Hessian approximations). Having potential for good convergence (up to superlinear), the DS-SD solves Example 1.3 only with linear convergence.

? *Any reason to explain such slow convergence?*

! *To answer this question, the reader may examine the structure of the exact Hessian that is probably not a diagonally dominant matrix.*

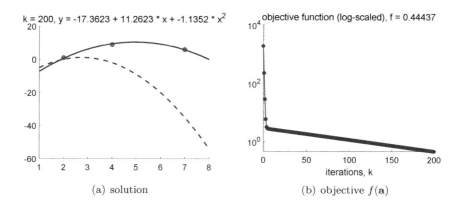

(a) solution   (b) objective $f(\mathbf{a})$

FIGURE 5.6
Results of applying the diagonally scaled SD (DS-SD) method.

Our next modification of Newton's approach, considered in Section 5.3.5, is the quasi-Newton BFGS method with proclaimed superlinear "Newton-like" convergence. As seen in Figure 5.7, BFGS competes pretty well with the pure and discretized Newton's methods converging to the optimal solution $\mathbf{a}^* = [-15\ 10\ -1]^T$ with $f \sim 10^{-29}$ in just five iterations. This fast convergence is supported by our investments into computations for the accurate step size α; refer to Figure 5.4(b). This figure also suggests that BFGS approximation D^k

of the inverted Hessian becomes accurate (close to $\left[\nabla^2 f(\mathbf{x}^k)\right]^{-1}$) starting at k = 2 when α^k is very close to 1. Then, it starts accumulating numerical errors as we do not use any restart strategies discussed in Section 5.3.5. However, this accumulation is not rapid as it does not prevent the method from converging accurately when $k \geq 3$.

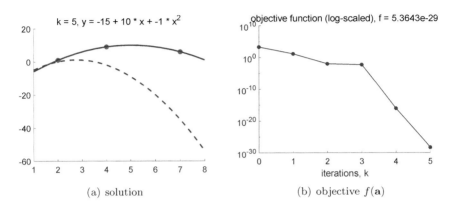

(a) solution   (b) objective $f(\mathbf{a})$

FIGURE 5.7
Results of applying the quasi-Newton BFGS method.

We have to make an additional comment about the performance of the BFGS method using data provided in Figure 5.5(c). The computational analysis for the convergence provides its coefficient $r = 6.6254$ that we may not trust as it is based on two apparent outliers. Unfortunately, removing them will lead to insufficient data – only two points that are very close to each other.

Finally, we apply the Gauss–Newton method, which uses approximations for the Hessians via Jacobians as described in Section 5.3.6. Figure 5.8 shows its convergence to the optimal solution $\mathbf{a}^*$ with $f \sim 10^{-26}$ just in two iterations.

? *Any reason to explain such fast convergence as, in general, the GN method converges linearly?*

! *The reader may refer to our discussion (Section 5.3.6) about the possibility of exhibiting quadratic convergence of the GN approach by considering the structure (quadratic) of the problem in Example 1.3 and checking the condition number of $\mathcal{J}^T \mathcal{J}$.*

The last method we explore here for the performance, see Figure 5.9, is the conjugate gradient. The provided solution is relatively accurate, $f \sim 10^{-8}$, and obtained after only nine iterations with no use of Hessians. The reader may notice the inability of the CG approach here to solve our 3D problem

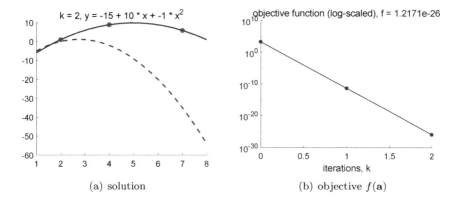

(a) solution   (b) objective $f(\mathbf{a})$

FIGURE 5.8
Results of applying the Gauss–Newton (GN) method.

of Example 1.3 in 3 iterations (or less) discussed in Section 5.4 (quadratic problems require n iterations). We advise the reader to explore our implementation of the method; see m-file `direction_search.m` on p. 115. For making it more computationally efficient, we use the practical approach for iterative updates (5.44)–(5.45) rather than the Gram-Schmidt procedure (5.43). As seen in the combined Figure 5.10 and discussed in Section 5.4, the CG approach cannot compete with Newton's and "Newton-like" methods (dN, BFGS, and GN). However, it shows the result much better than those demonstrated by the steepest descent (SD and DS-SD) methods. We also refer the reader back to Figure 5.2. Paring it with the combined performance results shown in Figures 5.5 and 5.10 adds more clarity to the concept of comparing various gradient-based methods by their accuracy, speed of convergence, and implementation simplicity.

We finalize our performance analysis here by making a final comment about the rate of convergence obtained computationally for the CG method. Figure 5.5(d) returns the value of $r = 3.8916$. It is also not trustful with the same arguments we made for a similar situation observed for the BFGS approach: one outlier located in the left bottom corner. However, here we have enough data to repeat the computational test for convergence after removing this outlier. As shown in Figure 5.11, the updated convergence test returns rate $r = 1.0071$ consistent with the typical behavior of the CG methods (linear or superlinear convergence, $1 \leq r < 2$) discussed in Section 5.4.

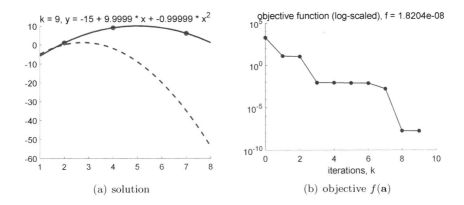

(a) solution   (b) objective $f(\mathbf{a})$

FIGURE 5.9
Results of applying the conjugate gradient (CG) method.

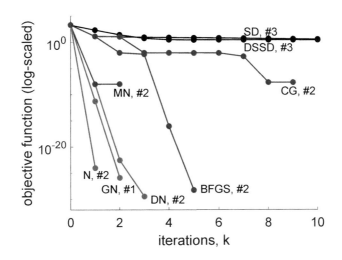

FIGURE 5.10
Combined graph illustrating the performance of all line search methods discussed in Chapter 5: (SD) steepest descent, (N) Newton's, (dN) discretized Newton's, (mN) modified Newton's, (DSSD) diagonally scaled SD, (BFGS) quasi-Newton BFGS, (GN) Gauss–Newton, and (CG) conjugate gradient. The numbers attached to method labels identify the termination conditions each approach used to complete the search; refer to m-file `params.m` on p. 117.

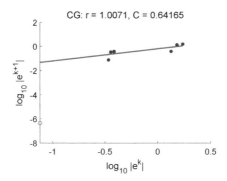

FIGURE 5.11
Convergence analysis for the CG methods similar to Figure 5.5(d) with one outlier (shown by the empty circle) removed.

5.6 Homework Problems

1. Suggest any ideas on optimal parallelization of discretized Newton's approach and evaluate the performance gain using units associated with the time necessary to evaluate objective function $f(\mathbf{x})$.

2. By assuming that modified Newton's method is chosen to update Hessians periodically during the optimization process, suggest the practical implementation of such updates based on some analysis for their necessity.

3. Suggest any scheme of using combinations of pure, discretized, modified Newton's, and diagonally scaled SD methods. Guess on expected benefits and performance.

4. Derive 1-order Hessian approximation (5.23) using Taylor series expansion for multidimensional cases given in (5.3).

5. Derive the iteration algorithm given by (5.36) for solving optimization problem (5.35) of Example 1.3.

6. Using the definition of the Q-conjugate vectors (5.39) prove that these vectors are linearly independent.

7. Explore `MATLAB` code `Chapter_5_data_fit_by_line_search.m` for Example 1.3 with added (in addition to existing Steepest Descent (SD) and pure Newton's methods) the following line search approaches:

   (a) discretized Newton's (dN),

   (b) modified Newton's (mN),

(c) diagonally scaled SD (DS-SD),

(d) quasi-Newton BFGS,

(e) Gauss–Newton (GN), and

(f) conjugate gradient (CG).

For all incorporated methods, check the convergence and approximate convergence parameters r and C. Rate these methods in terms of their rate of convergence shown for the current problem, and make a conclusion.

READ   Where to Read More

Bertsekas (2016), [4]
Chapter 1 (Unconstrained Optimization: Basic Methods), Chapter 2 (Unconstrained Optimization: Additional Methods)

Boyd (2004), [6]
Chapter 4 (Convex Optimization Problems), Chapter 9 (Unconstrained Minimization), Chapter 10 (Equality Constrained Minimization)

Griva (2009), [15]
Chapter 11 (Basics of Unconstrained Optimization), Chapter 12 (Methods of Unconstrained Optimization)

Nocedal (2006), [25]
Chapter 2 (Fundamentals of Unconstrained Optimization), Chapter 3 (Line Search Methods), Chapter 5 (Conjugate Gradient Methods), Chapter 6 (Quasi-Newton Methods), Chapter 7 (Large-Scale Unconstrained Optimization), Chapter 8 (Calculating Derivatives), Chapter 10 (Least-Squares Problems)

Press (2007), [26]
Chapter 10 (Minimization or Maximization of Functions)

RUN   MATLAB Codes for Chapter 5

- `Chapter_5_data_fit_by_line_search.m`

- `params.m`

- `initialize.m`

- `fn_eval_hess_approx.m`

- `fn_eval_g.m`

- `fn_eval_g_grad.m`

- `mode_OPT.m`

- `direction_search.m`
- `golden_section_search.m`
- `data_main.dat`
- `data_6pt.dat`
- `fn_convergence_sol_norm.m`
- `fn_eval_f.m`
- `fn_eval_f_grad.m`
- `fn_eval_hess.m`
- `mode_TEST.m`
- `kappa_test.m`
- `visualize.m`

5.7 Lab Assignment #2: Review Chapters 3–5

Problem 1: Consider Example 1.3 on p. 10 to find parameters a_1, a_2, a_3 in the quadratic model $y(x) = a_1 + a_2 x + a_3 x^2$, which optimally fits m data points (x_i, y_i), $i = 1, \ldots, m$, in the least-squares sense. Modify MATLAB code `Chapter_3_data_fit_by_gradient_ver_final.m` by adding the following derivative-free approaches:

1. Add *brute-force method* extended from 1D to 3D search. Discretize the search/control space $[a_1, a_2, a_3]$ uniformly within intervals $[-50, 50]$ for all three dimensions with steps (a) $h = 1$, (b) $h = 2$, (c) $h = 5$, and (d) $h = 10$. Compare the computational time and accuracy of found solutions and conclude.

2. Add *Monte Carlo method* extended from 1D to 3D search. Run the search by choosing sample solutions in each dimension using random numbers (a) uniformly distributed within interval $[-50, 50]$, (b) normally distributed with mean $\bar{a} = [0\ 0\ 0]^T$ and standard deviation $\sigma = 50/3$ (using "three-sigma" rule), and (c) normally distributed with mean $\bar{a} = [-15\ 10\ -1]^T$ and $\sigma = 50/3$. For each search (a-c) allow $n = 100$, $n = 10,000$, and $n = 1,000,000$ as the number of samples (f-evaluations). Compare the obtained solutions and conclude about the performance and applicability of cases (a-c).

Both methods, brute-force and Monte Carlo, should be incorporated into the same framework (use, e.g., variable `method` in `params_ver_final.m` to specify the approach). For all computations, use $m = 3$ and data from file `data_main.dat`. Check the completeness of the final version of your code by adding all necessary components.

! *Refer to Sections 7.5 and 7.6 to find some help in solving this problem.*

Problem 2: Consider the same problem as in Problem 1. Explore MATLAB code `Chapter_5_data_fit_by_line_search.m` with added (in addition to existing steepest descent (SD) and pure Newton's methods) the following line search approaches: discretized Newton's (dN), modified Newton's (mN), diagonally scaled SD (DS-SD), quasi-Newton BFGS (without restarts), Gauss–Newton (GN), and conjugate gradient (CG). For all incorporated methods, check the convergence and approximate convergence parameters r and C. Rate these methods in terms of their rate of convergence shown for the current problem and conclude. Explain the reasons for removing some methods from this "competition."

For all computations, use the following (refer to Section 5.5 for more details):

- $m = 3$ and data from file `data_main.dat`,

- initial guess $\mathbf{a}^0 = [-14 \ 11 \ -2]^T$,

- termination parameters: $\epsilon_1 = \epsilon_2 = 10^{-6}$ and $k_{max} = 200$; refer to (3.17)–(3.18) for explanations, and

- existing GS search method for finding step size α with bracketing interval $[0, 10]$ and termination tolerance $\epsilon_\alpha = 10^{-6}$.

In addition to this, consider

- approximated Hessians (whenever used) computed by FD approximations with $\Delta a = 10^{-4}$, and

- rate of convergence based on the solution $\mathbf{a}^{last}$ obtained at termination rather than on the exact solution $\mathbf{a}^*$ (practical approach discussed in Section 3.4).

6

Choosing Optimal Step Size

In this chapter, the focus is on various methods for performing an α-search to support optimal scaling of the gradient-based search directions. We consider simple algorithms (constant and diminishing steps) and methods for inexact (the Armijo rule with added Wolfe and Goldstein conditions) and exact (bracketing-Brent toolbox in `MATLAB`) line search by analyzing their performance, computational efficiency, and applicability based on the practical example developed in Chapter 3. We build our discussion around the concept of "proper communication" between different parts of a gradient-based optimization algorithm, namely d- and α-search methods, to boost the performance of their joint work, while solving problems of various complexities.

6.1 Overview

Before making our last step while exploring the implementation and performance of various line search methods (review Chapters 3–5 for more details), we refer the reader to its basic concept already discussed in Section 4.2.1. Any line search algorithm used to improve the solution $\mathbf{x}$ iteratively

$$\mathbf{x}^{k+1} = \mathbf{x}^k + \alpha^k \cdot \mathbf{d}^k, \quad k = 0, 1, \dots$$
$$\mathbf{x}^0 = \mathbf{x}_0,$$

$$(6.1)$$

consists of two steps made at every iteration k:

- *Step 1:* choose a search direction $\mathbf{d}^k$ (see Figure 4.2), then

- *Step 2:* search along this direction by choosing appropriate (optimal) step size α^k.

In Chapters 3 and 5, we experimented with different options for *Step 1* by playing with the array of methods within the generalized line search principle provided by gradient modifier D^k in (5.12). Although the choice of gradient-based methodology with accurate computing search directions $\mathbf{d}^k = -\boldsymbol{\nabla} f(\mathbf{x}^k)$ is crucial, we should not underestimate the contribution of optimal step size α^k to the overall performance. We already have some intuitive experience after solving the optimization problem of Example 1.3 using constant step

DOI: 10.1201/9781003275169-6

(Section 3.3) and then by applying the golden section search for finding its optimal size (Sections 3.7 and 5.5). In this chapter, we will explore this more systematically. Again, we will consider this aspect twofold: gaining more performance of (or, at least, not weakening) the line search methods with minimal changes to allow "optimal growth" of our existing optimization framework.

In the first place, we have to generalize the procedure of finding an optimal step size $\alpha^k > 0$, while solving n-dimensional unconstrained optimization problem

$$\min_{\mathbf{x} \in \mathbb{R}^n} f(\mathbf{x}), \tag{6.2}$$

by using iterative algorithm (6.1). This generalization involves the following assumption: if $\boldsymbol{\nabla} f(\mathbf{x}) \neq 0$, then there exists a small positive number, $\delta > 0$, such that

$$f(\mathbf{x} - \alpha \, \boldsymbol{\nabla} f(\mathbf{x})) < f(\mathbf{x}), \quad \forall \alpha \in (0, \delta). \tag{6.3}$$

The motivation behind this assumption is to enable the search for optimal step size α^k within the bounded 1D region, $0 < \alpha^k < \delta$, as a local minimizer of $f\left(\mathbf{x}^k - \alpha^k \, \boldsymbol{\nabla} f(\mathbf{x}^k)\right)$ by achieving an **adequate reduction** in $f(\mathbf{x})$ at **minimal cost**; refer to Figure 6.1 for this concept visualization.

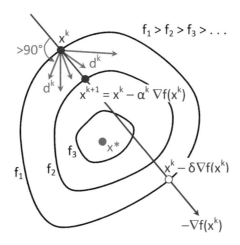

FIGURE 6.1
Schematic illustrating the concept of the optimal step size α^k search, while solving unconstrained optimization problem (6.2) using iterative line search algorithm (6.1).

To proceed, we consider two stages for this step search:

- *Stage 1 – Bracketing:* find interval containing desirable step length, and

- *Stage 2 – Bisection/interpolation:* compute a good step length within this interval.

According to methods chosen to represent each stage, there exist different algorithms with varied implementation complexity, expected accuracy, and computational costs. Among other existing classification approaches to estimate step size for the line search methods, here, we refer to a simplified one.

- *Simple algorithms:* function evaluations **not explicitly required**.

- *Inexact line search:* **moderate efforts** required to achieve **adequate** reduction in $f(\mathbf{x})$ at a **minimal cost**.

- *Exact line search:* enabled to find **accurate** and, therefore, **expensive** solutions for 1D optimization problem

$$\alpha^k = \underset{0<\alpha\leq\delta}{\operatorname{argmin}}\, f\left(\mathbf{x}^k + \alpha \cdot \mathbf{d}^k\right). \tag{6.4}$$

We will explore some examples for each type of the search approaches in the rest of this chapter.

6.2 Simple Approaches

Assigning a small constant value $\bar{\alpha}$ to step size α^k, i.e.,

$$\alpha^k = \bar{\alpha} = \text{const} > 0, \qquad k = 0, 1, 2, \ldots \tag{6.5}$$

kept unchanged during entire optimization is one of the simplest strategies. We refer the reader to our practical experiments back to Section 3.3, where the constant α approach exhibited **very slow convergence** or even **divergence** when chosen α^k did not properly align with the problem. However, we have to mention a positive feature – the method's simplicity in implementation that makes it suitable for initial experiments and checking (roughly) the framework functionality.

Another simple approach is *diminishing step size*

$$\alpha^k \to 0, \qquad k = 0, 1, 2, \ldots \tag{6.6}$$

provided α^k satisfies the *infinite travel condition*

$$\sum_{k=0}^{\infty} \alpha^k = \infty. \tag{6.7}$$

Although this method also shows an obvious simplicity in its practical implementation, it requires a "diminishing rule" that might be **problem-dependent**. Due to the latter, this approach may demonstrate **slow convergence** and may also diverge in case the chosen rule does not "fit" the problem.

? *Could you guess about reasons for divergence with any implementation of the diminishing step size algorithms?*

! *Look for the hint in the next section.*

6.3   Inexact Line Search

As an example of the inexact line search algorithms here we consider the Armijo rule with added Wolfe and Goldstein conditions.

6.3.1   Armijo Rule

Quite popular algorithms to perform inexact line search with a considerably simple computational scheme requiring moderate computational efforts uses the so-called *sufficient decrease condition* (also known as the *Armijo rule*)

$$f(\mathbf{x}^k + \alpha \, \mathbf{d}^k) \leq f(\mathbf{x}^k) + c_1 \, \alpha \, (\boldsymbol{\nabla} f(\mathbf{x}^k))^T \mathbf{d}^k, \tag{6.8}$$

where $0 < c_1 < 1$. This condition has a straightforward geometrical explanation illustrated in Figure 6.2. Here, a nonlinear function $\phi(\alpha) = f(\mathbf{x}^k + \alpha \, \mathbf{d}^k)$ represents the left side of inequality (6.8), the updated objective function $f(\mathbf{x})$ evaluated with precomputed search direction $\mathbf{d}^k$ and selected step size α. Similarly, a linear function $\mu(\alpha) = f(\mathbf{x}^k) + c_1 \, \alpha \, (\boldsymbol{\nabla} f(\mathbf{x}^k))^T \mathbf{d}^k$ with a negative slope due to descent condition (5.13) describes the right side of (6.8). Therefore, the selected α is acceptable if $\phi(\alpha) \leq \mu(\alpha)$. All computational steps to implement the Armijo rule are provided in Algorithm 6.1.

Algorithm 6.1 (Practical Implementation of the Armijo Rule)

1. Choose c_1 to be quite small, e.g., $c_1 = 10^{-4}$

2. Choose initial guess for α^k, e.g.,

$$\alpha^0 = \bar{\alpha} \tag{6.9}$$

3. Resize α^k until sufficient decrease condition (6.8) satisfies, e.g.,

$$\alpha^k = \beta^{m_k} \alpha^0, \qquad m_k = 0, 1, 2, \ldots, \quad 0 < \beta < 1 \tag{6.10}$$

Although the practical implementation of this rule is straightforward, it has a well-known drawback. Sufficient decrease condition (6.8) may not ensure the algorithm makes **reasonable progress**. However, the performance of this method could be further improved by pairing it with additional schemes (e.g., Wolfe or Goldstein conditions) to enhance the convergence.

Here, we could provide a hint for answering the question posed in the previous section. We assume using Algorithm 6.1 for the Armijo Rule without sufficient decrease condition (6.8). We may establish the "diminishing rule" for (6.6) using (6.10) (however, with infinite travel condition (6.7) neglected). Now, Figure 6.2 helps provide a clear explanation of why choosing $0 < \beta < 1$ cannot guarantee the reduction in $f(\mathbf{x})$ as expected by our generalized assumption (6.3).

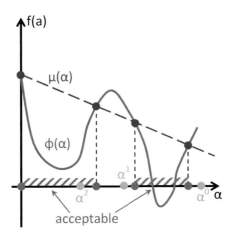

FIGURE 6.2
Schematic illustrating the concept of the Armijo rule or sufficient decrease condition (6.8).

6.3.2 Wolfe Conditions

One of the commonly used methods to improve the performance of the Armijo rule is adding a *curvature condition*

$$(\boldsymbol{\nabla} f(\mathbf{x}^k + \alpha \, \mathbf{d}^k))^T \mathbf{d}^k \geq c_2 \, (\boldsymbol{\nabla} f(\mathbf{x}^k))^T \mathbf{d}^k, \tag{6.11}$$

where $c_1 < c_2 < 1$. It also has a straightforward explanation in terms of its geometry as shown in Figure 6.3. Briefly, we are interested in the slope of $\phi(\alpha)$ being c_2 times greater than the gradient $\phi'(0)$. Otherwise, we cannot expect a **substantially more** decrease in $f(\mathbf{x}^k)$ along this direction. As it follows from Figures 6.2 and 6.3, both sufficient decrease (6.8) and curvature

(6.11) conditions make a pair for simultaneous use also known as the *Wolfe conditions*; refer to both figures for the intervals to define acceptable α with both conditions.

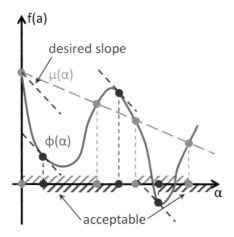

FIGURE 6.3
Schematic illustrating the concept of applying the Wolfe (combined sufficient decrease (6.8) and curvature (6.11)) conditions.

We may choose constant c_2 in (6.11), e.g., by using trial-and-error experiments for finding the best fit for the solved problem. However, the reader may find some recommendations in the literature related to the particular use of the Wolfe conditions coupled with methods for selecting the search direction $\mathbf{d}^k$. These recommendations may serve as good initial approximations for c_2 being adjusted later during the optimization runs and analysis of the computational performance. For example, a typical piece of advice is to set

- $c_2 = 0.9$ when $\mathbf{d}^k$ is chosen by Newton or any quasi-Newton methods, and

- $c_2 = 0.1$ when $\mathbf{d}^k$ is chosen by the conjugate gradient method.

We could move ahead and require even more from the curvature condition set in (6.11). For instance, the modified condition may involve forcing α to lie in at least a broad neighborhood of a local minimizer or stationary point of $\phi(\alpha)$

$$\left| (\boldsymbol{\nabla} f(\mathbf{x}^k + \alpha \, \mathbf{d}^k))^T \mathbf{d}^k \right| \geq c_2 \left| (\boldsymbol{\nabla} f(\mathbf{x}^k))^T \mathbf{d}^k \right|, \qquad (6.12)$$

with $0 < c_1 < c_2 < 1$. Both sufficient decrease (6.8) and modified (strong) curvature (6.12) conditions, when used together, are commonly referred to as the *strong Wolfe conditions*.

We have to mention that the performance of applying the Wolfe conditions in terms of the existence of α is proven analytically subject to some assumptions on function $f(\mathbf{x})$ in (6.2).

Lemma 6.1 Existence of Step Length

Let $f(\mathbf{x}) : \mathbb{R}^n \to \mathbb{R}$ be continuously differentiable. Suppose $\mathbf{d}^k$ is a descent direction at $\mathbf{x}^k$, and assume that $f(\mathbf{x})$ is bounded below along the ray $\{\mathbf{x}^k + \alpha \mathbf{d}^k : \alpha > 0\}$. Then if $0 < c_1 < c_2 < 1$, there exist intervals of step length satisfying the Wolfe conditions

$$
\begin{aligned}
f(\mathbf{x}^k + \alpha\,\mathbf{d}^k) &\leq f(\mathbf{x}^k) + c_1\,\alpha\,(\boldsymbol{\nabla} f(\mathbf{x}^k))^T \mathbf{d}^k, \\
(\boldsymbol{\nabla} f(\mathbf{x}^k + \alpha\,\mathbf{d}^k))^T \mathbf{d}^k &\geq c_2\,(\boldsymbol{\nabla} f(\mathbf{x}^k))^T \mathbf{d}^k,
\end{aligned}
\tag{6.13}
$$

and the strong Wolfe conditions

$$
\begin{aligned}
f(\mathbf{x}^k + \alpha\,\mathbf{d}^k) &\leq f(\mathbf{x}^k) + c_1\,\alpha\,(\boldsymbol{\nabla} f(\mathbf{x}^k))^T \mathbf{d}^k, \\
\left|(\boldsymbol{\nabla} f(\mathbf{x}^k + \alpha\,\mathbf{d}^k))^T \mathbf{d}^k\right| &\geq c_2\,\left|(\boldsymbol{\nabla} f(\mathbf{x}^k))^T \mathbf{d}^k\right|.
\end{aligned}
\tag{6.14}
$$

The application of the Wolfe conditions is also straightforward as they are added on top of the Armijo rule using the same Algorithm 6.1 with Step 3 modified by adding an extra condition, either (6.11) or (6.12). As the readers may see later in Section 6.4, the Wolfe conditions are **easily applicable** in most line search methods. Finally, the Wolfe conditions are **scale-invariant**. They do not alter if either $f(\mathbf{x})$ is multiplied by a constant, or $\mathbf{x}$ is changed by any affine transformation (linear mapping that preserves points, straight lines, and planes).

6.3.3 Goldstein Conditions

Another modification applied to the sufficient decrease condition (6.8) helps prevent α from being too small and, thus, increases the performance of the search for the optimal step size

$$
\begin{aligned}
f(\mathbf{x}^k) + (1-c)\,\alpha\,(\boldsymbol{\nabla} f(\mathbf{x}^k))^T \mathbf{d}^k &\leq f(\mathbf{x}^k + \alpha\,\mathbf{d}^k) \\
&\leq f(\mathbf{x}^k) + c\,\alpha\,(\boldsymbol{\nabla} f(\mathbf{x}^k))^T \mathbf{d}^k,
\end{aligned}
\tag{6.15}
$$

with $0 < c < \frac{1}{2}$. The two inequalities in (6.15) are usually referred to as the *Goldstein conditions*. Its brief analysis shows that the second (right) inequality is, in fact, the sufficient decrease condition (6.8), while the first (left) one controls the step length from below.

The interested readers are encouraged to read more on both Goldstein and Wolfe conditions to see that they are relatively close in their convergence theory. Here, we need to mention that although the Goldstein conditions are well suited for Newton-type methods, they are not so good for various quasi-Newton approaches.

? *Why may the quasi-Newton algorithms not show the same performance as Newton's and discretized Newton's methods?*

The main disadvantage of using the Goldstein conditions is the first inequality (kind of "lower bound" for α) may exclude all minimizers of $\phi(\alpha)$.

! *The reader may use this fact as a starting point for answering the question above.*

6.3.4   Backtracking Line Search

Here, we provide an algorithm for the practical implementation of the *backtracking line search (BLS)* via sufficient decrease condition (the Armijo rule) discussed in Section 6.3.1.

Algorithm 6.2 (Backtracking Line Search via Armijo Rule)

1. *Choose*

   - *initial step size:* $\bar{\alpha} > 0$
   - *contraction factor:* $\rho \in (0, 1)$
   - *constant c_1 in (6.8):* $c \in (0, 1)$

2. *Set* $\alpha = \bar{\alpha}$

3. *Check* **sufficient decrease condition**

$$f(\mathbf{x}^k + \alpha\, \mathbf{d}^k) \leq f(\mathbf{x}^k) + c\,\alpha\, (\boldsymbol{\nabla} f(\mathbf{x}^k))^T \mathbf{d}^k \tag{6.16}$$

4. *If (6.16) satisfies* → **TERMINATE** *with* $\alpha^k = \alpha$

5. *Improve α by setting it to $\rho\alpha$, i.e.,*

$$\alpha \leftarrow \rho\alpha \tag{6.17}$$

6. *Go to Step 3*

In Algorithm 6.2, we recommend setting initial step size $\bar{\alpha}$ to 1 in Newton's or quasi-Newton methods. For other approaches, proper initial values could be found, e.g., by trials. Contraction factor ρ is allowed to vary at each iteration k of line search. Our final comment is that this algorithm is relatively well-suited for Newton-like methods but may show limited performance for other approaches.

? *Why and when the performance of the BLS method may be limited?*

The reader is advised to review the entire Section 6.3 and check the computational results obtained in Section 6.4 to answer this question.

6.4 Step Size Performance Comparison

We similarly structurize this section as Section 5.5. There, we compared the performance of various line search algorithms fixing the methodology of choosing optimal step size α to the golden section search. Here, however, we perform a direction search via the Newton-like, namely BFGS, and conjugate gradient methods. The focus will be on changes seen in the performance of both approaches, while applying simple and inexact line search strategies discussed in Sections 6.2 and 6.3.

In a similar fashion, we invite the reader first to check back on our idea developed initially in Section 3.8 to allow the existing code (created and then modified, respectively, in Chapters 3 and 5 for solving Example 1.3) to grow up and include new methods by preserving the same structure (building concept) for our entire computational framework. We advise the reader to compare the version of the `MATLAB` code used in Chapter 5 (`Chapter_5_data_fit_by_line_search.m`) and the updated version named `Chapter_6_data_fit_by_line_search.m` containing various step size search methods considered in the current chapter; refer to Table 6.1 for a summary for all implemented changes.

element	implementation	software
main OPT	written manually	MATLAB
f-evaluator	m-function, analytically defined function $f(\mathbf{a})$	MATLAB
d-evaluator	m-code, various line search approaches	MATLAB
1D α search	**m-code for various step size searches**	MATLAB
visualizer	plain m-code	MATLAB

TABLE 6.1
Computational elements of `Chapter_6_data_fit_by_line_search.m`.

In particular, we advise the reader to check the following `MATLAB` scripts to verify the **minimal changes** in the structure required for adding new functionalities:

- computational facility to perform 1D search for optimal step size α has been moved from m-file `mode_OPT.m` to the new one, `stepsize_search.m`, and re-organized by using `MATLAB`'s `switch` block to allow easy additions of any other step search methods (by adding extra `case` statements)

- m-files `params.m` and `initialize.m` are updated accordingly, e.g., with new variable `methodAlpha` to specify the α-search method and associated parameters (`alphaConst`, `alphaIni`, `alphaRho`, etc.).

MATLAB: `stepsize_search.m`

```matlab
switch methodAlpha                                                        1
                                                                          2
  case 'CONST'                                                            3
    alphaOpt = alphaConst;                                                4
                                                                          5
  case 'GS'                                                               6
    golden_section_search;                                                7
                                                                          8
  case 'DIM'                                                              9
    if (k==0)                                                            10
      alphaOpt = alphaIni;                                               11
    else                                                                 12
      alphaOpt = alphaOpt * alphaRho;                                    13
    end                                                                  14
                                                                         15
  case 'BLS'                                                             16
    alphaOpt = alphaIni;                                                 17
    % sufficient decrease condition                                     18
    while(f(a+alphaOpt*d,data) > f(a,data)+alphaC*alphaOpt*...           19
        (grad_f(a,data))'*d)                                            20
      alphaOpt = alphaOpt*alphaRho;                                      21
    end                                                                  22
                                                                         23
  case 'BLSW'                                                            24
    alphaOpt = alphaIni;                                                 25
    % Wolfe conditions                                                   26
    while(f(a+alphaOpt*d,data) > f(a,data)+alphaC*alphaOpt*...           27
        (grad_f(a,data))'*d || (grad_f(a+alphaOpt*d,data))'*d ...       28
        < alphaCw*(grad_f(a,data))'*d)                                  29
      alphaOpt = alphaOpt*alphaRho;                                      30
    end                                                                  31
                                                                         32
  case 'BLSSW'                                                           33
    alphaOpt = alphaIni;                                                 34
    % strong Wolfe conditions                                           35
    while(f(a+alphaOpt*d,data) > f(a,data)+alphaC*alphaOpt*...           36
        (grad_f(a,data))'*d || abs((grad_f(a+alphaOpt*d,data))'...      37
        *d) < abs(alphaCw*(grad_f(a,data))'*d))                         38
      alphaOpt = alphaOpt*alphaRho;                                      39
    end                                                                  40
                                                                         41
  case 'BLSG'                                                            42
    alphaOpt = alphaIni;                                                 43
    % Goldstein conditions                                              44
    while(f(a+alphaOpt*d,data) > f(a,data)+alphaC*alphaOpt*...           45
        (grad_f(a,data))'*d || f(a+alphaOpt*d,data) < f(a,data)...      46
```

```
            +(1-alphaC)*alphaOpt*(grad_f(a,data))'*d)           47
        alphaOpt = alphaOpt*alphaRho;                           48
    end                                                         49
                                                                50
    otherwise                                                   51
        disp(['Unknown alpha-search method ' methodAlpha ' chosen!']); 52
        exCode = 0; % termination due to error                  53
        return;                                                 54
                                                                55
end                                                             56
```

The reader now could check the functionality of the updated framework by running it, e.g., with the BFGS and conjugate gradient methods (`method = 'BFGS'` and `method = 'CG'`) supplied with the golden section search for optimal α (`methodAlpha = 'GS'`) and the same parameters used in the previous version of `params.m` file; refer to p. 117 for details. Figure 6.4(a) shows the results of both runs consistent with the results obtained previously for Figure 5.10.

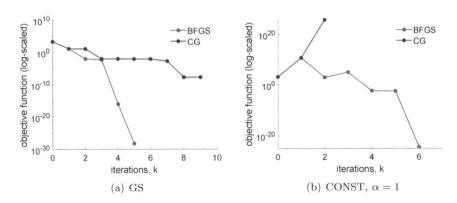

(a) GS

(b) CONST, $\alpha = 1$

FIGURE 6.4

(a) Excerpt graph from Figure 5.10 illustrating the performance of the BFGS and conjugate gradient methods both using the golden section search for α. (b) Performance of the BFGS and CG methods with constant $\alpha = 1$.

While the performance of other methods also shows sensitivity to the applied α-method (and the readers are encouraged to play with this), we have chosen the BFGS and CG ones as their convergence may vary significantly between linear and superlinear, as discussed in Sections 5.3.5 and 5.4. First, we would try the constant α (`methodAlpha = 'CONST'`) with a preset value of $\alpha = 1$. As we see it immediately, in Figure 6.4(b), the conjugate gradient method diverges as its search direction is not scaled. Even the BFGS approach experiences some difficulties not being able to improve objective function $f(\mathbf{x})$

at every iteration as it uses not the Hessian but its approximation; it becomes accurate after completing several steps. If we continue exploring the simplest possible (constant value) approach for choosing step size α (by playing with `alphaConst` variable), we might want to make various (tiny) steps to guarantee convergence. As seen in Figure 6.5, the outcomes are very method-dependent, making this strategy almost impossible for reliable predictions made for other methods and in applications to other problems. However, simplicity in implementation makes this method attractive as a valuable option for different stages in experimentation despite its slow convergence for varied α.

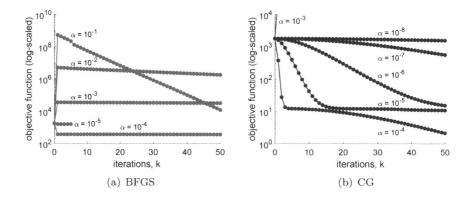

(a) BFGS   (b) CG

FIGURE 6.5
Performance of the (a) BFGS and (b) CG methods with various constant α values.

As the next step, we could move to another simple approach discussed in Section 6.2, diminishing step size via (6.6)–(6.7). As mentioned, it requires a "diminishing rule" that might be problem-dependent. As such, applying this method is not straightforward, especially if someone needs a systematic analysis of the performance while using it together with various line search approaches. We invite the reader to experiment with this method using the existing functionality (`methodAlpha = 'DIM'`) of our optimization framework: variables `alphaIni` for setting the initial $\bar{\alpha}$ value and `alphaRho` as the contraction factor ρ in the diminishing rule (6.17). We have to notice that the suggested rule borrowed from Step 5 in Algorithm 6.2 violates infinite travel condition (6.7). However, with $\rho \to 1$, this violation becomes less severe and setting $\rho = 1$ transforms the method to the constant step approach.

In an attempt to perform a systematic (probably, better to say pseudo-systematic) analysis, we vary the contraction factor ρ while using the BFGS ($0.9 \leq \rho \leq 1$) and CG ($0.1 \leq \rho \leq 1$) approaches. As initial $\bar{\alpha}$, we selected, respectively, 1 and 10^{-5} as values confirmed to be the good ones when the constant step is used; refer to Figures 6.4(b) and 6.5(b). Figure 6.6 presents the results that may be used for some analysis valid, again, only

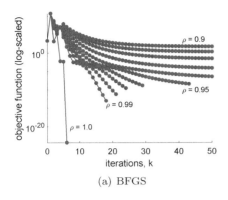

(a) BFGS

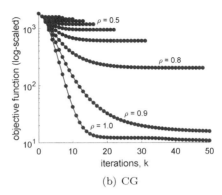

(b) CG

FIGURE 6.6
Performance of the (a) BFGS and (b) CG methods with applied diminishing step rule (6.17) with various contraction factor ρ.

within the frame of our current problem. For instance, our simple diminishing step rule $\alpha \leftarrow \rho\alpha$ cannot improve the results obtained with the constant steps. However, the created ρ-profiles may give the reader ideas about new diminishing rules with more potential leading to the improved convergence.

We could only expect a linear (sometimes, sublinear) computational convergence for both BFGS and CG methods, while applying constant or diminishing sizes of step α. The reader may also compare the results of the computational analysis obtained now (with $\rho = 0.99$ for BFGS and $\rho = 0.9$ for CG) and previously with the GS method; see, respectively, Figures 5.5(c,d) and 6.7 to confirm this conclusion.

At this point, we could stop our numerical experiments with the simple algorithms by concluding on their simplicity both toward implementation and computational efforts (no need in function evaluations). Now, we proceed with the group of inexact line search methods targeting to make the α-search less sensitive to the problem itself and the chosen method for the gradient-based direction search. As discussed in Section 6.3, these methods require moderate efforts to achieve adequate reduction in $f(\mathbf{x})$ at a minimal cost.

First, we apply the backtracking line search via the Armijo rule using Algorithm 6.2 of Section 6.3.4 with the same BFGS and CG methods. The performance of the BLS will change by varying three parameters: initial step size $\bar{\alpha}$ (same variable `alphaIni`), contraction factor ρ (same variable `alphaRho`), and slope parameter c (variable `alphaC`).

To make the analysis of our observations simple, we fix $\bar{\alpha} = 1$ for all experiments. By choosing c close to 1, we make the checked α values "less acceptable"; refer to Figure 6.2 for geometrical interpretation. Figure 6.8 shows the results of our experiments for two cases of contraction factor: $\rho = 0.75$ (smaller jumps) and $\rho = 0.25$ (bigger jumps). In the case of BFGS, $\rho = 0.75$

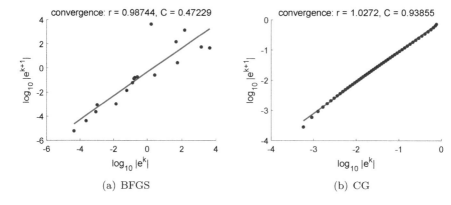

(a) BFGS   (b) CG

FIGURE 6.7
Computational convergence of the (a) BFGS ($\rho = 0.99$) and (b) CG ($\rho = 0.9$) methods with applied diminishing step rule (6.17).

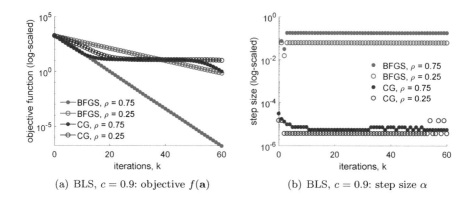

(a) BLS, $c = 0.9$: objective $f(\mathbf{a})$ (b) BLS, $c = 0.9$: step size α

FIGURE 6.8
Results of applying BLS approach for α-search with slope parameter $c = 0.9$ ("less acceptable").

makes more sense as it allows to find the optimal α^k within interval $[10^{-2}, 1]$ close to its initial value $\bar{\alpha} = 1$ when this method is expected to show its best performance. In its turn, the conjugate gradient makes more progress than previously: $\rho = 0.75$ allows better performance with optimal α spanning interval $[10^{-6}, 10^{-4}]$. The analysis of the computational convergence for $\rho = 0.75$ in Figure 6.9 now confirms its superlinear type, while the BFGS is still at linear one. Here, the reader may have a better understanding of the convergence. To reiterate, we consider it a **cumulative measure of the progress made at subsequent iterations** rather than a reference to **absolute numbers**, e.g., objective function $f(\mathbf{a})$ values; review formula (1.29) to clarify the rate of convergence concept.

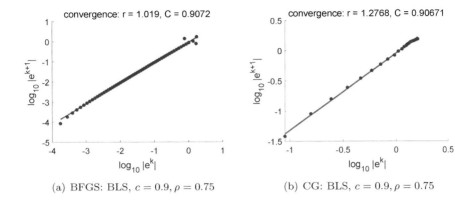

(a) BFGS: BLS, $c = 0.9, \rho = 0.75$ (b) CG: BLS, $c = 0.9, \rho = 0.75$

FIGURE 6.9
Computational convergence of the (a) BFGS and (b) CG methods paired with BLS approach using $c = 0.9$ and $\rho = 0.75$.

Next, we choose slope parameter c close to 0, e.g., $c = 0.1$, therefore, making the checked α values "more acceptable." Figure 6.10 shows the results of our experiments for the same contraction factors to allow smaller ($\rho = 0.75$) and bigger ($\rho = 0.25$) jumps. It permits BFGS to return to performance similar to that seen while applying the golden section method, see Figure 6.4(a). Here, the contraction factor plays an insignificant role as shortly after starting the optimization optimal steps α^k are allowed to take their initial values $\bar{\alpha} = 1$, see Figure 6.10(b) to boost the performance of the BFGS approach. The conjugate gradient also improves enabled to take larger steps, e.g., within $[10^{-4}, 10^{-2}]$ for $\rho = 0.75$.

The results obtained with "more acceptable" α ($c = 0.1$) are obviously more accurate than those obtained with "less acceptable" ones ($c = 0.9$). But how quickly do both methods converge to the optimal solution? We refer to Figure 6.11 for the analysis of the computational convergence. The BFGS approach now shows superlinear convergence, as expected. However, the conjugate gradient converges only linearly (for $\rho = 0.25$, we removed one outlier at the bottom; see Section 6.4 for discussion) by making about the same progress during subsequent iterations with no "accelerations" as seen, e.g., in case of the BFGS optimization. At this point, we assume the readers have enough data from our figures and their own numerical results to address the following two questions.

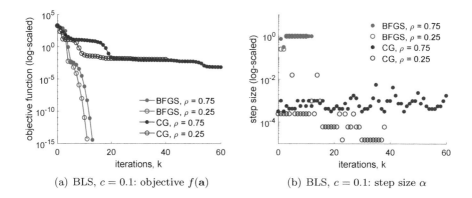

(a) BLS, $c = 0.1$: objective $f(\mathbf{a})$ (b) BLS, $c = 0.1$: step size α

FIGURE 6.10
Results of applying BLS approach for α-search with slope parameter $c = 0.1$ ("more acceptable").

> **?** *Could we further increase the performance of the conjugate gradient approach in terms of solution accuracy (by making $f(\mathbf{a})$ closer to 0)? What about convergence rate (by reaching it faster)?*

We may answer this question by performing a simple 2D search for optimal ρ and c in the backtracking line search method. So far, we experimented with two values of both contraction factor ρ (0.25 and 0.75) and slope parameter c (0.1 and 0.9). However, the reader may consider different strategies to conduct such a search within the interval $[0, 1]$ for both parameters. For instance, the performance of various pairs (ρ, c) may be examined by applying the brute force or Monte Carlo approaches discussed in Chapter 2 (with extended functionality from 1D to 2D). It may be also effective to improve one parameter, while keeping another fixed and then switching them in iterations (a modification of the *coordinate descent method*). We applied the latter to our problem to check how far we could improve the performance of the conjugate gradient method run with $c = 0.1$ and $\rho = 0.75$ in terms of solution accuracy.

Figure 6.12 shows the results of the first two steps (iterations). First, $c = 0.1$ is fixed, and optimization is running with various ρ taken from $[0.05, 0.95]$ discretized with step $\Delta\rho = 0.05$. The optimal value appears to be $\rho = 0.4$. Second, we fix this value and vary slope c also within the interval $[0.05, 0.95]$ discretized with step $\Delta c = 0.05$. As Figure 6.12(b) shows, we could consider 0.05, 0.1, 0.15, and 0.45 as optimal values for c (given fixed $\rho = 0.4$) where $c = 0.45$ provides a bit better accuracy.

The reader may proceed (in the same manner) to steps 3, 4, etc. However, we will stop at this point and make some comments. Although better parameters (ρ and c) are found, they are not unique. Even more, optimization using closer values of c (0.05 and 0.15) requires a different number of iterations to

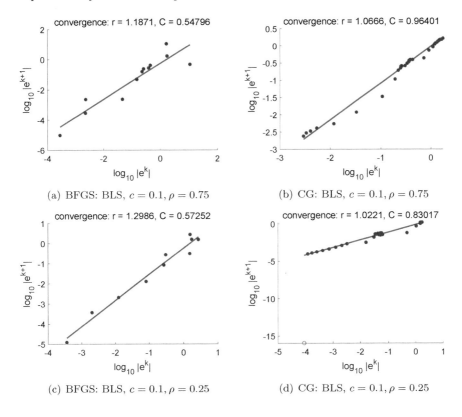

FIGURE 6.11
Computational convergence of the (a,c) BFGS and (b,d) CG methods paired
with BLS approach using $c = 0.1$ with (a,b) $\rho = 0.75$ and (c,d) $\rho = 0.25$.

complete. Far apart values (0.15 and 0.45) could lead to approximately the
same results (both accuracy and iteration count). In addition, the reader may
experiment more by changing, e.g., the initial step $\bar{\alpha}$, gradient-based method,
and the problem itself to notice how far the results of new "optimal" ρ and
c will deviate from those found here. Therefore, these parameters are very
problem-dependent, which is a well-known fact for parameter identification
algorithms. To conclude for future use, we will note the following:

(a) When choosing between various methods, consider finding a possible trade-
off between their accuracy and simplicity in tuning (adjusting parameters
for optimal performance).

(b) Check the performance and adjust necessary parameters after any, at least
significant, changes in the problem and solution method structures.

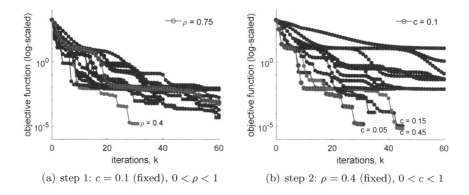

(a) step 1: $c = 0.1$ (fixed), $0 < \rho < 1$ (b) step 2: $\rho = 0.4$ (fixed), $0 < c < 1$

FIGURE 6.12
2-step search for optimal parameters $0 < \rho < 1$ and $0 < c < 1$ in the BLS method paired with the CG approach.

(c) Explore the sensitivity of the solution accuracy to all parameters in use. Fix the values of the less sensitive ones to reduce the dimensionality of the tuning process.

Now, for the current problem (Example 1.3) and the chosen gradient-based method (CG), we conclude on the (optimal?) parameters: $\rho = 0.4$ and $c = 0.45$. What about the convergence rate? Is it improved? We refer back to Figure 6.11(d) to find the old value, $r = 1.0221$ (linear convergence). Figure 6.13 shows the values $r = 1.274$ and $r = 1.3064$ after, respectively, steps 1 and 2 updates, both confirming faster (superlinear) convergence.

We leave it for the reader to explore more possibilities to enhance the performance (solution accuracy and convergence rate), while applying the Wolfe (`methodAlpha = 'BLSW'`) and strong Wolfe (`methodAlpha = 'BLSSW'`) conditions with one extra parameter c_2 (variable `alphaCw`). We will close the current section by comparing the results of applying the backtracking line search method for optimal α^k controlled by the Armijo (6.16) and Goldstein (6.15) conditions.

As presented in Figure 6.14, substituting the Armijo rule with the Goldstein conditions in Algorithm 6.2 provides the results of about the same accuracy (for $c = 0.15$ or $c = 0.05$) for $\rho = 0.4$ (fixed) and slope parameter c discretized within $[0.025, 0.475]$ using $\Delta c = 0.025$. While computational convergence appears a bit slower, $r = 1.186$, it is still superlinear and comparable with that obtained before. It might be possible to further improve it by adjusting the contraction factor ρ alone or performing the iterative search for optimal (ρ, c) as we did it before. The same analysis is applied to the results obtained with the Goldstein conditions (as well as the Wolfe and strong Wolfe conditions). It makes the process of parameter tuning just a technical routine.

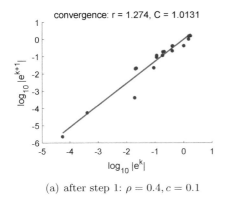

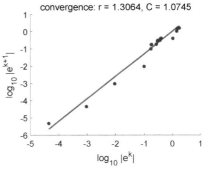

(a) after step 1: $\rho = 0.4, c = 0.1$ (b) after step 2: $\rho = 0.4, c = 0.45$

FIGURE 6.13
Computational convergence of the CG methods paired with BLS approach after (a) step 1 and (b) step 2 changes in ρ and c parameters.

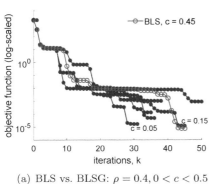

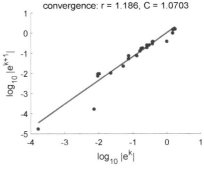

(a) BLS vs. BLSG: $\rho = 0.4, 0 < c < 0.5$ (b) BLSG: $\rho = 0.4, c = 0.15$

FIGURE 6.14
(a) Results of applying BLS approach for α-search based on the Goldstein (BLSG) conditions (6.15) with $\rho = 0.4$ and $0 < c < 0.5$. (b) Analysis for the computational convergence when $c = 0.15$.

> **?** *Setting initial step size $\bar{\alpha}$ and contraction factor ρ to values, respectively, bigger than and close to 1 would definitely allow more options for optimal α^k. How could this affect the computational load taking into account the necessity to evaluate objectives? How does the analysis we performed in this section help remove any redundant steps and avoid unnecessary efforts?*

! *Review the concept of inexact line search methods in Section 6.3 and re-*
fer to the results (step size α^k graphs) obtained in the current section by
applying various approaches supplied with different parameters.

6.5   Advanced Methods for 1D Search

In the previous section, we experimented extensively with various line search
algorithms for choosing optimal step size α by expanding the functionality of
our 1D search component in the existing computational framework. For doing
this, we added algorithms representing simple and inexact line search methods
and explored their performance in terms of both solution accuracy and convergence. The implemented approaches (CONST, DIM, GS, BLS, BLSW, BLSSW,
and BLSG) now provide quite a flexibility in choosing a proper one to satisfy
the need in 1D search applied to α^k. However, we conclude on the sensitivity
of all implemented methods to changes in their settings in particular and the
choice of the d-evaluator in general. Therefore, we are interested in finding a
suitable approach to minimize this sensitivity. At the same time, we will continue our discussion on the choice and implementation of the proper software
(see Section 3.2.2 for review) to further check the ability of our framework
to provide efficient communication and data processing functionality to any
added methods.

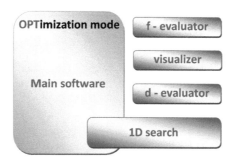

FIGURE 6.15
Generalized optimization framework: main (core) software part.

To restate the discussion above: our goal is to enhance our 1D search
module (currently structured by the stepsize_search.m file) by requiring
the following capabilities.

(a) Supporting the main optimization process by finding optimal step size α^k
at each iteration k.

(b) Being relatively **independent** of the nature of solved problems (linearity, convexity, constraints, etc.).

(c) Communicating effectively with f- and d-evaluators.

This goal is achievable using an *exact line search* approach to find the balance between **accuracy** and **computational cost** while solving 1D optimization problem (6.4). We selected the *bracketing-Brent* (BB) method written in MATLAB as a toolbox to perform computations in two steps:

- *Step 1:* **Bracketing** (finding interval containing the desirable step length by employing the (modified) golden section search).

- *Step 2:* **Interpolation** (computing a good step length within this interval by applying *Brent's method*).

We advise the reader to review the main concepts for the golden section search (Section 2.3 and Algorithm 2.2) used to support Step 1 in the BB approach. In the next sections, we discuss some theory for Step 2, practical implementation, and extended flexibility through added parameters, "floating" endpoints for search intervals, etc.

6.6 Brent's Method

Here, we briefly overview the concept used as a central point in *Brent's method*, namely *inverse parabolic interpolation*. Given values a, b, and c, the abscissa d that is the minimum (vertex) of a parabola through three points (not collinear) $(a, f(a))$, $(b, f(b))$, and $(c, f(c))$ can be computed as follows

$$d = b - \frac{1}{2} \frac{(b-a)^2[f(b)-f(c)] - (b-c)^2[f(b)-f(a)]}{(b-a)[f(b)-f(c)] - (b-c)[f(b)-f(a)]}. \tag{6.18}$$

Figure 6.16 illustrates the concept of the solution update performed after completing the inverse parabolic interpolation. Briefly, the initial triplet of points (1, 2, and 3) defines a parabola $\{1, 2, 3\}$ with the abscissa of its vertex d computed using (6.18). Evaluating minimized function $f(x)$ at $x = d$ produces a new point $(d, f(d))$, point 4 on the graph. Therefore, function $f(x)$ is now approximated in the neighborhood of its minimum by a new parabola passing through new point 4. The choice of the rest two points depends on the location of d: $a < d < b$ or $b < d < c$. For example, if we assume the former, as shown in Figure 6.16, a new parabola to approximate $f(x)$ will be $\{1, 4, 2\}$. A vertex for this parabola is again defined using (6.18) and evaluating function $f(x)$ (point 5). Repeating this algorithm in iterations leads to finding the global minimum of $f(x)$ over the interval $[a, c]$ under certain assumptions: e.g., continuity and unimodality of $f(x)$ inside $[a, c]$.

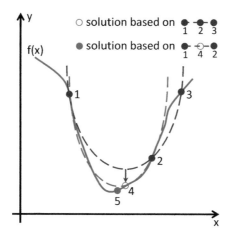

FIGURE 6.16
Schematic illustrating the solution update by the inverse parabolic interpolation of Brent's method.

The general framework for Brent's approach assumes switching between the golden section search and inverse parabolic interpolation whenever it has maximum benefits. Although the method itself may look pretty straightforward, its computational implementation is a nontrivial task due to:

- complicated housekeeping for **switching** between two methods to avoid unnecessary function evaluations,

- **accurate computations** when $f(x)$ is evaluated near to roundoff limit of the equation, and

- **robust scheme** to detect when switching should be performed.

We are not going to be deep inside all of these details. Instead, we will incorporate this algorithm into our existing optimization framework and play with its functionality similar to using any other "black-box" method. An interested reader could find more details on Brent's approach in Chapter 5 of the book of Richard P. Brent [7].

6.7   Bracketing-Brent Toolbox in `MATLAB`

Our bracketing-Brent toolbox in `MATLAB` consists of two m-files

- `fn_min_brack.m` for performing initial bracketing routine by using the (modified) golden section search, and

• `fn_brent.m` for minimizing $f(x)$ using line search by Brent's method called sequentially.

6.7.1 General Description

This set of routines was initially written in FORTRAN 77 and taken directly from Numerical Recipes [26]. The only slight modification is in the calling convention saving a few calls to function FUNC (if properly used). Two functions `fn_min_brack.m` and `fn_brent.m` achieve a line minimization of the user-supplied function FUNC, which is presumed to be a function of a single variable $f(x)$. It means that all appropriate parameters for the computation of FUNC must be passed to it.

MATLAB: `fn_min_brack.m` & `fn_brent.m`

Input:

• *initial guesses for* x: `AX, BX`

• *corresponding values of* FUNC: `FA, FB`

Output:

• *minimum point of* $f(x)$: `BX`

• *minimum value of* $f(x)$ *at the minimum point* $x =$ `BX`: `FB`

While a curious reader is encouraged to read more and look into the implementation of all methods used to run the bracketing-Brent toolbox, here, we will experiment with a new skill for adding and tuning the performance of a ready-to-use software. Again, this practical experience is particularly useful as we see it twofold. As previously, we want our computational framework to grow with no limitations for including new modules written by us from scratch as well as "third-party" codes assumed to work correctly. Thus, we will get to the next level of flexibility and ergonomic convenience by enhancing the modular structure of our framework with shared access to all existing functionalities. At the same time, working with the black-box tools requires some competence and prior research on at least the following:

(a) general suitability for the problems to be solved,

(b) structure of the input data, parameters, and initialization procedures,

(c) starting (calling) procedures,

(d) debugging and tuning the performance, and

(e) analysis of warning/error messages and unexpected outcomes.

In the rest of Section 6.7, we provide the reader with some details found to describe the functionality of the bracketing-Brent toolbox that may serve as a draft plan for the same type of exercises related to other software planned for future addition. We begin with Algorithm 6.3 to describe the general use of both `fn_min_brack.m` and `fn_brent.m` added to the existing code.

Algorithm 6.3 (Bracketing-Brent Toolbox)

1. Set up initial bracketing guess:

- `AX = AXini`
- `BX = BXini`
- `FA = FUNC(AX)`
- `FB = FUNC(BX)`

2. Call bracketing routine:

- `[AX, FA, BX, FB, CX, FC] = fn_min_brack(AX, BX, FA, FB, FUNC, MAXITER, GLIMIT)`

3. Call minimization routine:

- `[BX, FB] = fn_brent(AX, BX, CX, FA, FB, FC, FUNC, TOL, ITMAX)`

4. Expected output:

- `FB` *containing the line minimized value of* `FUNC`
- `BX` *containing the argument of* `FUNC`

achieved with tolerance `TOL`

As desired, the changes to the existing code are **minimal** and include

- added `fn_min_brack.m` and `fn_brent.m` files mentioned before with a new user-supplied function `FUNC` described in Section 6.7.2,

- extra `case` statement (`'BB'`) added to MATLAB's `switch` block in `stepsize_search.m`, and

- `params.m` file updated with the parameters (`AXini`, `BXini`, `MAXITER`, `GLIMIT`, `TOL`, `ITMAX`) to tune up the BB procedure.

MATLAB: 'BB' case added to stepsize_search.m

```
case 'BB'                                                            1
    % initial bracketing guess                                      2
    AX = AXini; BX = BXini;                                         3
    FA = FUNC(AX,a,d,f,data);                                       4
    FB = FUNC(BX,a,d,f,data);                                       5
    % calling bracketing routine                                    6
    [AX,FA,BX,FB,CX,FC] = fn_min_brack(AX,BX,FA,FB,@(alphaCurr) ...  7
        FUNC(alphaCurr,a,d,f,data),MAXITER,GLIMIT);                 8
    % calling minimization routine                                  9
    [BX,FB] = fn_brent(AX,BX,CX,FA,FB,FC,@(alphaCurr) ...           10
        FUNC(alphaCurr,a,d,f,data),TOL,ITMAX);                      11
    alphaOpt = BX;                                                  12
    % using BX from previous iterations                            13
    BXini = alphaOpt;                                              14
```

6.7.2 Initial Bracketing with fn_min_brack.m

As outlined in Section 6.5, Step 1 in the BB procedure (fn_min_brack.m) performs initial bracketing for a minimum of $f(x)$ by finding an interval containing the desirable step length α. To avoid any confusion among various bracketing techniques we will compare two of them, namely

(a) *bracketing a* **root** *of a function* $f(x)$ by finding interval $[a, b]$ (pair of points a and b) such that $f(a)f(b) < 0$ meaning $f(x)$ has opposite sign at those two points, and

(b) *bracketing a* **minimum** *of a function* $f(x)$ by finding a **triplet of points** $a < b < c$ such that $f(b) < f(a)$ and $f(b) < f(c)$ (recall that any smooth function has a minimum in the interval $[a, c]$.

Pursuing (b) target for bracketing, fn_min_brack.m implements this procedure with the following requirements.

- Making steps **downhill**.

- Taking steps as large as possible and **increasing step size** whenever possible using, e.g., **constant factor** or **parabolic extrapolation** of the preceding points. Incorporating this requirement, in fact, allows searching to the right side of the initial interval (a, c) to achieve better performance.

- Making necessary **alterations** to the interval (a, c) if stepping **uphill**.

At the beginning of our discussion on the BB structure, we mention the bracketing stage as being implemented using the modified golden section method. Briefly, if it moves inside the provided interval by making steps down-hill, it utilizes a regular golden section search, as described in Section 2.3. Otherwise, it attempts to alter this interval, assuming a better solution exists for $x > c$. We identify two parameters available for tuning to control this search

- `MAXITER` – maximum allowed number of iterations, and

- `GLIMIT` – maximum magnification allowed for parabolic-fit step used for altering the initial interval

with the following syntaxis and structure of the input/output data.

MATLAB: fn_min_brack.m

Input:

- *initial guesses for x to bracket the minimum of $f(x)$*: `AX, BX`

- *initial function values at points* `AX` *and* `BX`: `FA, FB`

- *full description of function $f(x)$ to be minimized*: `FUNC`

- *search parameters*: `MAXITER, GLIMIT`

Output:

- *bracket for the minimum point*: `AX, BX, CX`

- *with the respective function values*: `FA, FB, FC`

Syntaxis: see Section 6.7.4 for details

```
[AX,FA,BX,FB,CX,FC]
= fn_min_brack(AX,BX,FA,FB,FUNC,MAXITER,GLIMIT)
```

Finally, running the bracketing procedure may be followed by two self-explanatory warning messages:

- *Bracketing: proceeding uphill - will use a safer bracketing procedure!*

- *Bracketing: MAXITER exceeded, could not bracket the minimum!*

6.7.3   Brent Minimization with `fn_brent.m`

Next, Step 2 in the BB procedure (`fn_brent.m`) performs line minimization of $f(x)$ using Brent's method briefly described in Section 6.6. Here, we also identify two tuning parameters available to control this method

- TOL – tolerance for accuracy of the final answer, and

- ITMAX – maximum number of iterations

with the following syntaxis and structure of the input/output data.

MATLAB: fn_brent.m

Input:

- *bracketing triplet of points (from* fn_min_brack.m*)*: AX, BX, CX

- *initial function values at points* AX, BX, FC: FA, FB, FC

- *full description of function $f(x)$ to be minimized*: FUNC

- *search parameters*: TOL, ITMAX

Output:

- *minimum point of $f(x)$*: BX

- *minimum value of $f(x)$ at the minimum point $x = $* BX: FB

Syntaxis: see Section 6.7.4 for details

[BX,FB] = fn_brent(AX,BX,CX,FA,FB,FC,FUNC,TOL,ITMAX)

As before, running this procedure may be followed by a warning message in case the number of iterations exceeds the number provided by the ITMAX parameter:

- *Brent: no convergence in ... iterations!*

6.7.4 Technicalities for MATLAB Implementation

Although running the bracketing-Brent toolbox in MATLAB in application to various optimization problems may raise multiple questions and concerns for proper implementation and tuning procedures for the best performance, we will focus here on three questions.

? *How to construct the user-supplied function* FUNC *and pass it to both routines (*fn_min_brack.m *and* fn_brent.m*) as a function of* **one** *argument?*

We remind the reader that for solving Example 1.3 we currently use the function fn_eval_f(a, data) defined in m-file fn_eval_f.m to evaluate objective $f(\mathbf{a})$ by providing the current value of control $\mathbf{a}$ and data (x_i, y_i). To achieve maximum flexibility in these function calls, we also created a function handler f = @fn_eval_f placed in initialize.m.

Having a unified handler (f) allows easy changes in the evaluated function being processed without any corrections in the code where this functionality (f-evaluation) is requested. However, we recall that if employing the *exact line search*, we no longer need evaluating function $f(\mathbf{a})$. Instead, solving the following 1D optimization problem

$$\alpha^k = \underset{0<\alpha\leq\delta}{\mathrm{argmin}}\ f\left(\mathbf{a}^k + \alpha \cdot \mathbf{d}^k\right) \tag{6.19}$$

requires, in fact, multiple evaluations of function $f(\alpha) = f\left(\mathbf{a}^k + \alpha \cdot \mathbf{d}^k\right)$, which is a function of one variable, current α, and multiple parameters, such as current control $\mathbf{a}^k$, search direction $\mathbf{d}^k$, data (x_i, y_i), and, eventually, function $f(\mathbf{a})$ itself. Therefore, we continue in the same fashion to support the ergonomic convenience of our framework by creating another function FUNC used exclusively by the exact line search BB method locating it in the separate file FUNC.m.

MATLAB: function $f\left(\mathbf{a}^k + \alpha \cdot \mathbf{d}^k\right)$ by FUNC.m

```
function f_FUNC = FUNC(alphaCurr,a,d,f,data)              1
                                                         2
   f_FUNC = f(a+alphaCurr*d,data);                        3
                                                         4
return                                                   5
```

The reader now could review the structure of the new **case** statement 'BB' added to MATLAB's **switch** block in stepsize_search.m on p. 153 to note the calling concept for single variable function $f(\alpha)$:

@(alphaCurr) FUNC(alphaCurr,a,d,f,data)

Calling FUNC from another function (e.g., fn_min_brack.m) is even simpler

FC = FUNC(CX);

as all the rest parameters (...,a,d,f,data) are fixed and kept inside the FUNC definition.

? *How to prevent divergence due to the* MAXITER *condition?*

We could recommend the following.

- Keep tracking the *"Bracketing: MAXITER exceeded, could not bracket the minimum!"* message.

- Increase the value of `MAXITER` for the current iteration or entire process.

- Deactivate or re-define `MAXITER` condition.

Despite being very general, some readers will find this advice very helpful. And, finally,

 *How to speedup computations?*

Here, we also could provide some advice with the potential for practical implementation.

- Use `BX` values "inherited" from previous iterations (already implemented in the current version, see p. 153 for details).

- Decrease value of `ITMAX` in Brent's method for the current iteration or entire process (of course, at the expense of lower accuracy).

- Deactivate Brent's method by setting `ITMAX` to 0 and substitute it with any simple interpolation approach.

6.8   Bracketing-Brent Toolbox Performance

In the last section of this chapter, we explore the performance of the bracketing-Brent toolbox added to the existing optimization framework by comparing it with the pure golden section search method and discussing the use of its various settings. We will use the current version of the MATLAB code, `Chapter_6_data_fit_by_line_search.m`, with added files `fn_min_brack.m`, `fn_brent.m`, and `FUNC.m`, and updated files `stepsize_search.m` and `params.m` as discussed in Section 6.7; refer to Table 6.2 for a summary of implemented changes.

Before proceeding to new experiments with the BB method functionality, we could run the updated framework with multiple gradient-based line search methods, e.g., the BFGS, CG, and DS-SD methods (`method = 'BFGS'`, `method = 'CG'`, and `method = 'DSSD'`) supplied with the golden section search for optimal α (`methodAlpha = 'GS'`) and the same parameters used previously for the GS method; refer to `params.m` file on pp. 117 and 158 for details. Figure 6.17 shows the results of these three runs consistent with the results obtained previously for Figure 5.10.

TABLE 6.2
Computational elements of `Chapter_6_data_fit_by_line_search.m` upgraded
with added functionality of the bracketing-Brent method.

element	implementation	software
main OPT	written manually	MATLAB
f-evaluator	m-function, analytically defined function $f(\mathbf{a})$	MATLAB
d-evaluator	m-code, various line search approaches	MATLAB
1D α search	**m-code, various methods including BB**	MATLAB
visualizer	plain m-code	MATLAB

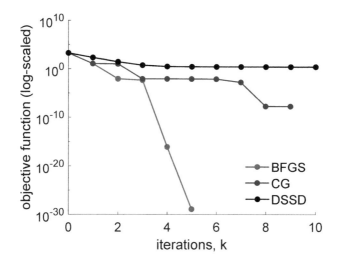

FIGURE 6.17
Excerpt graph from Figure 5.10 illustrating the performance of the BFGS,
CG, and DS-SD methods both using the golden section search for α.

Now, we could check the functionality of the BB method discussed in Sections 6.6 and 6.7 by running updated `Chapter_6_data_fit_by_line_search.m`
with the following settings (to maintain consistency with the results already
obtained in Chapters 5 and 6).

MATLAB: params.m

Settings for GS vs. BB comparison:

- *line search method:* method = 'BFGS'; choosing from the list: BFGS, CG, and DSSD

- *input data:* dataFile = 'data_main.dat';

- *initial guess:* aini = [-14 11 -2];

- *FD approximation step (for Hessian):* stepH = 1e-4;

- *step size search method:* methodAlpha = 'GS'; or methodAlpha = 'BB';

- *GS-search interval:* alphaA = 0; alphaB = 10; choosing also alphaB = 1; or alphaB = 0.1;

- *GS-search termination:* alphaEps = 1e-6;

- *BB-search interval:* AXini = 0; BXini = 0.1;

- *BB-search bracketing parameters:* MAXITER = 20; GLIMIT = 100.0;

- *BB-search Brent parameters:* TOL = 1e-6; ITMAX = 2;

- *termination #1:* epsilonF = 1e-6; for ϵ_1, refer to (3.18)

- *termination #2:* epsilonA = 1e-6; for ϵ_2, refer to (3.17)

- *termination #3:* kMax = 200;

We previously ran quasi-Newton's BFGS paired with the golden section method for choosing optimal step size α within the interval $[0, 10]$. It allowed some flexibility resulting in convergence using only 5 iterations. However, whenever $\alpha^k \ll 1$ starting with such a large interval leads to unnecessary objective function evaluations to complete the search. We could adjust this interval, say to $[0, 1]$ (alphaB = 1), by assuming that BFGS will use α values of at most 1. As Figures 6.18 and 6.19(a) show, this assumption works only for the first two iterations. Then the performance deteriorates due to the inability to set $\alpha^k > 1$, leading to the overall convergence rate of about 1.3 (just superlinear convergence instead of expected quadratic; refer to Figure 5.5(c)).

The current version of the bracketing-Brent toolbox does not have such a drawback. As mentioned in Section 6.7, the current implementation of the BB approach adjusts the right boundary of the search interval using values of optimal α "inherited" from previous iterations (BXini = alphaOpt; in line 14 of stepsize_search.m, see p. 153). It is based on the assumption that closer proximity to the optimal solution requires smaller steps in the search

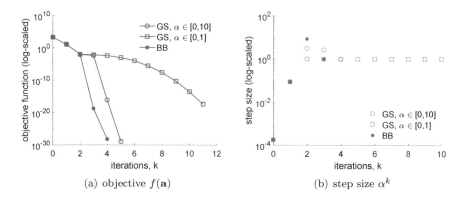

(a) objective $f(\mathbf{a})$ (b) step size α^k

FIGURE 6.18
Results of applying the BFGS method paired with the GS and BB approaches
for finding optimal step size α^k.

direction. Also, the search interval is altered (by allowing bigger steps) if stepping uphill within this interval is detected. The combination of these two features makes the BB approach (based on the same GS algorithm) more robust
and less dependent on the solved problem and paired method for selecting the
search direction. Here, for example, we choose the initial interval $[0, 0.1]$ to
be relatively small (`AXini = 0` and `BXini = 0.1`) and limiting parameter to
"magnify" the right boundary quite big (`GLIMIT = 100.0`) to pursue better
performance with minimal computational costs.

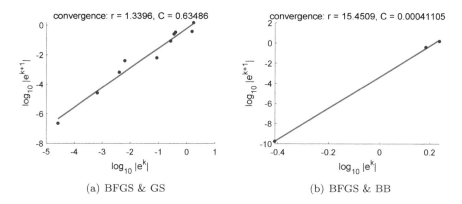

(a) BFGS & GS   (b) BFGS & BB

FIGURE 6.19
Computational convergence for the BFGS method paired with (a) the golden
section search for $\alpha \in [0, 1]$ and (b) the BB toolbox.

Figure 6.18 shows the results of applying the BB approach as a method for α-search. It finalizes in 4 iterations with the same solution accuracy as the GS method used with $\alpha \in [0,10]$. However, the former takes about 40 objective function $f(\mathbf{a})$ evaluations, while the latter completes the search with 180 evaluations. The convergence analysis for the BB returns a fake value of 15.5, see Figure 6.19(b), for the rate explained by the small number of points used for its evaluation. We could assume the convergence to be quadratic (or even better) as the BFGS obviously uses its full power.

As our next step, we could shift from the BFGS to the conjugate gradient approach (`method = 'CG'`) to compare the performance of both methods, the pure GS and BB, applied for α-search. As seen in Figure 6.17, a previous run of the CG paired with the GS supplied with the initial interval $[0,10]$ converged nicely in 9 iterations showing linear/superliner convergence according to analysis from Figures 5.5(d) and 5.11. However, the count for the objective function $f(\mathbf{a})$ evaluations returns the value of 325. Like for the case of using the BFGS, we could also adjust the search interval, say to $[0,0.1]$ (`alphaB = 0.1`), by assuming that CG will use α values of at most 0.1. We base this decision on the analysis of the step size α^k history seen in Figure 6.20(b), where all steps (empty circles) are below 0.1 except two (which are close to 1). This run finalizes now in 17 iterations and with about 440 objective evaluations. What about the accuracy of the obtained solution? It is a bit worse than in the case of using $[0,10]$ interval; see Figure 6.20(a). However, the computational convergence is still superlinear; refer to Figure 6.21(a).

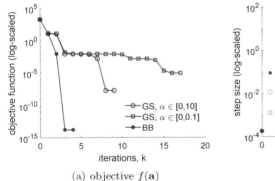
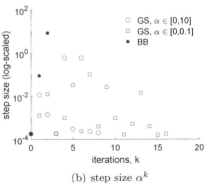

(a) objective $f(\mathbf{a})$ (b) step size α^k

FIGURE 6.20
Results of applying the CG method paired with the GS and BB approaches for finding optimal step size α^k.

Next, we apply the bracketing-Brent toolbox with the same settings as shown in the `params.m` file on p. 158. The idea is to minimize the changes in the settings once we have to switch between different methods for the search direction $\mathbf{d}^k$. Running this optimization returns much better results in terms of the solution accuracy (see Figure 6.20(a) for the comparison) and computational efficacy confirmed by just four iterations and 45 objective evaluations. Starting from the initial search interval $[0, 10^{-1}]$ does not prevent the BB toolbox from finding optimal steps of various orders, namely within $[10^{-4}, 10^2]$ as seen in Figure 6.20(b). This boost in the computational performance is supported by the superlinear/quadratic (maybe even better) convergence; refer to Figure 6.21(b). Without a doubt, applying the BB approach helps release the hidden potential of conjugate gradients by their optimal scaling.

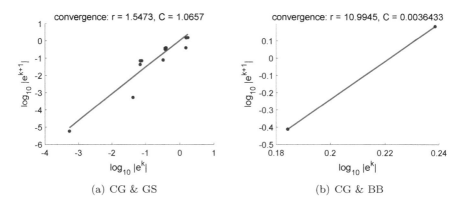

(a) CG & GS   (b) CG & BB

FIGURE 6.21
Computational convergence for the CG method paired with (a) the golden section search for $\alpha \in [0, 0.1]$ and (b) the BB toolbox.

Before proceeding to the last numerical experiment in this section, we must provide a "disclaimer" about the results obtained using the BB toolbox. The reader should avoid concluding that setting parameters BXini and GLIMIT to, respectively, very small and very large values will inevitably lead to superior performance as observed in our recent experiments. Although the BB approach evidently shows less sensitivity to changes in its parameters, it still requires some (moderate) efforts for tuning. The reader is advised to play more with all five parameters (BXini, MAXITER, GLIMIT, TOL, ITMAX) driving the BB functionality to explore this issue and gain practical experience to adjust the BB for the best possible performance. Our last numerical experiment is an example of such examinations.

Here, we change the conjugate gradient approach to a much slower diagonally scaled steepest descent (method = 'DSSD') to experiment with just one parameter, namely MAXITER, responsible for terminating the bracketing stage in the BB toolbox. We could explore the idea of the possibility to make our computations "cheaper" by limiting the number of iterations for proper bracketing. As seen in Figure 6.22, our initial setting (MAXITER = 20) allows rather large steps within the interval $[0.1, 10]$, resulting in good progress made, especially during the first (4 or 5) iterations and overall performance. Reducing

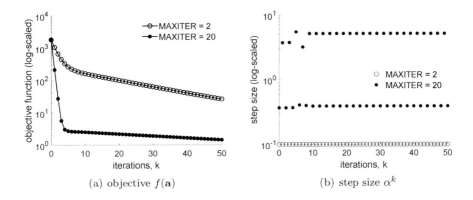

(a) objective $f(\mathbf{a})$ (b) step size α^k

FIGURE 6.22
Results of applying the DS-SD method paired with the BB approach for finding optimal step size α^k.

the value of `MAXITER` to 2 deteriorates the performance by, e.g., disabling the search outside the initial interval $[0, 0.1]$. It is also confirmed by the (slightly) reduced (from 1.04 to 1.02) convergence rate analyzed in Figure 6.23. The easiness of computations is also questionable: the count for the objective function $f(\mathbf{a})$ evaluations during the first five iterations returns, respectively, 172 (`MAXITER = 20`) and 168 (`MAXITER = 2`). We leave the complete analysis of these results to the reader.

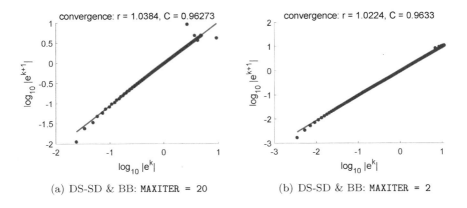

(a) DS-SD & BB: `MAXITER = 20`   (b) DS-SD & BB: `MAXITER = 2`

FIGURE 6.23
Computational convergence for the DS-SD method paired with the BB toolbox with parameter (a) `MAXITER = 20` and (b) `MAXITER = 2`.

! *Think about multiple issues playing together, such as*

- *inaccurate bracketing, in general, passed through consequent iterations,*
- *disabled search outside the provided interval (mentioned above),*
- *performance of the subsequent Brent algorithm based on inaccurate bracketing, etc.*

Finally, we conclude that many other interesting experiments leading to better performance are possible. However, we close this chapter at this point in hopes that the reader now has enough experience and practical skills to fully understand the concept of "proper communication" between different parts of a gradient-based optimization algorithm, namely d- and α-search methods. As discussed before, this concept adds a substantial portion to the big picture illustrating the structure of the efficient computational framework capable to solve problems of different complexities.

6.9   Homework Problems

1. Suggest any ideas on the diminishing step rules to apply in (6.6)–(6.7) and check the performance by implementing the new rule(s) within the existing optimization framework. Conclude on the choice of initial $\bar{\alpha}$, any parameters used, and the line search method selected for running the optimization.

2. Explore MATLAB code Chapter_6_data_fit_by_line_search.m for Example 1.3 with added (in addition to the existing constant and golden section search methods) the following approaches to find step size α^k:

   (a) diminishing step size,
   (b) backtracking line search with the Armijo rule,
   (c) BLS with the Wolfe and strong Wolfe conditions, and
   (d) BLS with the Goldstein conditions.

   Choose any existing approach for computing search directions and perform optimization with all α-search methods. Check the convergence and approximate convergence parameters r and C for every case. Rate these methods in terms of their rate of convergence and conclude.

3. Research the possibilities of enhancing the performance (solution accuracy and convergence rate), while applying the Wolfe and strong Wolfe conditions by finding the optimal set of parameters (ρ, c, c_2). Compare the results with those obtained using the Armijo rule and Goldstein conditions and conclude.

4. Explore `MATLAB` code `Chapter_6_data_fit_by_line_search.m` for Example 1.3 with added bracketing-Brent toolbox and incorporate any suggestions for avoiding divergence and speeding up computations by using strategies considered in Section 6.7.4.

5. Apply bracketing-Brent toolbox for finding optimal step size α^k together with all available methods for search directions (SD, NEWTON, DN, MN, DSSD, BFGS, GN, and CG). Play with bracketing and Brent parameters (`BXini, MAXITER, GLIMIT, TOL, ITMAX`) for each method to find an "optimal set" by checking the convergence and approximating convergence parameters r and C for every case. Check if this set has to be updated in case you change the initial guess $\mathbf{a}^0$ or the problem itself.

6. Similarly, check the performance of the bracketing-Brent toolbox with active/inactive "flexible" right boundary `BX` for the search interval and conclude.

7. Update `MATLAB` code `Chapter_6_data_fit_by_line_search.m` by adding extra functionality for counting the total number of objective function $f(\mathbf{a})$ evaluations. Repeat experiments with various methods for d- and α-search and conclude on the found trade-offs between the solution quality (accuracy) and required computational costs.

| READ | **Where to Read More**

Bertsekas (2016), [4]
Chapter 1 (Unconstrained Optimization: Basic Methods)

Boyd (2004), [6]
Chapter 9 (Unconstrained Minimization),

Griva (2009), [15]
Chapter 11 (Basics of Unconstrained Optimization), Chapter 12 (Methods of Unconstrained Optimization), Chapter 13 (Low-Storage Methods for Unconstrained Optimization)

Nocedal (2006), [25]
Chapter 3 (Line Search Methods)

Press (2007), [26]
Chapter 9 (Root Finding and Nonlinear Sets of Equations), Chapter 10 (Minimization or Maximization of Functions)

| RUN | **MATLAB Codes for Chapter 6**

- `Chapter_6_data_fit_by_line_search.m`

- `params.m`

- data_main.dat
- data_6pt.dat
- initialize.m
- mode_OPT.m
- fn_eval_f.m
- fn_eval_f_grad.m
- fn_eval_g.m
- fn_eval_g_grad.m
- fn_eval_hess.m
- fn_eval_hess_approx.m
- direction_search.m
- golden_section_search.m
- fn_convergence_sol_norm.m
- mode_TEST.m
- kappa_test.m
- visualize.m
- stepsize_search.m
- fn_min_brack.m
- fn_brent.m
- FUNC.m

7

Trust Region and Derivative-Free Methods

In this chapter, we broaden the discussion about the possibility for various approaches to optimally coexist in the same computational framework by sharing its functionality. We add the trust region method supplied with the simple dogleg algorithm as an alternative to the line search methods discussed and incorporated into computations in previous chapters. We also explore the derivative-free options by expanding the brute-force and Monte-Carlo approaches from applications in 1D (in Chapter 2) to our current 3D optimization problem.

7.1 Trust Region Outline

After practicing a lot with the practical implementation of various gradient-based line search algorithms in Chapter 5 paired with methods for choosing optimal step sizes in Chapter 6, we would like the reader to revisit Chapter 4 and Algorithm 1.1, discussing the general concept of optimization performed in iterations. This concept includes generating a sequence of suboptimal solutions or iterates $\{\mathbf{x}^0, \mathbf{x}^1, \mathbf{x}^2, \ldots, \mathbf{x}^k, \mathbf{x}^{k+1}, \ldots\}$ (see also Figure 4.1 illustrating the concept), while solving the following unconstrained optimization problem

$$\min_{\mathbf{x} \in \mathbb{R}^n} f(\mathbf{x}) \tag{7.1}$$

under the assumption $f(\mathbf{x}^{k+1}) < f(\mathbf{x}^k)$, facilitating monotonicity in decreased objective function $f(\mathbf{x})$ satisfied for every k.

In Section 4.2, we compared two fundamental strategies both based on the evaluated gradients, namely line search and trust region. Our focus was on the conceptual difference to help understand the ability and specificity of running both approaches within the same computational framework. We also discussed the practical implementation of line search methods in Chapter 5. In the current chapter, we consider the trust region approach more in-depth: technicalities of incorporating into the current framework by utilizing its existing functionalities, tuning steps, and the performance analysis while funding a solution for our Example 1.3.

DOI: 10.1201/9781003275169-7

In a nutshell, the general concept of any *trust region method* consists of two steps: (1) constructing a model function (often defined to be a quadratic function for easy minimization) to approximate function $f(\mathbf{x})$ near $\mathbf{x}^k$, e.g.,

$$m^k(\mathbf{p}) = m^k(\mathbf{x}^k + \mathbf{p}) = f(\mathbf{x}^k) + \mathbf{p}^T \nabla f(\mathbf{x}^k) + \frac{1}{2}\mathbf{p}^T B^k \mathbf{p}, \qquad (7.2)$$

and (2) minimizing m^k within some (trust) region around $\mathbf{x}^k$ by solving optimization problem

$$\mathbf{p}^k = \operatorname*{argmin}_{\|\mathbf{p}\|_2 \leq \Delta^k} m^k\left(\mathbf{x}^k + \mathbf{p}\right) \qquad (7.3)$$

to update the solution with the candidate step $\mathbf{p}^k$, i.e.,

$$\mathbf{x}^{k+1} = \mathbf{x}^k + \mathbf{p}^k. \qquad (7.4)$$

(7.3) represents, in fact, a constrained optimization problem as $\mathbf{x}^k + \mathbf{p}$ should be inside the trust region that could be a ball (also ellipse, box, etc.), i.e., $\|\mathbf{p}\|_2 \leq \Delta^k$, where $\Delta^k > 0$ is a **trust-region radius** and should be adjusted to ensure a sufficient decrease in $f(\mathbf{x})$. Finally, direction $\mathbf{p}^k$ and its scaler (step size) have to be chosen **simultaneously** to find new minimizer $\mathbf{x}^{k+1}$; refer to Figure 4.3 and Section 4.2.2 for a more detailed review.

To make this review complete, we need to mention a few known facts regarding the optimization problem defined in (7.2)–(7.3). First of all, objective $m^k(\mathbf{p})$ and constraint $\|\mathbf{p}\|_2^2 = \mathbf{p}^T\mathbf{p} \leq (\Delta^k)^2$ are both **quadratic** functions. Based on that, if B^k in (7.2) is a positive definite matrix, i.e.,

$$B^k \succ 0 \qquad (7.5)$$

and

$$\|[B^k]^{-1}\nabla f(\mathbf{x}^k)\|_2 \leq \Delta^k, \qquad (7.6)$$

the solution of (7.2)–(7.3) is unconstrained minimum

$$\mathbf{p}_B^k = -[B^k]^{-1}\nabla f(\mathbf{x}^k), \qquad (7.7)$$

which is also called a **full step**. If both conditions (7.5) and (7.6) are not met, the trust region method may use an approximate solution to obtain convergence and good practical behavior.

Finally, given step $\mathbf{p}^k$, the trust-region radius Δ^k may be defined, e.g., by computing the ratio between the **actual reduction** $f(\mathbf{x}^k) - f(\mathbf{x}^k + \mathbf{p}^k)$ and the **predicted reduction** $m^k(0) - m^k(\mathbf{p}^k) > 0$

$$\rho^k = \frac{f(\mathbf{x}^k) - f(\mathbf{x}^k + \mathbf{p}^k)}{m^k(\mathbf{0}) - m^k(\mathbf{p}^k)} \qquad (7.8)$$

followed by the decision analysis for Δ^k:

- $\rho^k < 0$: step Δ^k is **rejected** due to violation $f(\mathbf{x}^k + \mathbf{p}^k) < f(\mathbf{x}^k)$,

- $\rho^k \to 1$: there is a good agreement between m^k and $f(\mathbf{x})$, leading to **expanding/keeping the same** Δ^k, and

- $\rho^k \to 0$: bad agreement means Δ^k has to be **altered**.

7.2 General Algorithm

Using the discussion results of the previous section, we summarize the general procedure for applying the trust region method in Algorithm 7.1.

Algorithm 7.1 (General Algorithm for Trust Region Method)

1. *Choose parameters:*

$$\bar{\Delta} > 0, \qquad \Delta^0 \in (0, \bar{\Delta}), \qquad \eta \in \left[0, \frac{1}{4}\right]$$

2. *Set $k = 0$*

3. *Obtain $\mathbf{p}^k$ by solving (approximately)*

$$\min_{\|\mathbf{p}\|_2 \leq \Delta^k} m^k(\mathbf{p}) = f(\mathbf{x}^k) + \mathbf{p}^T \nabla f(\mathbf{x}^k) + \frac{1}{2} \mathbf{p}^T B^k \mathbf{p} \qquad (7.9)$$

4. *Evaluate ratio*

$$\rho^k = \frac{f(\mathbf{x}^k) - f(\mathbf{x}^k + \mathbf{p}^k)}{m^k(\mathbf{0}) - m^k(\mathbf{p}^k)}$$

5. *If $\rho^k < \dfrac{1}{4}$*

$$\Delta^{k+1} = \frac{1}{4} \|\mathbf{p}^k\|_2$$

else

- *if $\rho^k > \dfrac{3}{4}$ and $\|\mathbf{p}^k\|_2 = \Delta^k$*

$$\Delta^{k+1} = \min\left(2\Delta^k, \bar{\Delta}\right)$$

- *else*

$$\Delta^{k+1} = \Delta^k$$

6. *If $\rho^k > \eta$*

$$\mathbf{x}^{k+1} = \mathbf{x}^k + \mathbf{p}^k$$

else

$$\mathbf{x}^{k+1} = \mathbf{x}^k$$

7. *Set $k \leftarrow k + 1$ and go to Step 3*

? *Implementing Algorithm 7.1 is fairly straightforward, but how to solve approximately optimization subproblem (7.9) to make it **practical**?*

One of many other suggestions is approximating solutions using the *Cauchy point*, a minimizer of m^k along the steepest descent direction $-\nabla f(\mathbf{x}^k)$ subject to trust-region bound. Based on the structure of the computed Hessians B^k, we may perform computations for the Cauchy point by various methods.

- *Dogleg method:* it is good when Hessian $B^k \succ 0$.

- *Subspace minimization:* it could be applied when B^k is indefinite, see [12].

- *Steihaug method:* it is appropriate when B^k is the exact (large and sparse) Hessian $\nabla^2 f(\mathbf{x}^k)$; refer, e.g., to [27] for details.

In addition to the Cauchy point, the reader could explore many other methods, e.g., "nearly exact" solution due to Moré and Sorensen in [24], etc. In the rest of this chapter, we will discuss the Cauchy point calculation techniques and practical use of the dogleg method for improving this point.

7.3   Cauchy Point Calculation

In this section, we outline briefly the procedure for calculating the *Cauchy point* $\mathbf{p}_C^k$ – a minimizer of m^k along the steepest descent direction $-\nabla f(\mathbf{x}^k)$ subject to trust-region bound $\|\mathbf{p}\|_2 \le \Delta^k$ (refer to Algorithm 7.2 for details).

Algorithm 7.2 (Cauchy Point Calculation)

1. *Find vector $\mathbf{p}_s^k$ which solves linearized subproblem*

$$\mathbf{p}_s^k = \underset{\|\mathbf{p}\|_2 \le \Delta^k}{\operatorname{argmin}} \; f(\mathbf{x}^k) + \mathbf{p}^T \nabla f(\mathbf{x}^k) \qquad (7.10)$$

2. *Find minimizer τ^k of $m^k(\tau \mathbf{p}_s^k)$*

$$\tau^k = \underset{\tau > 0}{\operatorname{argmin}} \; m^k(\tau \mathbf{p}_s^k) \qquad (7.11)$$

   subject to the trust-region bound condition

$$\|\tau \mathbf{p}_s^k\|_2 \le \Delta^k$$

3. *Setup the Cauchy point $\mathbf{p}_C^k$ as*

$$\mathbf{p}_C^k = \tau^k \mathbf{p}_s^k$$

We have to note that computations provided in Algorithm 7.2 have a closed-form expression:

$$\mathbf{p}_C^k = -\tau^k \frac{\Delta^k}{\|\nabla f(\mathbf{x}^k)\|_2} \nabla f(\mathbf{x}^k), \qquad (7.12)$$

where

$$\tau^k = \begin{cases} 1, & (\nabla f(\mathbf{x}^k))^T B^k \nabla f(\mathbf{x}^k) \leq 0 \\ \min\left(\dfrac{\|\nabla f(\mathbf{x}^k)\|_2^3}{\Delta^k (\nabla f(\mathbf{x}^k))^T B^k \nabla f(\mathbf{x}^k)}, 1\right), & \text{otherwise} \end{cases}$$

$$(7.13)$$

We leave the derivation of (7.12)–(7.13) to the reader.

? *As Δ^k serves as a trust-region bound in the Cauchy point calculations, does the choice of this parameter have any influence on the performance of the trust region method?*

We will explore this issue in the next section.

7.4 Improving Cauchy Point by Dogleg Method

To answer the question posed in the previous section, we first examine the effect of Δ^k on solution $\mathbf{p}^k$. We advise the reader to refer to Figure 7.1 which illustrates the three cases discussed below.

- *Case A:* when Δ^k is significantly **large** (and $B^k \succ 0$), full step (7.7) is good and worth computing $[B^k]^{-1}$.

- *Case B:* for **tiny** Δ^k quadratic term $\frac{1}{2}\mathbf{p}^T B^k \mathbf{p}$ in (7.9) has little effect. Therefore, an unmodified Cauchy point, computed following the procedure of Algorithm 7.2, is good.

- *Case C:* for **intermediate** values of Δ^k, $\mathbf{p}^k(\Delta^k)$ follows the **optimal** (curved) trajectory; refer to Figure 7.1.

As such, focusing on *Case C* situation, our interest is in finding an approximate solution $\tilde{\mathbf{p}}^k$ by replacing the optimal (curved) trajectory with a path consisting of two line segments (known as the *dogleg method*):

#1: from origin to unconstrained minimizer along steepest descent direction

$$\mathbf{p}_U^k = -\frac{(\nabla f(\mathbf{x}^k))^T \nabla f(\mathbf{x}^k)}{(\nabla f(\mathbf{x}^k))^T B^k \nabla f(\mathbf{x}^k)} \nabla f(\mathbf{x}^k), \qquad (7.14)$$

#2: from $\mathbf{p}_U^k$ to $\mathbf{p}_B^k$.

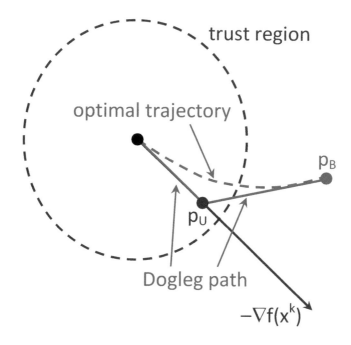

FIGURE 7.1
Schematic illustrating the concept of the dogleg method.

While adding this analysis to practical computations, this trajectory could be easily parameterized with $\tau \in [0, 2]$

$$\tilde{\mathbf{p}}^k(\tau) = \begin{cases} \tau \mathbf{p}_U^k, & 0 \leq \tau \leq 1 \\ \mathbf{p}_U^k + (\tau - 1)(\mathbf{p}_B^k - \mathbf{p}_U^k) & 1 \leq \tau \leq 2 \end{cases} \tag{7.15}$$

and the following lemma establishes the properties of path (approximate solution) $\tilde{\mathbf{p}}^k$.

Lemma 7.1 *Dogleg Path*
Let matrix B be positive definite. Then

- $\|\tilde{\mathbf{p}}^k(\tau)\|_2$ *is an increasing function of τ, and*

- $m^k(\tilde{\mathbf{p}}^k(\tau))$ *is a decreasing function of τ.*

This lemma has a very straightforward interpretation. Path $\tilde{\mathbf{p}}^k(\tau)$ intersects trust-region boundary $\|\mathbf{p}^k\|_2 = \Delta^k$ at exactly **one point** if $\|\mathbf{p}_B^k\|_2 \geq \Delta^k$, and nowhere otherwise. Having that, we could arrive at the practical implementation of the dogleg algorithm by taking into account *Cases A, B,* and *C* above, and also parameterization parameter $\tau \in [0, 2]$. Similarly, we consider three possible situations:

- $\|\mathbf{p}_B^k\|_2 \leq \Delta^k$ implies *Case A* and we could use $\mathbf{p}_B^k$ in (7.7) directly,

- $\|\mathbf{p}_B^k\|_2 \geq \Delta^k$ and $\|\mathbf{p}_U^k\|_2 \geq \Delta^k$ define *Case B* when $0 \leq \tau \leq 1$, and

- $\|\mathbf{p}_B^k\|_2 \geq \Delta^k$ and $\|\mathbf{p}_U^k\|_2 \leq \Delta^k$ entail *Case C* with $1 \leq \tau \leq 2$.

We complete this discussion by providing Algorithm 7.3 with step-by-step instructions for applying the dogleg method in practical computations. In Section 7.5, we discuss adding Algorithms 7.3 and 7.1 to our existing optimization framework and applications of the trust region method accompanied by the dogleg method (for Δ^k size analysis) for solving Example 1.3.

Algorithm 7.3 (Dogleg Method)

1. *Find full step*

$$\mathbf{p}_B^k = -[B^k]^{-1} \boldsymbol{\nabla} f(\mathbf{x}^k)$$

2. *If* $\|\mathbf{p}_B^k\|_2 \leq \Delta^k$

$$\mathbf{p}^k = \mathbf{p}_B^k$$

and go to Step 6

3. *Find*

$$\mathbf{p}_U^k = -\frac{(\boldsymbol{\nabla} f(\mathbf{x}^k))^T \boldsymbol{\nabla} f(\mathbf{x}^k)}{(\boldsymbol{\nabla} f(\mathbf{x}^k))^T B^k \boldsymbol{\nabla} f(\mathbf{x}^k)} \boldsymbol{\nabla} f(\mathbf{x}^k)$$

4. *If* $\|\mathbf{p}_U^k\|_2 \geq \Delta^k$, *approximate* $\mathbf{p}^k$ *by* $\tilde{\mathbf{p}}^k(\tau)$ *at the point of intersection of the* **first dogleg segment** *and trust-region boundary by computing appropriate* τ *($0 \leq \tau \leq 1$) as a solution of*

$$\tau \|\mathbf{p}_U^k\|_2 = \Delta^k \tag{7.16}$$

and go to Step 6

5. *Otherwise, approximate* $\mathbf{p}^k$ *by* $\tilde{\mathbf{p}}^k(\tau)$ *at the point of intersection of the* **second dogleg segment** *and trust-region boundary by computing appropriate* τ *($1 \leq \tau \leq 2$) as a solution of*

$$\|\mathbf{p}_U^k + (\tau - 1)(\mathbf{p}_B^k - \mathbf{p}_U^k)\|_2^2 = (\Delta^k)^2 \tag{7.17}$$

6. *Use found solution for* $\mathbf{p}^k$ *to proceed to Step 4 in Algorithm 7.1*

? *The solution of linear equation* (7.16) *for* τ *is pretty straightforward* $\left(\tau = \frac{\Delta^k}{\|\mathbf{p}_U^k\|_2} \right)$, *but* (7.17) *appears to be a scalar quadratic equation (again, in* τ). *While adding a numerical method for solving this equation is fairly easy, could we derive the solution analytically to reduce computational time and associated errors?*

! *Look into the practical implementation of Algorithm 7.3 by* MATLAB *file* trust_region_method.m *on p. 177 for help.*

7.5   Checking and Tuning Performance

Now we are ready to experiment with the trust region and dogleg (TRDL) methods. As we did before, whenever our existing optimization framework required any updates, we have to discuss performing these updates most optimally. We advise the reader to refer back to Section 3.8 and our discussion on the **modular** structure of our current framework. At the moment, it has

- two **modes:** OPT, TEST;

- eight **methods:** SD, NEWTON, DN, MN, DSSD, BFGS, GN, CG;

- eight functions for α-**search:** CONST, GS, DIM, BLS, BLSW, BLSSW, BLSG, BB;

with the unique and self-explanatory names used within the entire code (like global variables) to support proper communication between its parts as shown schematically in Figure 3.18.

Currently, we have to add another gradient-based search method (trust region) and, even more, consider this framework for applications to more complicated problems with, probably, a necessity to add more functionality. When we left the code in Chapter 6, it contained more than 20 files, and it keeps growing. A feasible solution for easy maintenance could be creating a modular structure (now at the "physical level") applied to all files by organizing them in several folders. Thus, starting with this chapter, we introduce

- *root folder* containing just two files required for running the framework: main file Chapter_7_data_fit_comprehensive.m to start optimization and params.m for changing its parameters,

- folder *data* with data-files data_main.dat and data_6pt.dat,

- folder *algorithms* containing main methods, e.g., d- and α-search algorithms, initialization, testing, visualization, etc.,

- folder *functions* with m-functions performing evaluation of objectives, gradients, Hessians, convergence analysis, etc.,

- folder *output* containing final output products, and

- folder *temp* reserved for future needs to store temporarily data needed to communicate between MATLAB and other software added to the framework.

Again, such re-organization at this time requires minimal changes by adding (in the main file) paths to two new folders containing algorithms and functions, i.e.,

- addpath algorithms;

- addpath functions;

and a path (in params.m) to the new location of all data, e.g.,

- dataFile = 'data/data_main.dat';

However, deferring this step to a later time will make the process obviously tougher due to the increased number of issues involved.

Finally, adding a final output procedure (to the *algorithms* folder) file_output.m may be a good idea to enable storing the history of objective function $f(\mathbf{x})$ values and corresponding solutions (controls $\mathbf{x}$) obtained throughout the entire optimization at every iteration. It allows visualizing some information with no need to re-run the optimization as it may be **time-consuming** especially for big problems. Think about running the code for an hour, a day, or a week. Ideally, the output of any optimization run will be not a collection of saved images, but all the source (or output) files from which these images are easily redrawn. As before, we suggest the reader to explore the structure of the re-organized MATLAB code to be used for future experiments and further development.

MATLAB: file_output.m

```
% core part of file name                                              1
fileName = ['_with_' method];                                        2
if (~isempty(methodAlpha))                                           3
  fileName = [fileName '_and_' methodAlpha];                         4
end                                                                   5
                                                                      6
% storing objective history                                          7
fileID = fopen(['output/opt_obj' fileName '.txt'], 'w');            8
fprintf(fileID, '%1.10e\r\n', obj);                                 9
fclose(fileID);                                                      10
                                                                      11
```

```
% storing iterative solutions                                      12
fileID = fopen(['output/opt_sol' fileName '.txt'], 'w');          13
fprintf(fileID, '%1.10e   %1.10e   %1.10e\r\n', aHist');          14
fclose(fileID);                                                    15
```

After completing the code re-organization, it is time to add the new functionality related to the trust region methodology (`method = 'TRDL'`) discussed in this chapter. As a final touch, we also add two derivative-free methods, namely the brute-force (`method = 'BF'`) and Monte Carlo (`method = 'MC'`) methods, based on earlier discussions for 1D applications in Chapter 2. Problem 1 in Lab Assignment #2 (Section 5.7) suggested adding these two approaches to the existing framework by extending their search capability from 1D to 3D. Now, the reader can check and compare how this functionality is added to our current version (`Chapter_7_data_fit_comprehensive.m`); refer to Table 7.1 for a summary of implemented changes. As continuously discussed throughout all chapters, we try to minimize the code structure updates to make gradient-based and derivative-free methods work together by sharing required functionalities, e.g., f-evaluator, visualizer, etc.

element	implementation	software
main OPT	**upgraded by added gradient-based &** **derivative-free methods**	MATLAB
f-evaluator	m-function, analytically defined function $f(\mathbf{a})$	MATLAB
d-**evaluator**	**m-code, various gradient-based approaches**	MATLAB
1D α search	m-code, various step size searches	MATLAB
visualizer	plain m-code	MATLAB

TABLE 7.1
Computational elements of `Chapter_7_data_fit_comprehensive.m` upgraded with added functionality of the trust region and derivative-free methods.

We advise the reader to review all changes incorporated in the following files.

- A new m-file `trust_region_method.m` has been added to folder *algorithms*. The reader may want to check the practical implementation of the trust region and dogleg approaches as provided in Algorithms 7.1 and 7.3. We also added `brute_force_method.m` and `monte_carlo_method.m` to the same folder. However, we will discuss the derivative-free methods in Section 7.6.

- Files `params.m`, `initialize.m`, `mode_OPT.m`, and `visualize.m` are also updated by adding correspondent parts for all three methods.

We also note that the new structure of our computational framework assumes all benefits from any possible additions to expand the current collection of line search, trust region, and derivative-free methods. The reader can do it easily by adding extra **case** statements in the **switch** blocks inside mode_OPT.m and trust_region_method.m files.

MATLAB: trust_region_method.m

```matlab
switch method                                                      1
                                                                   2
  case 'TRDL' % Dogleg method                                      3
    delta = deltaTR(end);                                          4
                                                                   5
    % search for p                                                 6
    grad = grad_f(a,data); % gradient (for multiple use)           7
    B = hess(a,data);       % Hessian (for multiple use)           8
    pB = -inv(B)*grad;                                             9
    if (norm(pB,2)<delta)                                          10
      p = pB;                                                      11
    else                                                           12
      pU = -((grad'*grad)/(grad'*B*grad))*grad;                    13
      % search for tau                                             14
      if (norm(pU,2)>=delta)                                       15
        tau = delta/norm(pU,2);                                    16
      else                                                         17
        % solving scalar quadratic equation                        18
        TRa = (pB-pU)'*(pB-pU);                                    19
        TRb = 2 * pU' * (pB-pU);                                   20
        TRc = pU'*pU - delta^2;                                    21
        tau = (-TRb + sqrt(TRb^2-4*TRa*TRc))/(2*TRa)+1;            22
      end                                                          23
      % computing p                                                24
      if (tau <= 1)                                                25
        p = tau*pU;                                                26
      else                                                         27
        p = pU + (tau-1)*(pB-pU);                                  28
      end                                                          29
    end                                                            30
                                                                   31
    % update for delta                                             32
    rho = (f(a,data)-f(a+p,data))/(-p'*grad+0.5*p'*B*p);          33
    if (rho<0.25)                                                  34
      delta = 0.25*norm(p,2);                                      35
    elseif ( (rho>0.75) && (norm(p,2)==delta) )                   36
      delta = min(2*delta,deltaTRmax);                            37
    end                                                            38
    deltaTR = [deltaTR delta];                                     39
                                                                   40
    if (rho<=paramTR)                                              41
      p = zeros(size(p));                                          42
    end                                                            43
```

```
                                                                          44
   otherwise                                                              45
      disp(['Unknown trust region method ' method ' is chosen!']);        46
      exCode = 0; % termination due to error                              47
      return;                                                             48
                                                                          49
   end                                                                    50
```

Now, we could check the functionality of the added trust region method by running `Chapter_7_data_fit_comprehensive.m` with the following settings (we list only those related closely to the TRDL functionality).

MATLAB: `params.m`

Settings for checking TRDL performance:

- *gradient-based method:* `method = 'TRDL';`

- *input data:* `dataFile = 'data/data_main.dat';`

- *initial guess:* `aini = [-14 11 -2];`

- *max size (radius) of trust region:* `deltaTRmax = 0.1;`

- *trust-region parameter η:* `paramTR = 1/8;`

- *termination #1:* `epsilonF = 1e-6;` for ϵ_1, refer to (3.18)

- *termination #2:* `epsilonA = 1e-6;` for ϵ_2, refer to (3.17)

- *termination #3:* `kMax = 200;`

To compare the results with the best ones obtained so far in Chapter 6, we combined them (BFGS and CG methods with the added pure Newton's approach supported by the BB-based α-search using settings as provided on p. 158) in Figure 7.2.

First, we refer to Figure 7.3 depicting four plots updated dynamically during optimization: (a) solution curves with data points, (b) objective $f(\mathbf{a})$ as a function of iteration count k, (c) structure of the current trust-region step direction $\mathbf{p}^k$, and (d) the history of trust-region radius Δ^k changes with k. The MATLAB's visual output is provided for $k = 10$ with (c) and (d) plots modified (previously showing gradient $\mathbf{d} = -\nabla_{\mathbf{a}} f(\mathbf{a})$ structure and α-search history) subject to associated components of the trust region algorithm.

As we did it previously with various line search methods, we can play with some of the trust-region parameters to check how the method's performance is

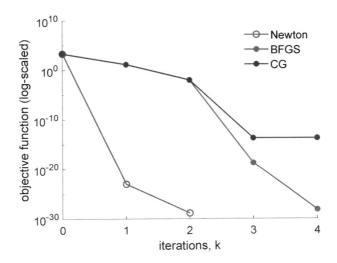

FIGURE 7.2
Excerpt graph from Figures 6.18(a) and 6.20(a) illustrating the performance of the BFGS and GS (with added pure Newton's) methods all using the BB search for α.

sensitive to these changes. For example, we could fix initial trust-region radius Δ^0 and improvement acceptance η, i.e.,

$$\Delta^0 = \frac{1}{2}\bar{\Delta}, \qquad \eta = \frac{1}{8},$$

and play with different values of the maximum size of the trust region $\bar{\Delta}$. Figure 7.4 shows the results obtained with $\bar{\Delta} = 0.1, 1.0$, and 10.

As expected from any gradient-based approach based on accurately computed Hessians B^k (either exact or approximated), the TRDL's performance is comparable to those seen if applying line search methods (both pure Newton's and BFGS). This result also nicely reflects the general concept of the trust region: solutions discoverable by the 2-order methods will be found in a few iterations if located within this region (Steps 1 and 2 in Algorithm 7.3). Otherwise, the number of iterations required to follow Steps 3–5 will depend on the user's strategy for choosing optimal bounds $\bar{\Delta}$ for the size (radius) of the trust region.

We refer the reader to Figure 7.5, showing the history of trust-region radius Δ^k changes and the results of the convergence analysis when $\bar{\Delta} = 0.1$ and 1.0. Choosing a higher bound $\bar{\Delta}$ while solving the current Example 1.3 is obvious: the TRDL sets Δ^k very close to $\bar{\Delta}$ for almost all iterations (except the last step when the solution is in the small neighborhood of the optimal one). It also leads to faster convergence: superlinear ($\bar{\Delta} = 0.1$) and quadratic ($\bar{\Delta} = 1.0$

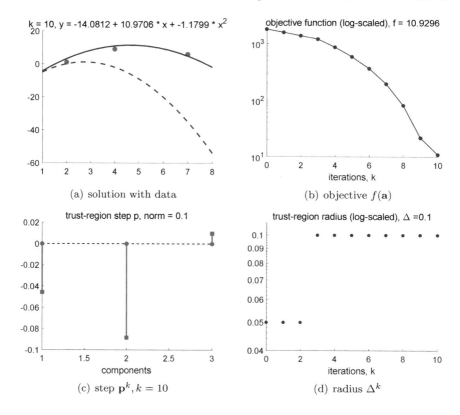

(a) solution with data   (b) objective $f(\mathbf{a})$

(c) step $\mathbf{p}^k, k = 10$ (d) radius Δ^k

FIGURE 7.3
MATLAB window with four plots updated dynamically while solving Example 1.3 by the trust region and dogleg (TRDL) methods.

or 10). Unfortunately, the same strategy cannot be successfully applied to all problems.

 *Could you guess or explain why?*

Our current Example 1.3 is an unconstrained quadratic optimization problem having a nice unique solution. More generally, (highly) nonlinear problems with possible constraints and without easy access to accurate gradients and Hessians will require careful consideration for trust-region bounds set to fixed values or changed during the optimization as determined by a particular implementation of the trust region approach.

 *Assuming no access to accurately computed (and inverted) Hessians, could you suggest any modifications to the TRDL approach using other*

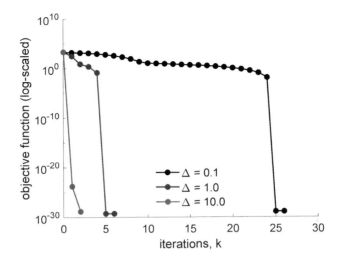

FIGURE 7.4
Results of applying the TRDL method with $\bar{\Delta} = 0.1, 1.0$, and 10.

gradient-based facilities existing in the current computational framework?
What would be their expected performance?

We close this section by inviting the reader to experiment more with the rest TRDL parameters (e.g., initial trust-region radius Δ^0 and improvement acceptance η) in a way we did it in Chapter 6. The reader may also be interested in getting more insights into the trust region's functionality, performance, and applicability by exploring its various modifications and adding them to the existing framework.

7.6 Exploring Derivative-Free Options

As a final touch, in the last section of this chapter, we have more experiments with the "comprehensive" version of the current optimization framework that now includes derivative-free approaches for solving Example 1.3, as discussed in Section 7.5. We also advise the reader to revisit Sections 2.5 and 2.6 for reviewing concepts used, respectively, by the brute-force and Monte-Carlo approaches for solving 1D optimization problems.

We start with the *brute-force method* expanded to perform an exhaustive search over three dimensions (related to the components of control $\mathbf{a} = [a_1 \, a_2 \, a_3]^T$) each discretized uniformly within intervals $[a_{\min}, a_{\max}]$ with

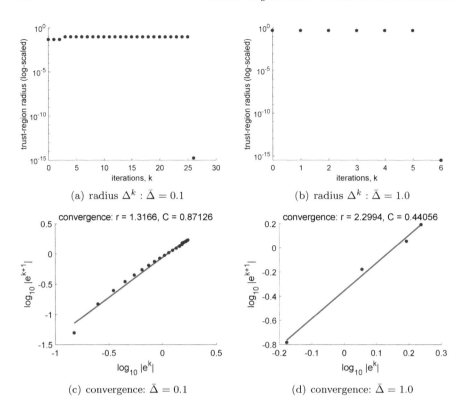

(a) radius $\Delta^k : \bar{\Delta} = 0.1$

(b) radius $\Delta^k : \bar{\Delta} = 1.0$

(c) convergence: $\bar{\Delta} = 0.1$

(d) convergence: $\bar{\Delta} = 1.0$

FIGURE 7.5
(a-b) History of trust-region radius Δ^k changes, and (c-d) computational convergence for the TRDL when (a,c) $\bar{\Delta} = 0.1$ and (b,d) $\bar{\Delta} = 1.0$.

step h. To avoid high computational loads, we set $a_{\min} = -50$, $a_{\max} = 50$, and $h = 2$; refer to the associated settings in `params.m` file (other parameters are not involved in computations and, as such, ignored).

MATLAB: `brute_force_method.m`

```matlab
% initial state                                                    1
a = aMin; fOpt = f(a,data);                                        2
                                                                   3
% brute-force loop                                                 4
for a1 = aMin(1):bfStep(1):aMax(1)                                 5
  for a2 = aMin(2):bfStep(2):aMax(2)                               6
    for a3 = aMin(3):bfStep(3):aMax(3)                             7
```

```
    k = k + 1;                                              8
    % current solution                                      9
    aCurr = [a1 a2 a3];                                    10
    fCurr = f(aCurr,data);                                 11
    if fCurr < fOpt % if better solution found             12
      a = aCurr; fOpt = fCurr; % update                    13
      obj = [obj fOpt];                                    14
      visualize;                                           15
    end                                                    16
  end                                                      17
 end                                                       18
end                                                        19
```

MATLAB: params.m

Settings for BF-based optimization:

- *solution method:* method = 'BF';

- *input data:* dataFile = 'data/data_main.dat';

- *search direction lower limits:* aMin = [-50 -50 -50];

- *search direction upper limits:* aMax = [50 50 50];

- *step size for each direction h:* bfStep = [2 2 2];

The result is shown in Figure 7.6, with the graphical output limited to two figures only: (a) solution curves with data points and (b) objective $f(\mathbf{a})$ as a function of solution updates (accepted whenever the solution is better than the stored one) counter n. The applied discretization implies $k = 51^3 = 132,651$ objective function evaluations with only $n = 61$ solution updates leading to an optimal (approximate) solution $\mathbf{a}^n = [-36 \; 20 \; -2]^T$ with $f(\mathbf{a}^n) = 34$. An exact solution $\mathbf{a}_{ex} = [-15 \; 10 \; -1]^T$ is obviously "missing" from the control space as values -15 and -1 are not "aligned" with the used discretization. Setting $h = 1$ will definitely resolve this problem and allow finding an exact solution at the increased cost of $k = 101^3 = 1,030,301$ objective evaluations.

In the next experiment, we use a random search by the *Monte-Carlo method* to create the "candidate" (sample) solutions from random numbers in each dimension. These numbers are selected, e.g., by using the normal (or Gaussian) distribution generator running it with mean $\bar{\mathbf{a}} = [0 \; 0 \; 0]^T$ and standard deviation $\sigma = 50/3$ (using the so-called "three-sigma" rule). To compare (somehow) the results obtained by the BF and MC methods, we limit the number of the sample solutions in the MC runs to the same $k = 132,651$ used by the BF approach; refer to the associated settings in params.m file.

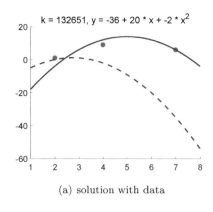

(a) solution with data

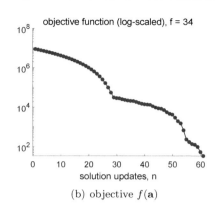

(b) objective $f(\mathbf{a})$

FIGURE 7.6
MATLAB window with two plots updated dynamically while solving Example 1.3
by the brute-force (BF) method.

MATLAB: params.m

Settings for MC-based optimization:

- *solution method:* method = 'MC';

- *input data:* dataFile = 'data/data_main.dat';

- *number of samples:* mcN = 132651;

- *distribution method:* mcDist = 'NORM'; for normal (or Gaussian)

- *mean values:* mcMean = [0 0 0];

- *standard deviations:* mcSigma = [50/3 50/3 50/3];

The result is shown in Figure 7.7, with the graphical output similarly limited to the same two figures (solution curves with data points and objective
$f(\mathbf{a})$ as a function of counter n for accepted solution updates). The MC approach returns a better solution ($f = 4.2$) "this time," meaning that it might
be different (either better or worse) when running the MC again and again.

Finally, to better distinguish the difference between deterministic approaches (e.g., line search or trust region) and stochastic methods (e.g., Monte-
Carlo), we can run the MC-based optimization multiple times to elaborate
more on the randomness of the results provided by this algorithm. For example, we can run it, say, 20 times with the same parameters as noted
in params.m file on p. 184 to assess the variation in the final value of the

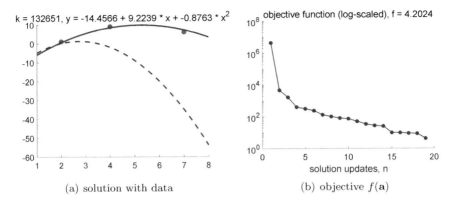

(a) solution with data (b) objective $f(\mathbf{a})$

FIGURE 7.7
MATLAB window with two plots updated dynamically while solving Example 1.3 by the Monte-Carlo (MC) method.

objective function $f(\mathbf{a})$. For the next set of 20 runs, we assume that we have "some" knowledge of the exact solution and could "navigate" optimization via changing parameters of the normal distribution (like making the focus on a particular solution), i.e., by setting mcMean = [-15 10 -1] (which is the exact solution) and mcSigma = [5 5 5]. For the next 20 runs, we could go even further and narrow down the area of random search by setting mcSigma = [0.5 0.5 0.5].

Figure 7.8 depicts how the average solution in each group (or solution span) reflects changes in the parameters defining the random structure of the solution process. Such analysis is often used to make an initial exploration of the solution space, generate approximate solutions or initial guesses used later in deterministic approaches, etc. The reader may repeat this experiment by changing: the number of samples (reduced or increased), means (closer to or farther from the exact solution), standard deviations (more or less focus on the provided means), or distribution types (e.g., uniform UNIF already implemented in our framework). The observational analysis may include the overall quality (accuracy) of obtained solutions, computational workloads in terms of the number of objective evaluations or elapsed computational time, exploring possibilities for parallelization with performance estimations, etc.

? *Suggest any approaches for coupling the BF and MC approaches for approximating the optimal solution by minimizing the computational workload during the exhaustive and randomized searches.*

Despite having different mechanisms to perform the search for the optimal solution, various approaches now coexist in the same computational framework sharing, whenever necessary, its functionality. It removes almost

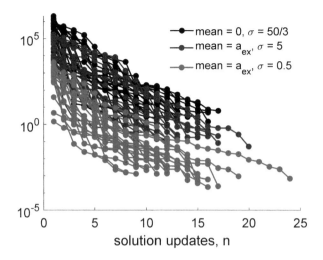

FIGURE 7.8
Results of applying the Monte-Carlo (MC) method for solving Example 1.3
with different parameters (mean and standard deviation) of the normal distribution.

all bounds in attempts to add new capabilities or modify existing utilities to
expand this framework and improve its performance. Our next step will be to
practice keeping the functionality of the working framework while changing
the optimization problem, from the simple 3D quadratic one to large-scale
constrained problems.

7.7 Homework Problems

1. Given $B^k \succ 0$ (positive definite) and $\|[B^k]^{-1}\nabla f(\mathbf{x}^k)\|_2 \leq \Delta^k$, prove that
   full step (7.7) is a solution for problem (7.9).

2. Derive the closed-form solution formula for the Cauchy point $\mathbf{p}_C^k$ in (7.12)–
   (7.13).

3. Derive an analytical solution of the scalar quadratic equation (7.17) in
   Algorithm 7.3 (look into the practical implementation of Algorithm 7.3 by
   MATLAB file trust_region_method.m on p. 177 if you need help).

4. Explore MATLAB code Chapter_7_data_fit_comprehensive.m for Example 1.3 with the added trust-region approach and dogleg method (TRDL).

Compare the performance of the trust region and line search methods (e.g., SD, Newton, CG, etc.) by checking their convergence and conclude.

5. Assuming no access to accurately computed (and inverted) Hessians, suggest any modifications to the TRDL approach using other gradient-based facilities existing in the current computational framework and conclude on the expected performance.

6. Suggest any approaches for coupling the brute-force and Monte-Carlo approaches for approximating the optimal solution by minimizing the computational workload during the exhaustive and randomized searches.

‖READ‖   Where to Read More

Griva (2009), [15]
Chapter 11 (Basics of Unconstrained Optimization)

Nocedal (2006), [25]
Chapter 4 (Trust-Region Methods), Chapter 9 (Derivative-Free Optimization)

Tarantola (2005), [28]
Chapter 2 (Monte Carlo Methods)

‖RUN‖   MATLAB Codes for Chapter 7

- **root** folder:

    - Chapter_7_data_fit_comprehensive.m

    - params.m

- folder algorithms:

    - brute_force_method.m

    - direction_search.m

    - file_output.m

    - golden_section_search.m

    - initialize.m

    - kappa_test.m

    - mode_OPT.m

    - mode_TEST.m

    - monte_carlo_method.m

    - stepsize_search.m

    - trust_region_method.m

- visualize.m

- folder data:

  - data_6pt.dat
  - data_main.dat

- folder functions:

  - fn_brent.m
  - fn_convergence_sol_norm.m
  - fn_eval_f.m
  - fn_eval_f_grad.m
  - fn_eval_g.m
  - fn_eval_g_grad.m
  - fn_eval_hess.m
  - fn_eval_hess_approx.m
  - fn_min_brack.m
  - FUNC.m

- folder output

- folder temp

7.8 Lab Assignment #3: Review Chapters 6–7

Problem 1: Consider Example 1.3 on p. 10 to find parameters a_1, a_2, a_3 in the quadratic model $y(x) = a_1 + a_2x + a_3x^2$, which optimally fits m data points (x_i, y_i), $i = 1, \ldots, m$, in the least-squares sense. Explore MATLAB code Chapter_6_data_fit_by_line_search.m or the more recent ("comprehensive") version Chapter_7_data_fit_comprehensive.m with added *backtracking line search* (BLS) supplied with sufficient decrease condition (Armijo rule), Wolfe (BLSW), strong Wolfe (BLSSW), and Goldstein (BLSG) conditions.

1. Perform optimization using the conjugate gradient (CG) method and golden section (GS) search for the step size run with search interval $[0, 1]$ and termination $\epsilon_\alpha = 10^{-6}$.

2. Perform now optimization using the same CG method and the BLS step size search using only the sufficient decrease condition (Armijo rule). Repeat this search every time changing $0 < \rho < 1$ and $0 < c_1 < 1$ parameters to identify the best performance.

3. Repeat the same experiments using now BLS step size search with added Wolfe conditions to identify $0 < \rho < 1$, $0 < c_1 < 1$ and $c_1 < c_2 < 1$ parameters which give you the best performance.

4. Repeat the same experiments with added strong Wolfe conditions and identify ρ, c_1 and c_2 parameters which give you the best performance.

5. Repeat the same experiments with added Goldstein conditions and identify $0 < \rho < 1$ and $0 < c < 1$ parameters which give you the best performance.

For all five cases, record the number of iterations necessary for convergence and optimal values of the objective function. Present the results (solution and objective function graphs) with the best performance. Approximate convergence parameters r and C. Compare the performance of all five approaches and conclude about using inexact vs. exact line search methods.

For all computations, use the following (refer to Section 6.4 for more details):

- $m = 3$ and data from file data_main.dat,

- initial guess $\mathbf{a}^0 = [-14\ 11\ -2]^T$,

- termination parameters: $\epsilon_1 = \epsilon_2 = 10^{-6}$ and $k_{max} = 200$; refer to (3.17)–(3.18) for explanations, and

- initial step size in all BLS-methods $\bar{\alpha} = 1$.

Problem 2: Consider the same problem as in Problem 1. Explore MATLAB code
Chapter_7_data_fit_comprehensive.m with the added *trust region* approach
based on the *dogleg method*.

1. Perform optimization and show the results for different values of the max-
   imum size of a trust region: (a) $\bar{\Delta} = 0.25$, (b) $\bar{\Delta} = 0.5$, (c) $\bar{\Delta} = 1.0$, and
   (d) $\bar{\Delta} = 5.0$.

2. Approximate convergence parameters r and C (whenever possible) and
   conclude on the relation between cases (a–d) performance and choice of
   $\bar{\Delta}$.

3. Find the simplest way to identify the minimum value of $\bar{\Delta}$ to get convergence in one (Newton) iteration (for the current problem).

For all computations, use the following (refer to Section 7.5 for more details):

- $m = 3$ and data from file data_main.dat,

- initial guess $\mathbf{a}^0 = [-14 \ 11 \ -2]^T$

- termination parameters: $\epsilon_1 = \epsilon_2 = 10^{-6}$ and $k_{max} = 200$; refer to (3.17)–
  (3.18) for explanations, and

- trust-region parameters: $\eta = \frac{1}{8}$ and $\Delta^0 = \frac{1}{2}\bar{\Delta}$.

For Problems 1 and 2, consider the rate of convergence based on the solution $\mathbf{a}^{last}$ obtained at termination rather than on the exact solution $\mathbf{a}^*$
(practical approach discussed in Section 3.4).

7.9   Midterm Assignment: Review Chapters 1–7

Problem: Consider the following system of nonlinear equations (adopted from [11])

$$3x_1 - \cos(x_2 x_3) - \frac{1}{2} = 0,$$

$$x_1^2 - 81(x_2 + 0.1)^2 + \sin x_3 + 1.06 = 0, \qquad (7.18)$$

$$e^{-x_1 x_2} + 20x_3 + \frac{10\pi - 3}{3} = 0.$$

Solve system (7.18) iteratively by applying optimization techniques mentioned below.

1. Reformulate problem (7.18) as a constrained optimization (feasible region) problem and show the method, which allows you to solve (7.18) within an unconstrained optimization framework.

2. Explore the structure and computational facilities of MATLAB code `Chapter_7_data_fit_comprehensive.m` ("comprehensive" version) and make all necessary adjustments to solve the problem in (7.18) following the **Coding Requirements** below.

3. Run the TEST mode for your optimization code to check the correctness of your gradient computations. Use vector $\mathbf{x} = [1\ 1\ 1]^T$ as an initial guess for the "kappa-test." Consider both "cheap" and "expensive" tests and create all three plots. Make a conclusion on the accuracy of the computed gradients and expected performance of gradient-based approaches.

4. Apply the brute-force (BF) method to approximate the solution for the problem in (7.18). Discretize the search/control space uniformly within intervals $[-50, 50]$ for all three dimensions with step $h = 5$. Use obtained approximate solution as the initial guess for your further computations in 5–7, and conclude.

5. Find the optimal solution for (7.18) by performing optimization using the Gauss–Newton (GN) method and compare performance by applying different methods for step size search: (a) backtracking line search (BLS), (b) golden section (GS) search, and (c) bracketing-Brent (BB) toolbox. Use the following parameters:

   - BLS: $\bar{\alpha} = 1.0$, $\rho = 0.25$, $c = 0.5$,
   - GS: bracketing interval $[0, 1]$, termination tolerance $\epsilon_\alpha = 10^{-6}$, and
   - BB: initial interval $[0, 1]$; bracketing: `MAXITER = 10`, `GLIMIT = 100.0`; Brent: `TOL = 10^{-6}`, `ITMAX = 2`.

Create four plots for each method (for the solution, objective function, step size, and convergence analysis) whenever possible and conclude for the performance.

6. Find the optimal solution for (7.18) by performing optimization using the BFGS method and compare performance by applying different methods for step size search: (a) backtracking line search (BLS), (b) golden section (GS) search, and (c) bracketing-Brent (BB) toolbox. Use the same parameters for BLS, GS, and BB as in 5. Create four plots for each method (for the solution, objective function, step size, and convergence analysis) whenever possible and conclude for the performance.

7. Find the optimal solution for (7.18) by performing optimization using the trust region approach based on the dogleg method (TRDL) with the following parameters: $\eta = \frac{1}{8}$ and $\bar{\Delta} = 0.5$. Create four plots related to this approach. Compare performance with line search methods (GN and BFGS) used in 5–6 and conclude.

8. Find the optimal solution for (7.18) by performing optimization using the Monte-Carlo (MC) method. Run the search by choosing sample solutions in each dimension using random numbers uniformly distributed within the interval $[-50, 50]$. Allow $n = 1,000,000$ as number of samples (f-evaluations). Repeat this numerical experiment 25 times and place all solutions on the same plot. Compare obtained results and conclude about the performance and applicability of stochastic approaches to problem (7.18).

For all computations, use the following termination parameters: $\epsilon_1 = \epsilon_2 = 10^{-6}$ and $k_{max} = 200$; refer to (3.17)–(3.18) for explanations.

Coding Requirements:

- **Allow minimal changes** to the provided MATLAB code. For example, you may still use control vector $\mathbf{a} = [a_1\ a_2\ a_3]^T$ instead of $\mathbf{x} = [x_1\ x_2\ x_3]^T$ to minimize your code adjustments. You may also keep vector `data` (e.g., having it empty) to preserve the same structure of all m-functions in the code.

- **Do not use** symbolic differentiation to compute Jacobians and gradients.

- In the visualization procedure, adjust only the solution (top-left) subplot by showing (a) initial guess (3 dots), (b) current (optimal) solution (3 dots), and (c) $\mathbf{x} = [x_1\ x_2\ x_3]^T$ for the current (optimal) solution in the title of the subplot.

! *Before solving this problem, you may also read Section 8.2.1 to find extra help with conversion to the unconstrained optimization.*

8

Large-Scale and Constrained Optimization

This chapter sets a divider in our discussion for constructing a computational framework enabled to solving unconstrained and constrained optimization problems using various gradient-based and derivative-free approaches. We start with a very formal generalization to identify the notation of large-scale optimization and provide multiple examples of such optimization problems. As many LSOs are constrained, we briefly discuss possible practical techniques to implement various constraints exemplified by the Lagrange multiplier and penalization approaches. We also discuss the transition of our existing computational framework to the state with the constraint handling facilities added to the solution process. Finally, we include several examples to introduce the notation of the DE-based (DE-constrained) optimization in anticipation of the focus topics in the rest of the chapters.

8.1 Generalization of Large-Scale Optimization

In previous chapters, we discussed plenty of different algorithms setting up a computational framework to solve optimization problems (so far, unconstrained). We also modified its structure to create this framework as an open platform for multiple numerical experiments with the potential to grow both in the number of incorporated computing methodologies and the complexity of the optimization problems to be solved. In this chapter, we formally start a discussion on the latter – moving from the simple 3D quadratic problem (our Example 1.3) to large-scale constrained ones.

We start this discussion by revisiting some general notations discussed in Section 1.2 to exemplify their operation with large objects related to *large-scale optimization* (LSO) problems. Here, we re-iterate its commonly used and rather general form

$$\min / \max_{\mathbf{u} \in \mathbb{R}^n} \quad \mathcal{J}(\mathbf{x}; \mathbf{u})$$

$$\text{subject to} \quad \mathbf{g}(\mathbf{x}; \mathbf{u}) = \mathbf{0} \quad (8.1)$$

$$\mathbf{h}(\mathbf{x}; \mathbf{u}) \leq \mathbf{0}$$

In problem (8.1), we identify the following objects.

DOI: 10.1201/9781003275169-8

- m-component vector of *state variables* $\mathbf{x} \in \mathbb{R}^m$ included in the problem (and, possibly, the objective function) description, which are not control (optimization) variables.

- n-component vector of *optimization* (control, decision, design) *variables* $\mathbf{u} \in \mathbb{R}^n$.

- *Objective (cost) function(al)* $\mathcal{J}(\mathbf{x}; \mathbf{u}) : \mathbb{R}^m \times \mathbb{R}^n \to \mathbb{R}^k$ may be a vector of scalar functions if $k > 1$.

- Equality and/or inequality *constraints* given (in general) by vector functions $\mathbf{g}(\mathbf{x}; \mathbf{u})$ and $\mathbf{h}(\mathbf{x}; \mathbf{u})$. These constraints are also called as *governing equations* or *equations of states* (EOS).

Based on this generalized definition of the optimization problem, we could now establish a notation (again, in a rather broad context) of *large-scale optimization* – it is a problem whenever at least one of the following applies:

- high dimensionality of *control space*, i.e., n is big,

- high dimensionality of the problem itself, i.e., m is big,

- *multiobjective optimization* when $k > 1$ and considerably big; this may (or may not) require objective function *scalarization* as discussed briefly in Section 8.2.2,

- any *constrained optimization* if it faces difficulties while solving

  - nonlinear equations,
  - ODEs (also called *ODE-based optimization*),
  - PDEs (also called *PDE-based optimization*),

or any combinations (systems) of them.

8.2 LSO Examples

8.2.1 Solving Systems of Equations

As a motivation for moving forward to discussions about the practical approaches used to solve LSOs, we would like to consider several examples of optimization problems recently solved with success based on the techniques to construct an efficient computational framework discussed in this book. Here, we start with the problem of solving a system of equations

$$\mathbf{g}(\mathbf{x}) = \mathbf{0} \tag{8.2}$$

similar to that one suggested in Section 7.9. To consider this problem in the "LSO-format," we assume vector function $\mathbf{g}(\mathbf{x}) : \mathbb{R}^m \to \mathbb{R}^\ell$ of high dimensionality (i.e., both m and ℓ are significantly big numbers).

A simple preliminary analysis of (8.2) for the possible solution approaches returns the following. Vector $\mathbf{x}$ plays the role of the control and, at the same time, the state. The solution method depends obviously on the dimensionality of $\mathbf{x}$ and also on the number and complexity of function $\mathbf{g}$ components. Therefore, if (8.2) represents just a **square** system of **linear** equations $A\mathbf{x} - \mathbf{b} = \mathbf{0}$ with matrix A **invertible**, we could easily find a solution

$$\mathbf{x} = A^{-1}\mathbf{b} \qquad (8.3)$$

similar to Example 1.2. To distinguish this solution approach from others that require formal optimization, we could call it a *direct solver*. Otherwise, we reformulate (8.2) as a constrained optimization or *feasible region problem* (with fictitious objective equalized to any constant, e.g., 0)

$$
\begin{aligned}
\min_{\mathbf{x} \in \mathbb{R}^n} \quad & \mathcal{J}(\mathbf{x}) = 0 \\
\text{s.t.} \quad & \mathbf{g}(\mathbf{x}) = \mathbf{0}
\end{aligned}
\qquad (8.4)
$$

and solve it as the following unconstrained optimization problem

$$\min_{\mathbf{x} \in \mathbb{R}^n} \bar{\mathcal{J}}(\mathbf{x}) = \frac{1}{2}\|\mathbf{g}(\mathbf{x})\|^2. \qquad (8.5)$$

(8.5) also serves as a good exemplar of a simple *scalarization* technique. Now we can apply the framework developed so far to solve this problem iteratively: we will refer to this option as an *iterative solver*.

8.2.2 Space-Dependent Parameter Reconstruction

An excellent example of an optimization problem qualified as a large-scale one could be any problem that involves reconstructing numerous parameters in equations (usually ODEs or PDEs) that describe various physical processes. To illustrate the concept of LSO, we pick quite a rather challenging model often used in geoscience modeling (in particular, in petroleum reservoir simulations) called the *identification of space-dependent transport coefficients*. To describe the movement of all fluids and gases through the underground (porous) media, we consider the liquid-gas structure as an N_p-phase system with N_c components. For example, $N_p = 2$ means two phases (i.e., liquid and gas), and each component c ($c = 1, \ldots, N_c$) is considered (with some exceptions) to be in both phases. For simplicity, the reader may think of two components (water and oil) present in the liquid phase and also as vapors. The mass balance equation for each component c could be written in the following PDE form:

$$\frac{\partial}{\partial t}\left(\phi \sum_{p=1}^{N_p} y_{cp}\rho_p s_p\right) + \nabla \cdot \sum_{p=1}^{N_p} y_{cp}\rho_p \mathbf{v}_p + \sum_{j=1}^{N_{well}} \sum_{p=1}^{N_p} y_{cp}\rho_p q_{j,p} = 0. \qquad (8.6)$$

In (8.6), the flux $\mathbf{v}_p$ of phase p is given by Darcy's law

$$\mathbf{v}_p = -\mathbf{k}\frac{k_{rp}}{\mu_p}\left(\boldsymbol{\nabla}p_p - \rho_p g \boldsymbol{\nabla}D\right), \tag{8.7}$$

t is time, while y_{cp}, p_p, and s_p denote the state variables (respectively, a fraction of component c in phase p, phase pressure, and saturation); see, e.g., [3] for more details.

Given the knowledge of the model geometry (domain) and all other parameters present in (8.6)–(8.7), we could solve this system of N_c PDEs (isothermal oil-water flow problem) using any available (or favorite) numerical approaches and discretized structure of the domain. We usually call such a system of equations a *forward (direct) problem*. We also move all parameters used in (8.6)–(8.7) into two separate groups:

- **phase-dependent parameters:** density ρ_p, relative permeability k_{rp}, viscosity μ_p, and source term $q_{j,p}$ associated with the jth ($j = 1, \ldots, N_{well}$) well drilled, e.g., for injecting water or producing oil (or combination of water and oil);

- **space-dependent parameters:** porosity ϕ, absolute permeability $\mathbf{k}$, and vertical depth D.

We also should notice that it may be required to add several more (algebraic or differential) equations in addition to (8.6)–(8.7) to make this system closed: e.g., capillary pressure and thermodynamical relationships, mass conservation applied to saturation, equations governing the fluid flow inside the wells and accounting for frictional and other pressure losses, etc. In general, to model the flow in porous media, we could consider the PDE system (8.6)–(8.7) in the following form

$$\begin{aligned}\mathbf{g}(\mathbf{x}, t; \mathbf{u}) &= \mathbf{0},\\ \mathbf{x}(t_0) &= \mathbf{x}_0,\end{aligned} \tag{8.8}$$

where $\mathbf{x} = \{y_{cp}, p_p, s_p\}$ is a vector of state variables, and $\mathbf{u}$ are model parameters (optimization or control variables) to represent physical and geometrical properties of a specific reservoir. For example, without loss of generality, the reader may think of reconstructing just one unknown space-dependent parameter such as permeability $\mathbf{k}$. Similar to the least-squares data fitting used in Example 1.3, this reconstruction may be performed by considering an optimization problem, often referred to as an *inverse problem*, with an objective function based on data obtained from measurements of different types. For this particular problem, there are usually at least three available sources of such data:

- *production (injection) rates $\tilde{q}_{j,p}$ measured in wells* by amounts of fluid injected into the media or produced during the reservoir life,

- *pressure in wells $\tilde{p}_j^{BHP}$*, also referred to as the bottom hole pressure (BHP), and

- *seismic data* available, for example, as a result of the inversion to estimate phase saturation $\tilde{s}_{k,p}$ in the kth ($k = 1, \ldots, N_{block}$) block of the discretized reservoir.

These data are referred to as the observed history of petroleum reservoirs: therefore, the associated optimization problems of reconstructing any parameters are known as *history matching* problems. The general form of the objective function created for such problems may be given by the following formula

$$
\begin{aligned}
\mathcal{J}(\mathbf{u}) = \alpha_1 \sum_{j=1}^{N_{well}} \sum_{p=1}^{N_p} C_{j,p} \left(q_{j,p} - \tilde{q}_{j,p}\right)^2 \\
+ \alpha_2 \sum_{j=1}^{N_{well}} C_j \left(p_j^{BHP} - \tilde{p}_j^{BHP}\right)^2 \\
+ \alpha_3 \sum_{k=1}^{N_{block}} \sum_{p=1}^{N_p-1} C_p \left(s_{k,p} - \tilde{s}_{k,p}\right)^2,
\end{aligned}
\tag{8.9}
$$

where coefficients $C_{j,p}$, C_j, and C_p provide weights for data obtained for different phases and within various wells. Additional scaling between different data types is required because the sensitivity of the full (scalarized) objective $\mathcal{J}(\mathbf{u})$ with respect to control $\mathbf{u}$ may vary by orders of magnitude for contributions of individual data sources. In (8.9), this scaling is provided by coefficients α_1, α_2, and α_3.

> **?** *As the performance of the optimization problem and its solution quality may change while using different weights α_i, how could we optimally scale these coefficients for better (or the best) efficiency?*

In the presence of several sources for matching data in (8.9), this problem goes under the category of *multiobjective optimization problems* where every "individual" objective represents a particular type of data. A commonly used method for dealing with multiple objective functions is to apply *scalarization*. For instance, we could mention several scalarization methods to combine k objectives $\mathcal{J}_i$, $i = 1, \ldots, k$, in one:

- *SQPROD*. It allows objectives with different orders of magnitude to have similar significance

$$
\mathcal{J} = \prod_{i=1}^{k} \alpha_i \left(\mathcal{J}_i - \mathcal{J}_i^u\right)^p.
\tag{8.10}
$$

- *P-METRIC*. It reflects the relative importance of the objectives but guarantees only a weak Pareto optimality (used in our scalarization example above)

$$
\mathcal{J} = \sum_{i=1}^{k} \alpha_i \left(\mathcal{J}_i - \mathcal{J}_i^u\right)^p.
\tag{8.11}
$$

- *INF-METRIC.* It represents the weighted *min-max (Tchebycheff) method* (necessary and sufficient for establishing Pareto optimality)

$$\mathcal{J} = \max_i \alpha_i \left(\mathcal{J}_i - \mathcal{J}_i^u\right) + \varepsilon \sum_{i=1}^{k} \alpha_i \left(\mathcal{J}_i - \mathcal{J}_i^u\right). \tag{8.12}$$

As the reader may have already noticed, all methods (8.10)–(8.12) mentioned above require defining parameters for each objective included in the scalarization procedure, namely

- *utopian* (ideal) points $\mathcal{J}_i^u$, and

- weighting coefficients α_i and ε.

All these parameters may be found by establishing some reasonable assumptions or, of course, by trial and error. It makes scalarization, initially seen as a simple procedure, rather challenging in cases when k is big and multiple objectives $\mathcal{J}_i$ are *conflicting* (decrease in one of them may cause increase in others). In such a case, *Pareto fronts* may provide an effective tool to gain interesting insights into the problem of solving multiobjective optimization.

In general, Pareto fronts can be constructed for any optimization problem involving trade-offs between two or more objectives. If obtained in a convex form, they often exhibit a "knee" point where a significantly large sacrifice in one objective is required to gain a relatively small improvement in the other ones. Single points on the Pareto fronts appear after performing optimization utilizing objective function $\mathcal{J}$ constructed from the individual objectives using scalarization methods mentioned earlier in this section. For example, we could use the method of weighted sums (8.11) and construct a convex combination of normalized objective functions $\mathcal{J}_1$ and $\mathcal{J}_2$ by setting $k = 2$, $p = 1$, $\mathcal{J}_1^u = 0$, $\mathcal{J}_2^u = 0$, and

$$\alpha_1 = \frac{\alpha}{\mathcal{J}_1^* \mathcal{J}_1^0}, \qquad \alpha_2 = \frac{1-\alpha}{\mathcal{J}_2^* \mathcal{J}_2^0}, \tag{8.13}$$

to arrive eventually at the definition of the following bi-objective function:

$$\mathcal{J} = \frac{\alpha}{\mathcal{J}_1^* \mathcal{J}_1^0} \cdot \mathcal{J}_1 + \frac{1-\alpha}{\mathcal{J}_2^* \mathcal{J}_2^0} \cdot \mathcal{J}_2. \tag{8.14}$$

In (8.14), coefficients $\mathcal{J}_1^0$ and $\mathcal{J}_2^0$ are the values of the respected objectives evaluated at the initial guess, while $\mathcal{J}_1^*$ and $\mathcal{J}_2^*$ are the upper bounds of the normalized objective functions introduced to ensure that the linear combination is convex with $\alpha \in [0, 1]$. Here, we use the strategy outlined in [22] to estimate these bounds obtained under the assumption that objectives $\mathcal{J}_1$ and $\mathcal{J}_2$ are conflicting and reduce the negative impact of having very different magnitudes in their values. Figure 8.1 shows a typical convex Pareto front for a bi-objective minimization problem. The two anchor points located at $(\mathcal{J}_1^{\min}, \mathcal{J}_2^*)$ and $(\mathcal{J}_1^*, \mathcal{J}_2^{\min})$ are obtained by minimizing separately $\mathcal{J}_1 / \mathcal{J}_1^0$

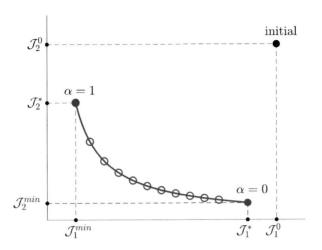

FIGURE 8.1
Schematic showing a convex Pareto front (solid line) with two anchor points (filled circles) for a typical minimization problem. The empty circles represent optimal solutions converged from the initial point $(\mathcal{J}_1^0, \mathcal{J}_2^0)$.

and $\mathcal{J}_2/\mathcal{J}_2^0$. These two points with other dots depict the *Pareto optimal solutions* that converged from the initial point $(\mathcal{J}_1^0, \mathcal{J}_2^0)$.

Finally, we draw the reader's attention to the point that objectives $\mathcal{J}_i$ could show their conflicting nature often due to being represented by data of significantly different sizes. For example, we may consider $\mathcal{J}_1$ and $\mathcal{J}_2$ the first and last terms in (8.9), representing, respectively, misfits for well production rates and seismic inversion data. The size of the latter is usually a way larger and depends on the number of (computational) blocks N_{block} in the discretized model for the simulated reservoirs. Figure 8.2 presents an example of such problems consistent with the definition of LSO by multiple criteria mentioned in Section 8.1.

This upscaled reservoir model [30] portrays the original Norne field (in Norwegian Sea) created by 388,080 total grid blocks ($49 \times 120 \times 66$) with 45,470 active ($N_{block} = 45,470$). The oil production uses 10 injecting (water) and 26 producing wells ($N_{well} = 36$) with individual open/closed operating schedules. The simulation time frame is 3,317 days (10 years), and the history matching data is available for the first 1,095 days (3 years). While solving a history matching problem (based on available production and seismic data) to identify one space-dependent transport coefficient (e.g., isotropic absolute permeability $\mathbf{k}$), a rough approximation returns the size of control vector $\mathbf{u}$ (i.e., $\mathbf{k}$) as of 45,470 components. The vector of states will be 2 or 3 times larger, depending on the $\mathbf{x}$ structure ($\mathbf{x} = \{p_p, s_p\}$ vs. $\mathbf{x} = \{y_{cp}, p_p, s_p\}$).

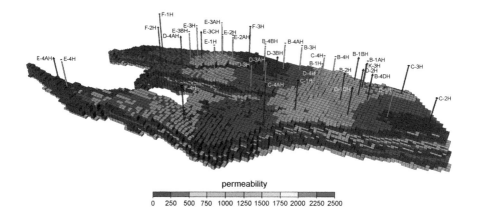

permeability

0   250 500 750 1000 1250 1500 1750 2000 2250 2500

FIGURE 8.2
3D reservoir structure, wells and permeability field for the modified (up-scaled) Norne field. Source: reprinted from Oleg Volkov, Vladislav Bukshtynov, Louis J. Durlofsky, and Khalid Aziz, "Gradient-based Pareto Optimal History Matching for Noisy Data of Multiple Types," Computational Geosciences 22, no. 6 (2018): 1465–1485.

Also, the seismic part of the objective function will be significantly enlarged if this data is obtained more than once. It obviously sets a necessity to apply some techniques for reducing dimensionality of control and, possibly, state and data spaces discussed briefly in the next section. We also refer the readers to [9, 30] once they are interested in getting a better perception of the solution approaches available for this particular problem using adjoint-gradient optimization and Pareto history matching with data of multiple types.

8.2.3   Parameter Identification – Another Example

In the previous section, we referred to the space-dependent parameter reconstruction problem based on the example of a petroleum reservoir. Its complicated computational model containing thousands of 3D blocks, supplied with massive historical data, and described by a system of PDEs leaves no doubt to consider the associated optimization (history matching) problem as the large-scale one. However, models of much smaller physical sizes may also satisfy the LSO criteria and challenge the solution process by complications related to reliable computations to accurately describe the complex nature of such models. We could easily exemplify it by considering another problem of space-dependent parameter identification in applications to imaging techniques used for cancer detection.

Electrical impedance tomography (EIT) recently gained popularity due to its noninvasive nature and proven safety for patients while screening for possible tumor growth. The underlying physical phenomenon is that the electrical properties (e.g., electrical conductivity $\sigma(x)$) of human body tissues are different depending on their health status (healthy vs. cancer-affected). It allows EIT to produce images of any biological tissues by interpreting their response to applied electrical voltages U or injected currents I. A commonly used mathematical model describes the distribution of electrical potential $u(x)$ by solving the following (*forward*) elliptic PDE problem:

$$\boldsymbol{\nabla} \cdot [\sigma(x)\boldsymbol{\nabla} u(x)] = 0, \qquad x \in \Omega$$

$$\frac{\partial u(x)}{\partial n} = 0, \qquad x \in \partial\Omega - \bigcup_{\ell=1}^{m} E_\ell \qquad (8.15)$$

$$u(x) + Z_\ell \sigma(x)\frac{\partial u(x)}{\partial n} = U_\ell, \qquad x \in E_\ell$$

In (8.15), Ω defines a set (usually in 2D or 3D) representing a part of the human body where m electrodes E_ℓ ($\ell = 1, \ldots, m$) with contact impedances Z_ℓ are attached to its periphery, $\partial\Omega$. If the so-called "voltage–to–current" model is used, constant voltages U_ℓ are applied to electrodes E_ℓ (see Figure 8.3 for clarification) to initiate electrical currents $\tilde{I}_\ell$ measured at the same electrodes. Boundary conditions set by the second and third equations in (8.15), respectively, represent the absence (no-flux) of electrical current through the portion of the boundary between the electrodes and the Ohms law applied to the areas where electrodes are placed. We may also compute electrical currents I_ℓ based on the current value of the control variable $\sigma(x)$ and the corresponding solution $u(x)$ of the forward problem (8.15), i.e.,

$$I_\ell = \int_{E_\ell} \sigma(x)\frac{\partial u(x)}{\partial n}\, ds. \qquad (8.16)$$

Now we could formulate the *inverse* EIT (conductivity) problem as a PDE-constrained optimization problem by considering the minimization of the following objective function

$$\mathcal{J}(\sigma) = \sum_{\ell=1}^{m} \left(I_\ell - \tilde{I}_\ell\right)^2. \qquad (8.17)$$

The reader may find more details on the description of this problem and some solutions available, e.g., here [1, 20, 2]. Being back to our discussion about the large-scale optimization, we have to point out that the dimensions of domain Ω considered in this example (for 2D or 3D cases) are pretty small – at the scale of the size of the human body parts (e.g., lungs, brain, breast, etc.). However, obtaining high resolution (and, as such, accurate) images requires solving this optimization problem on fine meshes with state and control spaces significantly expanded. For example, Figure 8.3(a) depicts the electrical conductivity $\sigma(x)$ with three areas (shown by dashed circles) identified as being suspicious of developing cancer. This image is obtained on a fine mesh, shown in Figure 8.3(b), containing about 8,000 elements (triangles) where discretized control σ takes distinct values at every spatial location, i.e., inside every finite element. Despite the potential for high resolution, optimization performed with such fine discretization may suffer from *overparameterization*. It often happens if the problem is *underdetermined*, i.e., the number of controls overweighs the size of available data (measurements).

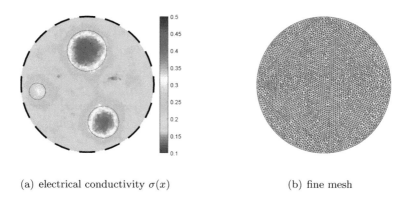

(a) electrical conductivity $\sigma(x)$ (b) fine mesh

FIGURE 8.3
(a) Electrical conductivity $\sigma(x)$ obtained by solving the inverse EIT problem on a fine mesh shown in (b). Line segments in (a) depicts 16-electrode model used for applying constant voltages U_ℓ and measuring electrical currents $\tilde{I}_\ell$.

A common approach to overcoming overparameterization of discretized $\sigma(x)$, along with the fact that the solutions obtained in due course of optimization should also honor any available prior information (e.g., available images, etc.), is implementing some types of *re-parameterization* of the control space. While a variety of approaches are available to re-define control vectors using a fewer number of components, here we consider widely used *principal component analysis* (PCA), also known as *proper orthogonal decomposition* (POD) or Karhunen–Loève expansion.

More specifically, PCA enables representing control $\sigma(x)$ with obviously correlated components (as a characterization of a spatial distribution) by a set of uncorrelated variables (components of new vector ξ). The initial control (σ) and the new one (ξ) are mapped by

$$\sigma = \Phi\,\xi + \bar{\sigma},$$
$$\xi = \hat{\Phi}^{-1}(\sigma - \bar{\sigma}),$$

$$(8.18)$$

where Φ is the basis (linear transformation) matrix, and $\hat{\Phi}^{-1}$ is the pseudo-inverse of Φ. Very often, the PCA transform matrix Φ is constructed using the *truncated singular value decomposition* (TSVD) of a centered matrix

$$M = \frac{1}{\sqrt{N_r - 1}} [\sigma_1 - \bar{\sigma} \ \ldots \ \sigma_{N_r} - \bar{\sigma}] \tag{8.19}$$

containing N_r sample solutions (realizations) σ_n ($n = 1, \ldots, N_r$) as its columns and the prior mean $\bar{\sigma} = \frac{1}{N_r} \sum_{n=1}^{N_r} \sigma_n$. The TSVD factorization provides

$$M \approx U_{N_\xi} \Sigma_{N_\xi} V_{N_\xi}^T, \tag{8.20}$$

where diagonal matrix Σ_{N_ξ} contains the singular values of M, and matrices U_{N_ξ} and $V_{N_\xi}^T$ consist of the left and right singular vectors of M. Matrix Σ_{N_ξ} is truncated to keep only the N_ξ largest singular values, and analogous truncations are applied to U_{N_ξ} and $V_{N_\xi}^T$. The PCA linear transformation is obtained by

$$\Phi = U_{N_\xi} \Sigma_{N_\xi}, \qquad N_\xi \leq N_r, \tag{8.21}$$

to project the initial σ-control space onto the reduced-dimensional ξ-space, containing only N_ξ principal components in ξ and employing mapping (8.18). Sample solutions σ_n may be represented by any available "images" of $\sigma(x)$ assuming their compatibility with the structure of discretized control σ. They could be also generated synthetically based on any available assumptions on the solution for $\sigma(x)$. For instance, if the cancerous spots are expected to be circular, the sample solutions may contain multiple circles of random size; refer to Figure 8.4 for some examples. The same technique applies to large models, e.g., to history matching in petroleum engineering as exemplified in Section 8.2.2. In such cases, more efficient algorithms could be used as the TSVD required to construct Φ can be quite time-consuming. We refer the reader to [9, 20] for more details on completing PCA representations for both examples.

Finally, re-parameterization of the control space allows restating the optimization problem in terms of the new model parameter ξ used in place of control $\sigma(x)$ (or, in general, $\mathbf{u}$ as used in the example of Section 8.2.2) as follows

$$\xi^* = \operatorname*{argmin}_{\xi} \mathcal{J}(\xi). \tag{8.22}$$

We also note the easiness of implementing gradient-based approaches aligned with the PCA-based space reduction (re-parameterization) techniques. New gradients $\nabla_\xi \mathcal{J}$ of objective $\mathcal{J}(\sigma)$ with respect to new control ξ can be expressed as

$$\nabla_\xi \mathcal{J} = \Phi^T \nabla_\sigma \mathcal{J} \tag{8.23}$$

to define the projection of original gradients $\nabla_\sigma \mathcal{J}$ from the initial (physical) σ-space onto the reduced-dimensional ξ-space.

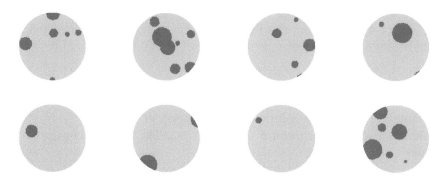

FIGURE 8.4
Sample solutions generated synthetically for solving the inverse EIT problem using PCA-based re-parameterization.

8.2.4 State-Dependent Parameter Reconstruction

In the previous sections, both examples taken from the fields of petroleum and biomedical engineering considered reconstructing various parameters as functions of space, i.e., solving optimization problems for space-dependent parameter identification. These examples referred to physical models (petroleum reservoir and cancer-affected biological tissue) of different sizes showing the same issue of both control and state spaces overparameterized due to applied spatial discretization. Such discretization is inevitable in case the models themselves are of large sizes, require high resolution in pursuit of accurate representation, or both. However, even pretty modest discretization applied to small-sized physical domains cannot guarantee the problem appears under the category of the hard-to-solve large-scale optimization. Here, our next model will provide the reader with such an example and the basic ideas about *reconstructing state-dependent parameters.*

Multiple questions and various practical concerns arise in the computational analysis and optimization applied to models where some material (physical) properties appear as functions of the state variables in the same system. In the automotive industry, e.g., new materials are often created after welding two or more metals. Thus, the manufacturers would like to know the physical properties of the weld pool, such as physical strength, plasticity, thermal conductivity, etc. Another example refers to new materials (e.g., polymers) created in a way (e.g., due to small sizes) preventing from applying laboratory procedures to examine their physical properties as results of direct measurements. As a model problem, here we consider the Navier-Stokes equation with temperature-dependent viscosity $\mu(T)$ coupled with the energy equation:

$$\frac{\partial}{\partial t}\mathbf{u} + \mathbf{u}\cdot\nabla\mathbf{u} + \nabla p - \nabla\cdot\left[\mu(T)[\nabla\mathbf{u} + (\nabla\mathbf{u})^T]\right] = 0, \qquad x\in\Omega$$

$$\nabla\cdot\mathbf{u} = 0, \qquad x\in\Omega \qquad (8.24)$$

$$\frac{\partial}{\partial t}T + \mathbf{u}\cdot\nabla T - \nabla\cdot[k\nabla T] = 0, \qquad x\in\Omega$$

subject to properly defined initial and boundary conditions.

In the PDE system (8.24), the state variables are the velocity of the moving fluid (e.g., the liquid metal alloy) $\mathbf{u}(x,t)$, pressure $p(x,t)$, and temperature $T(x,t)$. We could consider the heat conductivity coefficient k a space-dependent parameter or can even set it to a constant value. Our main focus here is on another parameter, namely viscosity $\mu(T)$, which is a function of another state variable, temperature T, governed by a separate (energy) equation. Consequently, we would like to reconstruct state-dependent viscosity $\mu(T)$ based on some available measurements, say temperature $\tilde{T}_i$, $i = 1,\ldots,M$, inside domain Ω. Similar to the least-squares data fitting used in both previous examples of the current chapter and also in Example 1.3, this reconstruction may be performed by considering an *inverse problem* formulated in terms of an optimization problem supplied with the objective function based on data obtained from the temperature measurements, i.e.,

$$\mathcal{J}(\mu) = \sum_{i=1}^{M}\left(T(x,t) - \tilde{T}_i\right)^2. \qquad (8.25)$$

Now, an obvious question comes after comparing the current problem with two previous ones.

? *If domain Ω assumes a pretty modest discretization, why do we consider this problem as a large-scale optimization?*

The forward problem (8.24) will require solving a system of PDEs, which is not a big deal in 2D, and even in 3D, when the discretized version of domain Ω contains not too many finite (spatial) elements. However, the solution space for state variable T requires additional discretization to serve as "spatial discretization" and support all computations for state-dependent function $\mu(T)$. Figure 8.5 provides a schematic illustration for this concept simplified by considering 1D spatial domain $\Omega = (-1, 1)$. Here, interval $I = [T_\alpha, T_\beta]$ represents the temperature range spanned by the solution $T(x)$ obtained, e.g., by solving forward problem (8.24); we will refer to this interval as the *identifiability interval*. As μ is a function of $T(x)$, interval I should have its own discretization (uniform or nonuniform) similar to domain Ω. Pursuing a proper quality for reconstructing $\mu(T)$ requires making T-discretization rather fine.

Now we ask the reader to think about the concept of reconstructing state-dependent parameters more broadly. Let us assume that some physical property ς is dependent on multiple (say, m) states s_i, i.e., $\varsigma = \varsigma(s_1, s_2, \ldots, s_m)$.

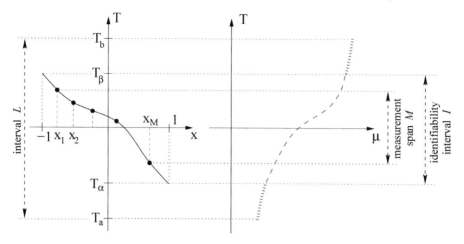

FIGURE 8.5
Schematic showing (left) the solution $T(x, \tau)$ at some fixed time τ and (right) the corresponding reconstruction of $\mu(T)$ defined over their respective domains, i.e., $\Omega = (-1, 1)$ and the identifiability region I. The thick dotted line represents an extension of $\mu(T)$ from I to the desired interval L. In the figure on the right, the horizontal axis serves as the ordinate. Source: reprinted from Vladislav Bukshtynov and Bartosz Protas, "Optimal Reconstruction of Material Properties in Complex Multiphysics Phenomena," Journal of Computational Physics 242, (2013): 889–914.

Computationally, it is a very hard problem to solve: once the problem is discretized (including solution spaces for states $s_1, s_2, \ldots s_m$), the ς-control space will be of size $N_1 \times N_2 \times \cdots \times N_m$, where N_i is the number of finite elements used to perform discretization for the s_i's solution space. With the increased number of such dependencies ($m > 1$), the complexity of the associated parameter identification problem will grow enormously. In computational practices, this phenomenon is well-known by its specific name – *"the curse of dimensionality."*

As a final remark, we need to mention that the state-dependent parameter reconstructions also pose a lot of other computational challenges. For example, the interval $L = [T_a, T_b]$ (refer back to Figure 8.5) is the temperature interval on which one could seek to obtain a reconstruction of the material property, e.g., viscosity $\mu(T)$. We have to note that, in general, this interval is larger than the identifiability interval I, and the identification will not progress outside this interval. Therefore, the computational framework designed to solve such a problem should have the methodology to deal with this issue. In addition, obtaining analytical expressions for the gradients is not straightforward. Analysis of (8.25) suggests that the objective $\mathcal{J}(\mu)$ does not contain $\mu(T)$ to allow explicit differentiation but is linked to it through the solution of the PDE system (8.24). Moreover, deriving feasible methodologies for computing gradients with respect to parameters of multiple states is still considered an open problem as the conventional approaches used so far are intractable due to the same problem caused by the "curse of dimensionality." To read more about computational aspects and available approaches to identify state-dependent parameters, we refer the reader to [10, 8].

8.3   Improving Performance of LSO

After reviewing an idea of a large-scale optimization in Section 8.1 and several examples in Section 8.2, here we briefly summarize some thoughts about improving the performance while running LSOs. As discussed multiple times before, we are particularly interested in finding a trade-off between the increased accuracy of the obtained solutions and minimized efforts invested into computations. On the one hand, to enhance the overall quality of the gradient-based optimization, we would think to improve accuracy while

- solving forward problem/equations of states,

- evaluating objectives,

- solving adjoint equations (if used for constrained optimization), and

- computing gradients.

It requires tuning up individual modules in the optimization framework; we refer the reader to Chapter 3 to review some methods and tips for making the tuning process less painful and more systematic.

On the other hand, we may consider reducing computational loads in several ways. Speeding up computations by dimensionality reduction for state variables may be reached by

- applying coarser discretization to space variables, or

- using various reduced-order (ROM) models.

Also, enhanced convergence of the optimization methods in use could be achieved by reducing the dimensionality of the control space if

- applying any parameterization techniques, e.g., PCA, or

- performing any types of upscaling, etc.

As found useful in practical computations, advancing, in general, the optimal solution search could be seen in using or combining different search techniques, e.g.,

- gradient-based,

- stochastic,

- combinatorial,

- metaheuristic, etc.

The reader may also think about other known solutions or developing the new approaches by keeping in mind that, in many cases, these solutions appear to be very problem-dependent.

8.4   General Theory for Constrained Optimization

As we observed in examples of Section 8.2, typical optimization problems based on the models taken from real-world applications usually come with at least one or several constraints. It is inevitable as these constraints, in particular, take responsibility for describing a model, while the objective function measures the optimality of solutions created for such a model. While before, in Sections 1.2 and 1.4, we discussed the connected notations of the *constrained optimization problem* and *feasibility* (*feasible region* or set of *feasible solutions*), so far, our focus was only on examples with no constraints. Now we have to move forward, and the second part of this chapter considers general theory on optimization problems supplied with various types of constraints (e.g., algebraic equations/inequalities or ODEs) to give the reader a flavor of their complexity based on pretty simple examples.

In pursuit of a very broad theory review, we consider the following general form of a constrained n-dimensional optimization problem:

$$\min_{\mathbf{x} \in \mathbb{R}^n} f(\mathbf{x})$$
$$\text{s.t. } \mathbf{x} \in \mathbb{S} \subset \mathbb{R}^n \tag{8.26}$$

It minimizes objective function $f(\mathbf{x}) : \mathbb{R}^n \to \mathbb{R}$ that hopefully is (however, sometimes, is not) continuously differentiable for any $\mathbf{x} \in \mathbb{S}$. Here, to focus on constraints, we also consider control and state variables as components of vector $\mathbf{x}$ and feasible region $\mathbb{S}$ as a set of points defined by a combination of all constraints that could be

- algebraic equations or inequalities,

- linear or nonlinear in their nature,

- represented by ODE or PDE systems, and

- consistent with any other requirements (e.g., functional space for $\mathbf{x}$).

Finally, we reiterate that, in general, an *infeasible solution* cannot be treated as a valid solution of (8.26) due to violating at least one constraint in the definition of $\mathbb{S}$. However, this violation is "negotiable" depending on the constraint type, e.g.,

- *hard constraints* require to be satisfied, but

- *soft constraints* allow some violations; if not satisfied, the objective function is penalized.

While creating a new computational framework, at some point, the readers should decide on the practical implementation of all constraints already imposed by the solved optimization problem: methods to be used and whether to add this functionality manually or by using any available external software. This decision is not straightforward as it depends on the existing preferences and problem complexity with expected growth in the future. We include the review of the Lagrange multiplier theory supplied with the simple example using our computational framework in the rest of this chapter once the readers choose to experiment on their own. To help them make this decision, we could briefly review some existing methods for adding constraints to computations.

We start with the *primal approach*, which means staying feasible at each iteration. For example, one could consider a variety of methods, such as

- the *feasible direction method* chooses search direction $\mathbf{d}^k$ such that for feasible current solution $\mathbf{x}^k$, new solution $\mathbf{x}^k + \alpha^k \mathbf{d}^k$ remains feasible,

- the *active set method* accounts continuously for active and inactive constraints,

- the *gradient projection method* projects all infeasible directions onto a feasible set,

- the *reduced gradient method* relates closely to the simplex method, and

- some possible variations of these approaches.

The *dual approach* considers a dual problem instead of the original primal problem. For instance,

- the *augmented Lagrangian (Lagrange multiplier) method* is claimed to be one of the most effective algorithms, and

- the *cutting plane method* to refine a feasible set through linear inequalities.

One of the very popular approaches, the *interior point method*, uses the so-called *primal-dual approach* when both problems (primal and dual) are solved simultaneously during all iterations. The last but not least group in our review is the penalization methods, e.g.,

- the *penalty function method* adds new terms to the objective to prescribe high costs for constraint violations, and

- the *barrier function method* also adds terms that favor points in the interior of the feasible region over those near the boundary.

We complete this brief review of methods for implementing constraints by noting that practical use may benefit from combining different approaches. For example, the augmented Lagrangian and penalty function methods are often used together as the *quadratic penalty function method* discussed later in Section 10.2.1.

8.5 Lagrange Multiplier Approach

Here, we limit our discussion to the basics of implementing the augmented Lagrangian or Lagrange multipliers to solve the following equality constrained optimization problem

$$
\min_{\mathbf{x} \in \mathbb{R}^n} f(\mathbf{x})
$$
$$
\text{s.t. } h_i(\mathbf{x}) = 0, \quad i = 1, \ldots, m \tag{8.27}
$$

under the necessary assumption that functions $f(\mathbf{x})$, $h_i(\mathbf{x}) : \mathbb{R}^n \to \mathbb{R}$ are **continuously differentiable**. We introduce the *Lagrangian (augmented objective) function* $\mathcal{L}(\mathbf{x}, \boldsymbol{\lambda}) : \mathbb{R}^{n+m} \to \mathbb{R}$

$$
\mathcal{L}(\mathbf{x}, \boldsymbol{\lambda}) = f(\mathbf{x}) + \sum_{i=1}^{m} \lambda_i \cdot h_i(\mathbf{x}), \tag{8.28}
$$

where scalars λ_i (components of vector $\boldsymbol{\lambda}$) are *Lagrange multipliers*. From (8.28), it is clear that the control space is augmented

$$
\mathbf{x} \quad \to \quad (\mathbf{x}, \boldsymbol{\lambda})
$$

by adding one multiplier λ_i per every constraint $h_i(\mathbf{x}) = 0$ in (8.27). The optimal solution for such updated controls becomes known after solving a new optimization problem

$$
(\mathbf{x}^*, \boldsymbol{\lambda}^*) = \operatorname*{argmin}_{\mathbf{x}, \boldsymbol{\lambda}} \mathcal{L}(\mathbf{x}, \boldsymbol{\lambda}). \tag{8.29}
$$

The Lagrange multiplier theory applicable to solving optimization problems (8.27) comes with two important theorems to provide necessary and sufficient *optimality conditions*.

Theorem 8.1 Necessary Optimality Conditions
Let $\mathbf{x}^*$ be a local minimizer of $f(\mathbf{x})$ subject to equality constraints $h_i(\mathbf{x}) = 0$, $i = 1, \ldots, m$. Assume that functions $f(\mathbf{x})$ and $h_i(\mathbf{x})$ are twice continuously

differentiable and constraint gradients $\nabla h_1(\mathbf{x}^*), \ldots, \nabla h_m(\mathbf{x}^*)$ *are linearly independent, then there exists a unique vector* $\boldsymbol{\lambda}^* = (\lambda_1^*, \ldots, \lambda_m^*)$ *such that the following conditions are true:*

- **1-order necessary**

$$\nabla_{\mathbf{x}}\mathcal{L}(\mathbf{x}^*, \boldsymbol{\lambda}^*) = \mathbf{0}, \qquad \nabla_{\boldsymbol{\lambda}}\mathcal{L}(\mathbf{x}^*, \boldsymbol{\lambda}^*) = \mathbf{0},$$

- **2-order necessary**

$$\nabla_{\mathbf{xx}}^2\mathcal{L}(\mathbf{x}^*, \boldsymbol{\lambda}^*) \succeq 0.$$

Theorem 8.2 Sufficient Optimality Conditions
Assume that functions $f(\mathbf{x})$ *and* $h_i(\mathbf{x})$, $i = 1, \ldots, m$, *are twice continuously differentiable and let* $\mathbf{x}^*$ *and* $\boldsymbol{\lambda}^*$ *satisfy both conditions*

- $\nabla_{\mathbf{x}}\mathcal{L}(\mathbf{x}^*, \boldsymbol{\lambda}^*) = \mathbf{0}, \qquad \nabla_{\boldsymbol{\lambda}}\mathcal{L}(\mathbf{x}^*, \boldsymbol{\lambda}^*) = \mathbf{0},$

- $\nabla_{\mathbf{xx}}^2\mathcal{L}(\mathbf{x}^*, \boldsymbol{\lambda}^*) \succ 0$ *(2-order sufficient).*

Then $\mathbf{x}^*$ *is a* **strict** *local minimizer of* $f(\mathbf{x})$ *subject to constraints* $h_i(\mathbf{x}) = 0$.

The proofs for both theorems may be easily found, e.g., in [4]. We also advise the reader to review Theorems 5.1 and 5.2 of Chapter 5 to examine how optimality conditions established for unconstrained and constrained (with the use of Lagrange multipliers) optimizers are related to each other.

8.6 Lagrange Multiplier vs. Penalization

We created our new example to be rather simple, with a quadratic objective and one linear equality constraint, to practice finding the analytical solution and adjusting the existing optimization framework to solve this problem computationally.

Example 8.1 Quadratic Objective + Linear Equality Constraint
Solve minimization problem

$$\min_{\mathbf{x} \in \mathbb{R}^4} \quad f(\mathbf{x}) = \frac{1}{2}\left(x_1^2 + x_2^2 + x_3^2 + x_4^2\right)$$
$$s.t. \quad x_1 + x_2 + x_3 + x_4 = 4 \tag{8.30}$$

First, we solve the optimization problem of Example 8.1 analytically by employing the Lagrange multiplier approach discussed in Section 8.5. We construct the augmented objective function (or Lagrangian)

$$
\begin{aligned}
\mathcal{L}(\mathbf{x}, \lambda) &= f(\mathbf{x}) + \lambda \cdot h(\mathbf{x}) \\
&= \frac{1}{2} \left(x_1^2 + x_2^2 + x_3^2 + x_4^2 \right) + \lambda \cdot (x_1 + x_2 + x_3 + x_4 - 4),
\end{aligned} \tag{8.31}
$$

where $\lambda \in \mathbb{R}$ is the Lagrange multiplier. Then the 1-order necessary conditions (in Theorems 8.1 and 8.2) $\nabla_{\mathbf{x}}\mathcal{L}(\mathbf{x}^*, \lambda^*) = \mathbf{0}$ and $\nabla_{\lambda}\mathcal{L}(\mathbf{x}^*, \lambda^*) = \mathbf{0}$ provide the following system of (linear) equations

$$
\begin{aligned}
x_1 + \lambda &= 0, \\
x_2 + \lambda &= 0, \\
x_3 + \lambda &= 0, \\
x_4 + \lambda &= 0, \\
x_1 + x_2 + x_3 + x_4 - 4 &= 0.
\end{aligned} \tag{8.32}
$$

Now we could find a *stationary point* $\mathbf{x}^* = [1\ 1\ 1\ 1]^T$, $\lambda^* = -1$ as a unique solution of system (8.32). We complete the analytical solution by making two conclusions:

(1) 2-order necessary condition (in Theorem 8.1) is also **sufficient** (in Theorem 8.2) as $\nabla_{\mathbf{xx}}^2 \mathcal{L} = \mathcal{I} \succ 0$. Therefore, optimal solution $\mathbf{x}^*$ is a **strict local minimizer** of $f(\mathbf{x})$.

(2) Functions $f(\mathbf{x})$ and $h(\mathbf{x}) = x_1 + x_2 + x_3 + x_4 - 4$ are both convex. We refer to Theorem 5.3 (Convex Cost Function) to confirm that $\mathbf{x}^*$ is also a **global minimizer** of $f(\mathbf{x})$.

A computational implementation of the Lagrange multiplier approach for solving the optimization problem of Example 8.1 differs slightly from the analytical solution method described above. Following the discussion in Section 8.5 and, more specifically, observing formula (8.29), it might be tempting to minimize Lagrangian (8.31) directly by considering the extended control set $\mathbf{u} = \{x_1, x_2, x_3, x_4, \lambda\}$, and computed gradients in the following form

$$
\nabla_{\mathbf{u}}\mathcal{L} = \begin{bmatrix} x_1 + \lambda \\ x_2 + \lambda \\ x_3 + \lambda \\ x_4 + \lambda \\ x_1 + x_2 + x_3 + x_4 - 4 \end{bmatrix}. \tag{8.33}
$$

The reader would gain valuable experience while giving it a try. The new framework will pass the kappa-test discussed in Section 3.5, but the solution will likely diverge even started very close to the optimal one.

? *Why? What is wrong?*

There is nothing wrong with the analytical solution leading directly to stationary point $\mathbf{x}^*$ – it is just intractable for applying in iterative algorithms when the initial guess may also be infeasible. By solving the last equation in (8.32), we explicitly require the solution to honor the constraint. However, the iterative solver will use correct gradients (8.33) and minimize objective (8.31) by highly likely setting $(x_1 + x_2 + x_3 + x_4 - 4)$ to a negative value as its "zero" target is not set anywhere, and, as such, unknown.

A computational procedure to apply Lagrange multipliers in practice requires a formal definition of three sets of variables: namely, controls, states, and multipliers. In this generalized situation, the variables in the last group are often called *adjoint (adjoined) variables*. We structurize the adjoint-based procedure for solving the optimization problem (8.27) in Algorithm 8.1.

Algorithm 8.1 (Adjoint-gradient Equality-constrained Optimization)

1. *Consider optimization problem:*

$$\min_{\mathbf{x} \in \mathbb{R}^n} f(\mathbf{x}; \mathbf{u})$$
$$s.t. \ \mathbf{h}(\mathbf{x}; \mathbf{u}) = \mathbf{0} \tag{8.34}$$

   with state $\mathbf{x}$ and control $\mathbf{u}$ variables

2. *Define Lagrangian*

$$\mathcal{L}(\mathbf{x}; \mathbf{u}, \boldsymbol{\lambda}) = f(\mathbf{x}; \mathbf{u}) + \boldsymbol{\lambda} \cdot \mathbf{h}(\mathbf{x}; \mathbf{u}) \tag{8.35}$$

   where $\boldsymbol{\lambda}$ is a vector of adjoint states (adjoint variable)

3. *Derive analytical expressions for gradients*

$$\nabla_{\mathbf{x}}\mathcal{L}, \quad \nabla_{\mathbf{u}}\mathcal{L}, \quad \nabla_{\boldsymbol{\lambda}}\mathcal{L} = \mathbf{h}(\mathbf{x}; \mathbf{u}) \tag{8.36}$$

4. *During each iteration*

   (a) solve **state equation**

$$\nabla_{\boldsymbol{\lambda}}\mathcal{L} = \mathbf{h}(\mathbf{x}; \mathbf{u}) = \mathbf{0} \tag{8.37}$$

   for current control $\mathbf{u}$ to obtain **new state $\mathbf{x}$**

   (b) solve **adjoint equation**

$$\nabla_{\mathbf{x}}\mathcal{L} = \mathbf{0} \tag{8.38}$$

   for current control $\mathbf{u}$ and updated state $\mathbf{x}$ to obtain **new adjoint $\boldsymbol{\lambda}$**

(c) evaluate analytical expression $\nabla_{\mathbf{u}}\mathcal{L}$ for **adjoint-based gradient** *using current control* **u** *and updated state* **x** *and adjoint* λ

5. *Utilize gradients $\nabla_{\mathbf{u}}\mathcal{L}$ to update control* **u** *using chosen schemes for gradient-based search directions and step sizes for solving* (8.34)

Now we apply this algorithm to solve the optimization problem of Example 8.1 computationally (iteratively). We still use the same Lagrangian as defined in (8.31) and consider three sets of variables with their respected gradients:

- control $\mathbf{u} = \{x_1, x_2, x_3\}$

$$\nabla_{\mathbf{u}}\mathcal{L} = \begin{bmatrix} x_1 + \lambda \\ x_2 + \lambda \\ x_3 + \lambda \end{bmatrix}, \tag{8.39}$$

- formally defined state $\mathbf{x} = \{x_4\}$ (the reader may use another choice of $\mathbf{x}$)

$$\nabla_{\mathbf{x}}\mathcal{L} = x_4 + \lambda, \tag{8.40}$$

- and adjoint λ

$$\nabla_{\lambda}\mathcal{L} = x_1 + x_2 + x_3 + x_4 - 4. \tag{8.41}$$

As outlined in Algorithm 8.1, we have to take three steps at each optimization iteration, i.e.,

(a) solve *state equation* $\nabla_{\lambda}\mathcal{L} = 0$ using (8.41) to obtain new "state" x_4

$$x_4 = 4 - (x_1 + x_2 + x_3), \tag{8.42}$$

(b) solve *adjoint equation* $\nabla_{\mathbf{x}}\mathcal{L} = 0$ using (8.40) to obtain new adjoint λ

$$\lambda = -x_4 = x_1 + x_2 + x_3 - 4, \tag{8.43}$$

(c) and finally update the *adjoint-based gradient* $\nabla_{\mathbf{u}}\mathcal{L}$ with respect to ("truncated") control $\mathbf{u} = \{x_1, x_2, x_3\}$ given by (8.39) using adjoint λ.

While solving more complicated problems, the adjoint-based procedure with steps (a-c) applies in full at every iteration. As we could see later, in general, it requires some effort to solve state and adjoint equations, e.g., given by systems of algebraic equations or ODEs/PDEs. However, in the current simple example, we could omit steps (a) and (b) by plugging (8.43) to (8.39) and arrive directly at the following shape of the gradient

$$\nabla_{\mathbf{u}}\mathcal{L} = \begin{bmatrix} 2x_1 + x_2 + x_3 - 4 \\ x_1 + 2x_2 + x_3 - 4 \\ x_1 + x_2 + 2x_3 - 4 \end{bmatrix} \quad (8.44)$$

to be used in practical computations. This gradient now "knows" the constraint $h(\mathbf{x}) = 0$ as it is derived using relation (8.42) to enforce (strictly) this condition. Finally, we could use (8.42) to obtain the optimal "state" value $\mathbf{x}^* = \{x_4^*\}$ after completing optimization and computing optimal control $\mathbf{u}^* = \{x_1^*, x_2^*, x_3^*\}$.

We also remark on the convexity of the constrained optimization problem of Example 8.1 solved by the adjoint-based gradient (8.44). The reader may easily check the positive definiteness of the Hessian

$$\nabla_{\mathbf{uu}}^2 \mathcal{L} = \begin{bmatrix} 2 & 1 & 1 \\ 1 & 2 & 1 \\ 1 & 1 & 2 \end{bmatrix} \quad (8.45)$$

by computing its eigenvalues: 1, 1, and 4 (all positive).

The simplicity of Example 8.1 allows us to illustrate the way of "adding" constraint $h(\mathbf{x}) = 0$ to the objective function $f(\mathbf{x})$ by comparing this augmentation with plain substitution. We express x_4 explicitly, e.g., as shown in (8.42), by using equality constraint in (8.30) and arrive at the new reduced-dimensional problem

$$\min_{\mathbf{x} \in \mathbb{R}^3} f(\mathbf{x}) = \frac{1}{2} \left[x_1^2 + x_2^2 + x_3^2 + (x_1 + x_2 + x_3 - 4)^2 \right]. \quad (8.46)$$

We advise the reader to check that the unconstrained optimization problem (8.46) is equivalent to the problem (8.30) solved by the Lagrange multiplier approach: it has the same gradient as in (8.44) and optimal solution $\mathbf{x}^* = [1\ 1\ 1]^T$.

Before proceeding to computations, it might be interesting to compare the results provided by the *hard-constraint* method (e.g., the Lagrange multiplier) with any *soft-constraint* one. As an example, here, we choose a simple *penalization* approach to solve the same problem in Example 8.1. Similar to Lagrangian, we create a new objective function $\bar{f}(\mathbf{x})$ by adding a penalization term $\beta h^2(\mathbf{x})$ to the objective $f(\mathbf{x})$ in (8.30), i.e.,

$$\min_{\mathbf{x} \in \mathbb{R}^4} \bar{f}(\mathbf{x}) = \frac{1}{2} \left(x_1^2 + x_2^2 + x_3^2 + x_4^2 \right) + \beta(x_1 + x_2 + x_3 + x_4 - 4)^2, \quad (8.47)$$

where $\beta \in \mathbb{R}_+$ is a positive constant. Solving this unconstrained optimization problem will require gradients

$$\nabla_{\mathbf{x}}\bar{f}(\mathbf{x}) = \begin{bmatrix} x_1 + 2\beta(x_1 + x_2 + x_3 + x_4 - 4) \\ x_2 + 2\beta(x_1 + x_2 + x_3 + x_4 - 4) \\ x_3 + 2\beta(x_1 + x_2 + x_3 + x_4 - 4) \\ x_4 + 2\beta(x_1 + x_2 + x_3 + x_4 - 4) \end{bmatrix} \quad (8.48)$$

and, if the use of the 2-order methods is planned, Hessians

$$
\nabla_{\mathbf{xx}}^2 \bar{f}(\mathbf{x}) = \begin{bmatrix} 2\beta + 1 & 2\beta & 2\beta & 2\beta \\ 2\beta & 2\beta + 1 & 2\beta & 2\beta \\ 2\beta & 2\beta & 2\beta + 1 & 2\beta \\ 2\beta & 2\beta & 2\beta & 2\beta + 1 \end{bmatrix} = 2\beta\mathcal{E} + \mathcal{I}, \qquad (8.49)
$$

where $\mathcal{E}$ and $\mathcal{I}$ are, respectively, unity and identity matrices. We may also prove the convexity of the problem (8.47) by showing that Hessian $H = 2\beta\mathcal{E} + \mathcal{I}$ is positive definite for any $\beta > 0$. We also advise the reader to obtain the analytical solution for this problem that will help explain our computational outcomes.

! *Solve a system of linear equations* $\nabla_{\mathbf{x}} \bar{f}(\mathbf{x}) = 0$ *and* (8.48) *to get*

$$
x_1^* = x_2^* = x_3^* = x_4^* = \frac{8\beta}{8\beta + 1}. \qquad (8.50)
$$

Now we are ready to experiment with the computational part of solving the constrained optimization in Example 8.1. We see this experimental part with manifold purposes. In Chapter 7, we completed the modular structure of our current optimization framework at

- "logical level" by establishing proper communication between its different parts, and

- "physical level" by organizing files within several folders; refer to Section 7.5 for more details on the current structure.

We left this framework with functionality to solve the unconstrained optimization problem of Example 1.3 computationally, including a range of gradient-based and derivative-free approaches. At this time, we have an opportunity to check the ergonomics of the current structure, while transitioning to a new problem of Example 8.1 that features

- a new objective,

- different number (four) of controls,

- an equality constraint, and

- various methods to incorporate this constraint.

We advise the readers to either perform this transition by themselves or compare the modified framework (`Chapter_8_opt_constrained.m`) with its version used previously in Chapter 7; refer to Table 8.1 for a summary of all modifications.

To check the minimized "invasion" in the structure of the existing code, we provide the reader with the complete list of all required changes.

element	implementation	software
main **OPT**	minimally modified to support solutions for constrained optimization	MATLAB
f-evaluator	m-function, analytically defined function (changed to functions $f(\mathbf{x})$ and $\bar{f}(\mathbf{x})$)	MATLAB
d-evaluator	m-code to keep **SD/CG/BFGS/NEWTON** (changed for new gradients & Hessians)	MATLAB
1D α search	m-code, various step size searches (no change)	MATLAB
visualizer	plain m-code (updated due to new solution structure)	MATLAB

TABLE 8.1

Computational elements of `Chapter_8_opt_constrained.m` modified from `Chapter_7_data_fit_comprehensive.m` to solve constrained optimization problem of Example 8.1.

- In the main file `Chapter_8_opt_constrained.m`, option to load data (`data = load(dataFile)`) is removed as not required for the new problem. However, variable `data` is preserved to pass some information (chosen method for handling constraint and penalization coefficient β) throughout the entire framework. The deactivated (not moved from the previous version) methods (DN, MN, DSSD, GN, TRDL, BF, MC) are commented (`<- deactivated`).

- The entire structure of the parameter file `params.m` is also preserved to allow the reader to experiment with activating (transferring from the previous version) gradient-based and derivative-free approaches (left as homework). Initial guess vector `aini` is extended by adding an extra component. We also added two more variables to select a constraint handling method (`exMethod`: 0 - Lagrange multiplier, 1 - penalization) and set penalization coefficient β (`penalCoeff`).

- Folder `data` is removed as not used; however, the readers may keep it if they plan to experiment more with this version.

- In folder `algorithms`, we modified only two files: `initialize.m` and `visualize.m`. A new version of the initialization file does not save the shape of the initial guess curve. Instead, it initializes two variables: `data = [exMethod penalCoeff]` and `aTrue = [1 1 1 1]`, which is self-explanatory. Changes to the visualization file are minimized to update the solution and objective function graphs.

- In folder `functions`, we modified three files (`fn_eval_f.m`, `fn_eval_f_grad.m`, and `fn_eval_hess.m`) to apply computations provided in (8.31), (8.44)–(8.45), and (8.47)–(8.49).

- Finally, we keep vector **a** (not **x**) for controls to avoid renaming this variable at each appearance in the entire code. We only notice that **a**(4) is supplied with the zero gradient component and updated manually by (8.42) whenever the Lagrange multiplier method is used (exMethod = 0).

Now, we could check the functionality of the modified computational framework to solve constraint optimization by running Chapter_8_opt_constrained.m with the following settings (we list only important ones and those related closely to the implemented changes).

MATLAB: params.m

Settings for solving constrained optimization:

- *mode:* mode = 'OPT'; or mode = 'TEST';

- *gradient-based method:* method = 'SD'; also NEWTON/BFGS/CG

- *initial guess:* aini = [5 -3 8 5];

- *constraint handling method:* exMethod = 0; for Lagrange multiplier (or 1 for penalization)

- *penalization coefficient:* penalCoeff = 1; will vary

- *step size search method:* methodAlpha = 'BB';

- *termination #1:* epsilonF = 1e-6; for ϵ_1, refer to (3.18)

- *termination #2:* epsilonA = 1e-6; for ϵ_2, refer to (3.17)

- *termination #3:* kMax = 200;

Following our discussions in Sections 3.5 and 3.6, we start with checking the correctness of implemented changes to the objective and gradient computations using the *kappa-test*. The results for both methods (the Lagrange multiplier and penalization) are shown in Figure 8.6.

Here, we see that the cheap test demonstrates that the gradients for both methods are computed correctly. The expensive test also confirms the correctness of all individual gradient components: three for the Lagrange multiplier approach (the fourth one is set to 0 as the fourth control is "moved" to the category of state variables) and four for the penalization. We advise the reader to review our discussion in Section 3.6 for the procedure and correct interpretation of the kappa-test results.

Next, we refer to Figure 8.7 depicting four plots updated dynamically during optimization finalized at $k = 8$: (a) solution (true, initial, and optimal),

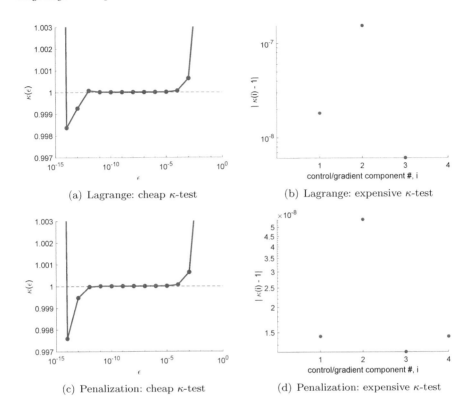

(a) Lagrange: cheap κ-test

(b) Lagrange: expensive κ-test

(c) Penalization: cheap κ-test

(d) Penalization: expensive κ-test

FIGURE 8.6
Results of (a,c) cheap and (b,d) expensive kappa-tests obtained, while applying (a,b) the Lagrange multiplier and (c,d) penalization approaches.

(b) objective $f(\mathbf{x})$ as a function of iteration count k, (c) structure of the current gradient-based search direction $\mathbf{d}^k$, and (d) the history of step size α changes with k.

The results shown in Figure 8.7 are obtained by applying the Lagrange multiplier approach (`exMethod = 0`) with the gradient-based steepest descent (`method = 'SD'`) paired with bracketing-Brent (`methodAlpha = 'BB'`) search for optimal step size α. It requires only $k = 8$ iterations to converge to the optimal solution $\mathbf{x}^* = [0.99996\ 0.99992\ 0.99997\ 1.0002]^T$ with $f^* = 2$ by satisfying the constraint in the strict sense (residual `res = 0` meaning $x_1 + x_2 + x_3 + x_4 - 4 = 0$). To conclude on the performance of other gradient-based approaches moved to the upgraded framework (i.e., NEWTON, BFGS, and CG), we run them all. The combined results are shown in Figure 8.8. All methods converge to the same optimal solution with speed expected based on our observations made for the unconstrained Example 1.3 (in fact, the

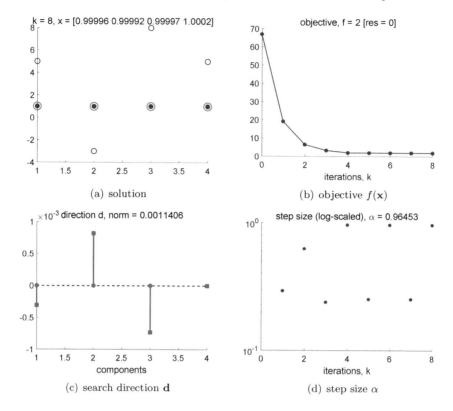

(a) solution   (b) objective $f(\mathbf{x})$

(c) search direction $\mathbf{d}$ (d) step size α

FIGURE 8.7

MATLAB window with four plots updated dynamically (finalized with SD at $k = 8$), while solving Example 8.1 by the Lagrange multiplier approach.

conjugate gradient method shows the performance comparable to the BFGS approach).

Now it is the turn to check the computational performance of the penalization approach. Being a soft-constraint method is considered one of the known drawbacks of applying penalization. Another disadvantage is the use of the parameter β in (8.47). Taking it small makes the penalization term in $\bar{f}(\mathbf{x})$ insignificant compared to the core objective part, i.e., $f(\mathbf{x})$ in (8.30): it leads to the constraint violation at higher levels. On the opposite, setting β to heavier values helps in satisfying the constraint but may disallow finding a better solution (lower value for the core objective part $f(\mathbf{x})$) due to early termination.

To give the reader some flavor of that phenomenon, we run the computational framework with multiple β values fixing the gradient-based method to the CG. First, we set β to a relatively small number, e.g., $\beta = 1$. As shown in

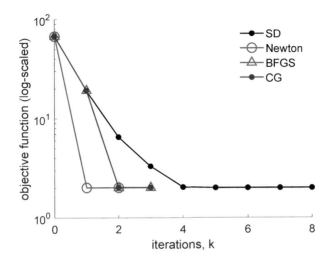

FIGURE 8.8
Performance of various gradient-based approaches (SD, NEWTON, BFGS, and CG), while solving Example 8.1 by the Lagrange multiplier approach.

Figure 8.9, the conjugate gradient method converges in just $k = 3$ iterations with optimal solution $\mathbf{x}^* = [0.88889\ 0.88889\ 0.88889\ 0.88889]^T$ and "better" objective value $\bar{f}^* = 1.7778$ at the expense of the severe constraint violation: $x_1 + x_2 + x_3 + x_4 - 4 = 0.44444$.

Next, we repeat the computations by only changing the penalization parameter, e.g., $\beta = 10, 100, 1000$; refer to Figure 8.10. Although heavier β implies higher values for the initial state of the objective function $\bar{f}(\mathbf{x})$, the final solution tends to satisfy the constraint as β becomes sufficiently large. The solution behavior is also consistent with the results provided in (8.50)

$$x_i^* = \frac{8\beta}{8\beta + 1} \to 1, \quad \beta \to \infty. \tag{8.51}$$

The reader may also notice that some values (e.g., $\beta = 100$) may require a longer iteration process for convergence.

? *What might be considered a good strategy for choosing β in the penalization approach implemented computationally?*

! *The reader may think about proper scaling applied for the core and penalization parts of the augmented objective function and flexible schemes to change parameter β during the optimization process. We also invite the reader to experiment with these strategies in practice using the current version of the computational framework.*

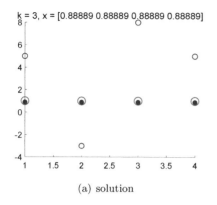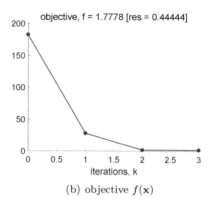

(a) solution   (b) objective $f(\mathbf{x})$

FIGURE 8.9
MATLAB window with two plots updated dynamically (finalized with CG at $k = 3$), while solving Example 8.1 by the penalization approach using $\beta = 1$.

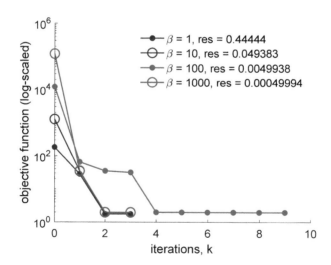

FIGURE 8.10
Performance of the CG approach while solving Example 8.1 by penalization with $\beta = 1, 10, 100, 1000$.

We complete the analysis part of this computational exercise by addressing the issue of choosing between these two methods. The simplicity of practical implementation for penalization is evident. However, constraints provided by too many equations will pose a problem of multiple penalization coefficients that may cause conflicts between different parts of the augmented objective if improperly selected. On the other hand, the Lagrange multiplier approach does not require any coefficients. However, the implementing methodology becomes rather sophisticated if the equalities are ODEs or PDEs; we will see these examples soon in the subsequent chapters.

We complete this section by revisiting two examples to reconsider them in terms of constrained optimization. The first one evolves from Example 1.3 for the least-square data fitting.

Example 8.2 Least-Squares Data Fitting Revisited
Consider the following constrained optimization problem

$$\min_{\mathbf{a} \in \mathbb{R}^3} f(y(x); \mathbf{a}) = \sum_{i=1}^{m} (y_i - y(x_i))^2 \tag{8.52}$$

$$\text{s.t. } y(x) = a_1 + a_2 x + a_3 x^2$$

where (x_i, y_i), $i = 1, \ldots, m$, *are given data* $(m \geq 3)$. *Unconstrained reformulation made with the Lagrange multipliers approach is fully equivalent to the simplest form* (1.6)–(1.7) *we used before, i.e.,*

$$\min_{\mathbf{a} \in \mathbb{R}^3} f(\mathbf{a}) = \sum_{i=1}^{m} \left(y_i - (a_1 + a_2 x_i + a_3 x_i^2) \right)^2 . \tag{8.53}$$

To prove the equivalency claimed in Example 8.2, we construct an augmented objective function (Lagrangian)

$$\mathcal{L}(\mathbf{y}; \mathbf{a}, \boldsymbol{\lambda}) = \sum_{i=1}^{m} (y_i - y(x_i))^2 + \sum_{i=1}^{m} \lambda_i \left(y(x_i) - (a_1 + a_2 x_i + a_3 x_i^2) \right). \tag{8.54}$$

Then the 1-order necessary condition (Theorem 8.1) $\nabla_{\boldsymbol{\lambda}} \mathcal{L}(\mathbf{y}^*; \mathbf{a}^*, \boldsymbol{\lambda}^*) = 0$ gives explicitly

$$y(x_i) = a_1 + a_2 x_i + a_3 x_i^2, \quad i = 1, \ldots, m \tag{8.55}$$

which reduces the dimensionality of the control space back to the size of control **a**, i.e.,

$$\mathcal{L}(\mathbf{y}; \mathbf{a}, \boldsymbol{\lambda}) \rightarrow f(\mathbf{a}),$$

by considering the unconstrained optimization problem (8.53).

The second example refers back to our discussion in Section 8.2.1 about solving (large) systems of (nonlinear) equations considered a form of large-scale optimization.

Example 8.3 Systems of (Nonlinear) Equations Revisited
Solve the system of equations

$$\mathbf{g}(\mathbf{x}) = \mathbf{0}, \tag{8.56}$$

where $\mathbf{g}(\mathbf{x}) : \mathbb{R}^n \to \mathbb{R}^m$, *by reformulating the problem as the constrained optimization (feasible region) problem*

$$\begin{aligned} \min_{\mathbf{x} \in \mathbb{R}^n} \quad & f(\mathbf{x}) = 0 \\ s.t. \quad & \mathbf{g}(\mathbf{x}) = \mathbf{0} \end{aligned} \tag{8.57}$$

Similar to the previous example, we could construct the augmented objective function (Lagrangian)

$$\mathcal{L}(\mathbf{x}, \boldsymbol{\lambda}) = 0 + \boldsymbol{\lambda} \cdot \mathbf{g}(\mathbf{x}) = \sum_{i=1}^{m} \lambda_i \cdot g_i(\mathbf{x}) \tag{8.58}$$

and apply 1-order necessary conditions $\nabla_{\mathbf{x}} \mathcal{L}(\mathbf{x}^*, \boldsymbol{\lambda}^*) = \mathbf{0}$ and $\nabla_{\boldsymbol{\lambda}} \mathcal{L}(\mathbf{x}^*, \boldsymbol{\lambda}^*) = \mathbf{0}$, i.e.,

$$\sum_{i=1}^{m} \lambda_i \nabla_{\mathbf{x}} g_i(\mathbf{x}) = \mathbf{0}, \tag{8.59}$$

$$\mathbf{g}(\mathbf{x}) = \mathbf{0}.$$

However, it still requires solving $\mathbf{g}(\mathbf{x}) = \mathbf{0}$ and moves the problem back to its initial formulation (8.56). Instead, we could consider a new unconstrained optimization problem

$$\min_{\mathbf{x} \in \mathbb{R}^n} \bar{f}(\mathbf{x}) = \frac{1}{2} \|\mathbf{g}(\mathbf{x})\|^2 \tag{8.60}$$

that has the same solution as (8.56) and (8.57) and may be solved, e.g., by the Gauss-Newton method. We advise the reader to review Section 5.3.6 for more details.

8.7 Extending Complexity – DE-constrained Optimization

In this section, we tend to motivate the readers to reconsider the complexity of optimization problems constrained by differential equations. Here, we review the general form (8.1) by focusing on the equality constrained optimization problem

$$\min_{\substack{\mathbf{u} \in \mathbb{R}^m}} \ \mathcal{J}(\mathbf{x}; \mathbf{u})$$
$$\text{s.t. } \mathbf{g}(\mathbf{x}; \mathbf{u}) = 0, \tag{8.61}$$

where $\mathbf{x} \in \mathbb{R}^n$ and $\mathbf{u} \in \mathbb{R}^m$. The reader will find a discussion on the practical approaches for solving such optimization problems (both ODE- and PDE-constrained) in the next chapters. Here, we give an overall idea of their structure and associated complexity using two simple examples of ODE-based problems.

Example 8.4 BVP-2 Parameter Identification
Identify parameters a_1, a_2, and a_3 in the 2-order boundary-value problem (BVP) by fitting available data $\tilde{y}$

$$\min_{\mathbf{a} \in \mathbb{R}^3} \ \mathcal{J}(y(t); \mathbf{a}) = \frac{1}{2} \int_{t_0}^{T} (y - \tilde{y})^2 \, dt$$
$$\text{s.t. } a_1 y'' + a_2 y' + a_3 y = f(t), \quad y(t_0) = c_1, \quad y(T) = c_2 \tag{8.62}$$

We notice that (8.62) provides a *continuous formulation* of the *ODE-based* (ODE-constrained) optimization problem. This formulation involves (both continuous) integration (in the objective function) and a description of BVP-2 with a solution (state variable) assumed to be a continuous function $y(t)$ defined over the interval $[t_0, T]$. If the solution to a such problem cannot be obtained analytically, we may proceed by applying numerical methods. It will require re-writing the problem (8.62) in the *discretized form* by

- discretizing time t with step $\Delta t = \dfrac{T - t_0}{n}$:

$$t_0, \ t_1 = t_0 + \Delta t, \ t_2 = t_0 + 2\Delta t, \ \ldots, \ t_n = T,$$

- discretizing state variable $y(t) \to \mathbf{y} \in \mathbb{R}^{n+1}$ and associated data $\tilde{\mathbf{y}}$, i.e.,

$$
\mathbf{y} = \begin{bmatrix} y_0 \\ y_1 \\ \cdots \\ y_n \end{bmatrix} = \begin{bmatrix} y(t_0) \\ y(t_1) \\ \cdots \\ y(T) \end{bmatrix}, \qquad \tilde{\mathbf{y}} = \begin{bmatrix} \tilde{y}_1 \\ \tilde{y}_2 \\ \cdots \\ \tilde{y}_{n-1} \end{bmatrix},
$$

- defining a discretized form of the objective function $\mathcal{J}(y(t); \mathbf{a})$, e.g.,

$$
\mathcal{J}(\mathbf{a}) = \frac{1}{2} \sum_{i=1}^{n-1} (y_i(\mathbf{a}) - \tilde{y}_i)^2, \tag{8.63}
$$

- and, finally, solving BVP-2 in (8.62) using the obtained discretization.

We show an additional example to help the reader create a general picture for diverse problems with optimization supported by differential equations. In this example, we choose an objective with no data required, DE in the form of the *initial-value problem (IVP)* with the free (source) term function $f(t)$ to represent the control. The latter will also require discretization to solve the optimization problem numerically.

Example 8.5 IVP-2 Parameter Identification
Identify free term function $f(t)$ in the 2-order initial-value problem by minimizing the 2-norm of the solution $y(t)$, i.e.,

$$
\min_{f(t)} \mathcal{J}(y(t); f(t)) = \|y(t)\|_2^2 = \int_{t_0}^{T} y^2 \, dt \tag{8.64}
$$
$$
\text{s.t. } a_1 y'' + a_2 y' + a_3 y = f(t), \quad y(t_0) = c_1, \quad y'(t_0) = c_2
$$

Similar to Example 8.4, we notice that (8.64) provides a *continuous formulation* of the *ODE-based* (ODE-constrained) optimization problem. A numerical solution for this problem also requires discretization for

- time t,

- both state $y(t) \to \mathbf{y} \in \mathbb{R}^{n+1}$ and control $f(t) \to \mathbf{f} \in \mathbb{R}^{n+1}$ variables, i.e.,

$$
\mathbf{y} = \begin{bmatrix} y_0 \\ y_1 \\ \cdots \\ y_n \end{bmatrix} = \begin{bmatrix} y(t_0) \\ y(t_1) \\ \cdots \\ y(T) \end{bmatrix}, \qquad \mathbf{f} = \begin{bmatrix} f_0 \\ f_1 \\ \cdots \\ f_n \end{bmatrix} = \begin{bmatrix} f(t_0) \\ f(t_1) \\ \cdots \\ f(T) \end{bmatrix},
$$

- objective function $\mathcal{J}(y(t); f(t))$, e.g.,

$$\mathcal{J}(\mathbf{f}) = \sum_{i=1}^{n} [y_i(\mathbf{f})]^2, \tag{8.65}$$

- and, finally, for solving IVP-2 in (8.64) using the obtained discretization.

After changing the continuous formulation of optimization problems in Examples 8.4 and 8.5 to the discretized ones, their structures become similar and easily fitted into the shape of the computational framework we modified to work in this chapter. Generally speaking, there are two main approaches for proceeding with the numerical solutions, namely:

(a) *"optimize–then–discretize"* with optimality conditions applied to the problem, while it is still in its continuous form, and

(b) *"discretize–then–optimize"* with the same conditions applied directly to the already fully discretized problem.

? *How to compute gradients $\nabla_{\mathbf{a}}\mathcal{J}(\mathbf{a})$ and $\nabla_{\mathbf{f}}\mathcal{J}(\mathbf{f})$ if required?*

We will discuss both approaches in more detail and methods for computing associated gradients in Chapter 9.

8.8   KKT Optimality Conditions

In the last section of this chapter, we briefly discuss the theory of implementing inequality constraints by the same Lagrange multiplier approach. We refer to the equality constrained optimization problem (8.27). Here, we update it by adding r inequalities, i.e.,

$$\begin{aligned} \min_{\mathbf{x}\in\mathbb{R}^n} \quad & f(\mathbf{x}) \\ \text{s.t.} \quad & h_i(\mathbf{x}) = 0, \quad i = 1,\ldots,m \\ & g_j(\mathbf{x}) \leq 0, \quad j = 1,\ldots,r \end{aligned} \tag{8.66}$$

and also assume that functions $f(\mathbf{x})$, $h_i(\mathbf{x})$, $g_i(\mathbf{x}) : \mathbb{R}^n \to \mathbb{R}$ are **continuously differentiable**. A commonly used approach treats inequality constraints $g_j(\mathbf{x}) \leq 0$ as equality constraints $g_j(\mathbf{x}) = 0$ by defining a set of *active constraints*

$$\mathbb{A}(\mathbf{x}) = \{j : g_j(\mathbf{x}) = 0\} \tag{8.67}$$

and considering the following equivalent problem:

$$\min_{\mathbf{x} \in \mathbb{R}^n} \quad f(\mathbf{x})$$

$$\text{s.t.} \quad h_i(\mathbf{x}) = 0, \tag{8.68}$$

$$g_j(\mathbf{x}) = 0, \quad \forall j \in \mathbb{A}(\mathbf{x})$$

Similar to the discussed in Section 8.5, we introduce the *Lagrangian* (*augmented objective*) *function* $\mathcal{L}(\mathbf{x}, \boldsymbol{\lambda}, \boldsymbol{\mu}) : \mathbb{R}^{n+m+r} \to \mathbb{R}$

$$\mathcal{L}(\mathbf{x}, \boldsymbol{\lambda}, \boldsymbol{\mu}) = f(\mathbf{x}) + \sum_{i=1}^{m} \lambda_i \cdot h_i(\mathbf{x}) + \sum_{j=1}^{r} \mu_j \cdot g_j(\mathbf{x}), \tag{8.69}$$

where scalars λ_i (components of vector $\boldsymbol{\lambda}$) and μ_j (components of vector $\boldsymbol{\mu}$) are *Lagrange multipliers* assigned, respectively, to equality and inequality constraints in (8.66) and (8.68). Therefore, the control space is augmented, i.e.,

$$\mathbf{x} \quad \to \quad (\mathbf{x}, \boldsymbol{\lambda}, \boldsymbol{\mu})$$

by adding now $m + r$ new controls where $\mu_j \geq 0$ ($\mu_j = 0$ if jth constraint is not in the set $\mathbb{A}(\mathbf{x})$ for current solution $\mathbf{x}$).

Finally, we provide two fundamental theorems that make the Lagrange multiplier theory applicable to solving optimization problems with various types of constraints, namely *Karush–Kuhn–Tucker (KKT) optimality conditions*.

Theorem 8.3 KKT Necessary Optimality Conditions

Let $\mathbf{x}^*$ be a local minimizer of $f(\mathbf{x})$ subject to equality $h_i(\mathbf{x}) = 0$, $i = 1, \ldots, m$, and inequality $g_j(\mathbf{x}) \leq 0$, $j = 1, \ldots, r$ constraints. Assume that functions $f(\mathbf{x})$, $h_i(\mathbf{x})$, and $g_j(\mathbf{x})$ are twice continuously differentiable and constraint gradients $\nabla h_1(\mathbf{x}^*), \ldots, \nabla h_m(\mathbf{x}^*)$ and $\nabla g_1(\mathbf{x}^*), \ldots, \nabla g_r(\mathbf{x}^*)$ are linearly independent. Then there exist unique Lagrange multiplier vectors $\boldsymbol{\lambda}^* = (\lambda_1^*, \ldots, \lambda_m^*)$ and $\boldsymbol{\mu}^* = (\mu_1^*, \ldots, \mu_r^*)$ such that the following conditions are true:

- **1-order necessary**

$$\nabla_{\mathbf{x}} \mathcal{L}(\mathbf{x}^*, \boldsymbol{\lambda}^*, \boldsymbol{\mu}^*) = \mathbf{0}, \qquad \nabla_{\boldsymbol{\lambda}} \mathcal{L}(\mathbf{x}^*, \boldsymbol{\lambda}^*, \boldsymbol{\mu}^*) = \mathbf{0},$$

  where

$$\mu_j^* \geq 0, \quad j = 1, \ldots, r,$$

  and

$$\mu_j^* = 0, \quad \forall j \notin \mathbb{A}(\mathbf{x}^*),$$

- **2-order necessary**

$$\nabla_{\mathbf{xx}}^2 \mathcal{L}(\mathbf{x}^*, \boldsymbol{\lambda}^*, \boldsymbol{\mu}^*) \succeq 0.$$

Theorem 8.4 KKT Sufficient Optimality Conditions

Assume that functions $f(\mathbf{x})$, $h_i(\mathbf{x})$, and $g_j(\mathbf{x})$, $i = 1, \ldots, m$, $j = 1, \ldots, r$, are twice continuously differentiable and let $\mathbf{x}^* \in \mathbb{R}^n$, $\boldsymbol{\lambda}^* \in \mathbb{R}^m$, and $\boldsymbol{\mu}^* \in \mathbb{R}^r$ satisfy both conditions

- $\nabla_{\mathbf{x}} \mathcal{L}(\mathbf{x}^*, \boldsymbol{\lambda}^*, \boldsymbol{\mu}^*) = \mathbf{0}$, $\qquad \nabla_{\boldsymbol{\lambda}} \mathcal{L}(\mathbf{x}^*, \boldsymbol{\lambda}^*, \boldsymbol{\mu}^*) = \mathbf{0}$, $\qquad \mathbf{g}(\mathbf{x}^*) \leq \mathbf{0}$

$$\mu_j^* \geq 0, \quad j = 1, \ldots, r,$$
$$\mu_j^* = 0, \quad \forall j \notin \mathbb{A}(\mathbf{x}^*),$$

- $\nabla_{\mathbf{xx}}^2 \mathcal{L}(\mathbf{x}^*, \boldsymbol{\lambda}^*, \boldsymbol{\mu}^*) \succ 0$ *(2-order sufficient)*.

*Assume also that $\mu_j^* > 0$, $\forall j \in \mathbb{A}(\mathbf{x}^*)$. Then $\mathbf{x}^*$ is a **strict** local minimizer of $f(\mathbf{x})$ subject to constraints $h_i(\mathbf{x}) = 0$ and $g_j(\mathbf{x}) \leq 0$.*

The proofs for both theorems may be easily found, e.g., in [4]. We also advise the reader to review Theorems 5.1 and 5.2 of Chapter 5 and Theorems 8.1 and 8.2 of the current chapter to examine how optimality conditions established for unconstrained and equality/inequality constrained optimizers with the use of Lagrange multipliers are related to each other.

8.9 Homework Problems

1. Review MATLAB code `Chapter_7_data_fit_comprehensive.m` for solving the unconstrained optimization problem of Example 1.3 and suggest the optimal ways for adding extra functionalities discussed in this chapter, e.g.,

   (a) multiobjective optimization via scalarization,
   (b) Pareto optimal solutions,
   (c) principal component analysis (PCA), etc.

2. Modify MATLAB code `Chapter_7_data_fit_comprehensive.m` by adding a new model problem of your choice to be solved by unconstrained optimization. Apply any available methods to create an initial guess and solve the problem by choosing appropriate settings. Whenever possible, compare the performance of the chosen approaches in terms of

   (a) accuracy of the obtained solution,
   (b) elapsed computational time, and
   (c) rate of convergence by approximating convergence parameters r and C.

   Make a conclusion.

3. Solve the optimization problem of Example 8.1 by applying the penalization approach.

4. Modify `MATLAB` code `Chapter_8_opt_constrained.m` by "activating" all methods (both gradient-based and derivative-free) previously existing in `Chapter_7_data_fit_comprehensive.m`. Apply all available approaches to the constrained optimization problem(s) of your choice. Compare the performance in terms of accuracy of the obtained solutions, computational time, and rate of convergence, and conclude.

5. Suggest any strategies for choosing penalization approach coefficients β and discuss their applicability. Implement the selected methods computationally, compare the performance, and conclude.

6. Modify `MATLAB` code `Chapter_8_opt_constrained.m` to solve

$$\min_{\mathbf{x} \in \mathbb{R}^4} f(\mathbf{x}) = -(x_1 x_2 x_3 + x_2 x_3 x_4 + x_1 x_3 x_4 + x_1 x_2 x_4)$$

$$\text{s.t. } x_1 + x_2 + x_3 + x_4 = 4$$

by the Lagrange multiplier and/or penalization approach. Compare the performance and conclude.

7. Modify `MATLAB` code `Chapter_8_opt_constrained.m` to solve

$$\min_{\mathbf{x} \in \mathbb{R}^4} f(\mathbf{x}) = \frac{1}{2} \left(x_1^2 + x_2^2 + x_3^2 + x_4^2 \right)$$

$$\text{s.t. } x_1 + x_2 + x_3 + x_4 \leq -4$$

by the Lagrange multiplier approach. Compare the performance and conclude.

READ   **Where to Read More**

Bertsekas (2016), [4]
Chapter 3 (Optimization Over a Convex Set), Chapter 4 (Lagrange Multiplier Theory)

Boyd (2004), [6]
Chapter 10 (Equality Constrained Minimization), Chapter 11 (Interior-Point Methods)

Griva (2009), [15]
Chapter 14 (Optimality Conditions for Constrained Problems), Chapter 15 (Feasible-Point Methods)

Nocedal (2006), [25]
Chapter 12 (Theory of Constrained Optimization), Chapter 15 (Fundamentals of Algorithms for Nonlinear Constrained Optimization)

| RUN | MATLAB **Codes for Chapter 8** |

- **root** folder:

  − Chapter_8_opt_constrained.m

  − params.m

- folder algorithms:

  − direction_search.m

  − file_output.m

  − golden_section_search.m

  − initialize.m

  − kappa_test.m

  − mode_OPT.m

  − mode_TEST.m

  − stepsize_search.m

  − visualize.m

- folder functions:

  − fn_brent.m

  − fn_convergence_sol_norm.m

  − fn_eval_f.m

  − fn_eval_f_grad.m

  − fn_eval_hess.m

  − fn_min_brack.m

  − FUNC.m

- folder output

- folder temp

9

ODE-based Optimization

In this chapter, we set the focus on solving computationally constrained optimization problems supported by one or several differential equations, e.g., ODEs. We start with reviewing the simplified notations of the directional derivative, Fréchet (Gâteaux) differential, and the Riesz Representation Theorem used in deriving the analytical formulas for gradients based on the adjoint (Lagrange multiplier) variables obtained as solutions of the adjoint problems. We practice this derivation within the optimize–then–discretize approach using the example modified from the ODE-based optimization problems discussed in Chapter 8. We also briefly discuss the computational algorithm to implement the discretize–then–optimize technique to compare the benefits and drawbacks of both methods. We devote the second part of this chapter to solving the optimization problem of reconstructing a parameter function in the Lotka–Volterra predator–prey model, a nonlinear ODE-based problem to describe dynamical processes in various biological systems. We discuss the quality of the solutions obtained within the optimize–then–discretize framework employing the discretized gradients, synthetic measurements, and the built-in MATLAB's functionality to solve systems of ODEs representing the forward and adjoint problems. Finally, we identified a range of issues related to the accuracy and regularity of the obtained solutions to be addressed later in Chapter 10.

9.1 Fitting Data by ODE-based Optimization

In the first part of this chapter, we will answer the question posed on p. 227 and go through the process of deriving gradients to serve iterations, while solving optimization problems containing constraints such as ODEs. We leave solving Examples 8.4 and 8.5 to the reader and apply our analysis to a new problem with continuous formulation modified from both examples.

Example 9.1 Fitting Data by ODE-based Optimization
Identify parameter $f(t)$ in the BVP-2 by fitting data $(\tilde{t}, \tilde{y})$ available

DOI: 10.1201/9781003275169-9   233 *continuously over interval* $[t_0, T]$

$$
\min_{f(t)} \mathcal{J}(y(t); f(t)) = \frac{1}{2} \int_{t_0}^{T} (y - \tilde{y})^2 \, dt \tag{9.1}
$$

$$
s.t. \ a_1 y'' + a_2 y' + a_3 y = f(t), \quad y(t_0) = c_1, \quad y(T) = c_2,
$$

where $a_1, a_2, a_3, c_1,$ *and* c_2 *are given constants.*

Following our previous discussions, we plan to solve this problem using the *adjoint-based gradient* method (Lagrange multiplier approach). First, as shown in Section 9.3, we derive gradients by the *"optimize–then–discretize"* approach with the optimality conditions and optimization algorithm based on the *continuous formulation* of the problem in Example 9.1. Next, as discussed further in Section 9.4, we look into finding gradients by the *"discretize–then–optimize"* approach where the same use, instead, the *discrete formulation*. Another alternative method would use objective function gradients approximated by any finite difference (FD) schemes. We have to note that the last approach may be very expensive in case of fine discretization produced in time t.

First, we review the main notations that apply to the continuous formulation of any ODE-based optimization problem used for deriving gradients. Therefore, we define the main (core) objective function

$$
\mathcal{J}(y(t); f(t)) = \frac{1}{2} \int_{t_0}^{T} (y - \tilde{y})^2 \, dt \tag{9.2}
$$

and the *forward problem* (also, *governing* or *state equation*) given by the BVP-2 constraint in (9.1)

$$
\begin{aligned}
a_1 y'' + a_2 y' + a_3 y &= f(t), \\
y(t_0) = c_1, \quad y(T) &= c_2.
\end{aligned} \tag{9.3}
$$

As discussed in Section 8.5, we obtain the Lagrangian

$$
\mathcal{L}(y(t); f(t), \psi(t)) = \mathcal{J}(y(t); f(t)) + \langle a_1 y'' + a_2 y' + a_3 y - f(t), \ \psi(t) \rangle_\chi \tag{9.4}
$$

by augmenting the objective function (9.2). In (9.4), functions $y(t)$, $f(t)$, and $\psi(t)$ play the roles of the *state*, *optimization (control)*, and *adjoint* (Lagrange multiplier) variables, respectively. We also note that we define the inner product, $\langle \cdot \, , \, \cdot \rangle_\chi$, in a broad context regarding functional space χ. For example, if $\chi = L_2$ (a space of square-integrable functions), then

$$
\langle f_1(t), \ f_2(t) \rangle_{\chi=L_2} = \int_{t_0}^{T} f_1 \cdot f_2 \, dt. \tag{9.5}
$$

Thus, the Lagrangian in (9.4) now takes the following form used throughout the "optimize–then–discretize" approach

$$\mathcal{L}(y(t); f(t), \psi(t)) = \frac{1}{2} \int_{t_0}^{T} (y - \tilde{y})^2 \, dt$$

$$+ \int_{t_0}^{T} [a_1 y'' + a_2 y' + a_3 y - f(t)] \, \psi(t) \, dt. \tag{9.6}$$

9.2 Derivative vs. Directional Derivative: Review

Here, we briefly review the notations of the derivative function and the directional derivatives. So, a *derivative* of a real variable function $y = f(x)$ is also a function

$$f'(x) = \frac{df}{dx} = \lim_{\Delta x \to 0} \frac{\Delta f}{\Delta x} = \lim_{\Delta x \to 0} \frac{f(x + \Delta x) - f(x)}{\Delta x}. \tag{9.7}$$

The reader may refer to Figure 9.1 for notations used in definition (9.7).

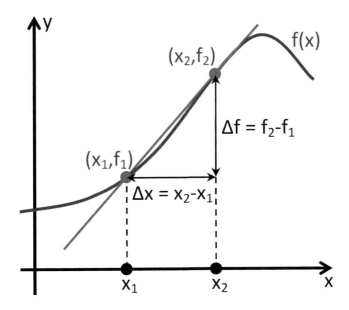

FIGURE 9.1
Schematic showing the concept of the derivative $f'(x)$ in (9.7).

There are several meanings to interpreting the results provided by the derivative function $f'(x)$:

- a number (if exists),

- an instantaneous rate of change by ratio

$$\frac{df}{dx} = \frac{\Delta f}{\Delta x}\bigg|_{\Delta x \to 0},$$

- a slope of the tangent line to the graph of the function $f(x)$ at point x.

Using (9.7), we also arrive at the notation of the *differential*

$$df(x) = f'(x)\,dx. \tag{9.8}$$

Next, we define the *directional derivative* of a scalar function $f(x_1, x_2, \ldots, x_n)$ along vector $v = (v_1, v_2, \ldots, v_n)$

$$\frac{\partial f(x)}{\partial v} = \boldsymbol{\nabla}_v f(x) = \lim_{\epsilon \to 0} \frac{f(x + \epsilon v) - f(x)}{\epsilon} = \boldsymbol{\nabla} f(x) \cdot v \tag{9.9}$$

and elaborate just a bit on its meaning; it is

- a number (if exists),

- an instantaneous rate of change of $f(x)$ moving through x with a velocity specified by v, and

- a special case of the Gâteaux derivative.

An eager reader may read more about the last meaning in the calculus of variations or perturbation analysis (also referencing the Fréchet derivative). Here, we focus only on the basics of the notation of the *Gâteaux derivative (or differential)* using the generalized concept of directional derivative in differential calculus, i.e.,

$$\frac{\delta F(u)}{\delta \psi} = \boldsymbol{\nabla}_\psi F(u) = \lim_{\epsilon \to 0} \frac{F(u + \epsilon \psi) - F(u)}{\epsilon}, \tag{9.10}$$

where $U \subset X$, $F : X \to Y$, $u \in U$, $\psi \in X$.

Our closing remark for this brief review section is about the notations and used definitions. We understand that removing some properties and simplifying the definition structures make those definitions (more specifically, for directional and Gâteaux derivatives) incomplete. However, we assume such simplification will not prevent us from applying the related analysis correctly and help the reader focus on the technical process of deriving gradients.

9.3 Optimize–then–Discretize

We split the process of solving the optimization problem of Example 9.1 using the "optimize–then–discretize" approach into two phases: (1) deriving the analytical expression for the adjoint-based gradients ("optimize" phase), and (2) solving the problem in the discretized settings ("discretize" phase) by using the discretized versions of those gradients.

9.3.1 Deriving Gradient

We start deriving gradients by referencing the *Riesz Representation Theorem*. Its simplified interpretation tells that the 1-order *total variation* for Lagrangian $\mathcal{L}(y(t); f(t), \psi(t))$ may be expressed by the linear part only, i.e.,

$$\delta\mathcal{L}(y(t); f(t), \psi(t)) = \langle \boldsymbol{\nabla}_f \mathcal{L}, \ \delta f \rangle_\chi + \langle \boldsymbol{\nabla}_y \mathcal{L}, \ \delta y \rangle_\chi + \langle \boldsymbol{\nabla}_\psi \mathcal{L}, \ \delta\psi \rangle_\chi + \dots, \quad (9.11)$$

where other (smaller than linear) terms denoted by "..." in (9.11) are neglected. Other terms in (9.11) are *variations* (*perturbations*) for control, state, and adjoint variables given, respectively, by δf, δy, and $\delta \psi$ (arbitrary chosen functions). There are also gradients with respect to control, state, and adjoint variables, namely $\boldsymbol{\nabla}_f \mathcal{L}$, $\boldsymbol{\nabla}_y \mathcal{L}$, and $\boldsymbol{\nabla}_\psi \mathcal{L}$.

Setting $\delta\mathcal{L}$ in (9.11) to zero will minimize the Lagrangian $\mathcal{L}(y(t); f(t), \psi(t))$ due to the following.

- Setting $\langle \boldsymbol{\nabla}_\psi \mathcal{L}, \ \delta\psi \rangle_\chi$ to 0 is natural as it represents a forward problem, e.g., as given by (9.3).

- Setting $\langle \boldsymbol{\nabla}_y \mathcal{L}, \ \delta y \rangle_\chi$ to 0 defines adjoint equation(s).

- Finally, *Fréchet differential* $\langle \boldsymbol{\nabla}_f \mathcal{L}, \ \delta f \rangle_\chi$ gives an expression for the gradients in the sense of *Fréchet* or *Gâteaux*. Then these gradients will be reduced (hopefully to 0) while running optimization iterations.

Therefore, we move forward to derive the 1-order total variation for Lagrangian given in (9.6) by keeping $\chi = L_2$ and obtain the following expression

$$\begin{aligned}
\delta\mathcal{L}(y(t); f(t), \psi(t)) = &\int_{t_0}^{T} (y - \tilde{y})\, \delta y\, dt \\
&+ \int_{t_0}^{T} [a_1(\delta y)'' + a_2(\delta y)' + a_3\delta y - \delta f]\, \psi(t)\, dt \quad (9.12) \\
&+ \int_{t_0}^{T} [a_1 y'' + a_2 y' + a_3 y - f(t)]\, \delta\psi\, dt
\end{aligned}$$

to be further modified to make it consistent with the Riesz Representation

Theorem (9.11). We remind the reader that our final goal is to set $\delta\mathcal{L}$ in (9.12) to 0 to satisfy the KKT optimality conditions discussed in Chapter 8. After examining the right side of (9.12), we make two conclusions:

(1) its second term is not consistent with the theorem: we will integrate it by parts to get rid of $(\delta y)''$ and $(\delta y)'$, and

(2) the third term $\int_{t_0}^{T} [a_1 y'' + a_2 y' + a_3 y - f(t)]\,\delta\psi\,dt = 0$ due to forward problem (9.3).

Both updates lead us to the new form, i.e.,

$$
\begin{aligned}
\delta\mathcal{L} = &\int_{t_0}^{T} (y - \tilde{y})\,\delta y\,dt + [a_1 \psi(\delta y)']_{t_0}^{T} - \int_{t_0}^{T} (a_1\psi)'(\delta y)'\,dt \\
&+ [a_2\psi\delta y]_{t_0}^{T} - \int_{t_0}^{T} (a_2\psi)'\delta y\,dt + \int_{t_0}^{T} a_3\psi\delta y\,dt - \int_{t_0}^{T} \psi\,\delta f\,dt.
\end{aligned}
\tag{9.13}
$$

In (9.13), the third term $\int_{t_0}^{T}(a_1\psi)'(\delta y)'\,dt$ is still not consistent with the theorem, and we integrate by parts one more time:

$$
\begin{aligned}
\delta\mathcal{L} = &\int_{t_0}^{T} (y - \tilde{y})\,\delta y\,dt + [a_1\psi(\delta y)']_{t_0}^{T} - [(a_1\psi)'\delta y]_{t_0}^{T} + \int_{t_0}^{T} (a_1\psi)''\delta y\,dt \\
&+ [a_2\psi\delta y]_{t_0}^{T} - \int_{t_0}^{T} (a_2\psi)'\delta y\,dt + \int_{t_0}^{T} a_3\psi\delta y\,dt - \int_{t_0}^{T} \psi\,\delta f\,dt.
\end{aligned}
\tag{9.14}
$$

Boundary terms $[(a_1\psi)'\delta y]_{t_0}^{T}$ and $[a_2\psi\delta y]_{t_0}^{T}$ in (9.14) are both zeros due to the perturbation system (more precisely, its boundary conditions)

$$
\begin{aligned}
a_1(\delta y)'' + a_2(\delta y)' + a_3\delta y &= \delta f, \\
\delta y(t_0) = 0, \quad \delta y(T) &= 0,
\end{aligned}
\tag{9.15}
$$

obtained by perturbing forward problem (9.3). It is also known as the variational form of the forward problem. And finally, we make the form fully "Riesz-consistent":

$$
\begin{aligned}
\delta\mathcal{L} &= \int_{t_0}^{T} (-\psi)\,\delta f\,dt \quad + [a_1\psi(\delta y)']_{t_0}^{T} \\
&\qquad\qquad\qquad\quad + \int_{t_0}^{T} [(y - \tilde{y}) + (a_1\psi)'' - (a_2\psi)' + a_3\psi]\,\delta y\,dt \\
&= \langle \boldsymbol{\nabla}_f\mathcal{L},\ \delta f\rangle_{L_2} \quad + \langle \boldsymbol{\nabla}_y\mathcal{L},\ \delta y\rangle_{L_2} \\
&= \int_{t_0}^{T} \boldsymbol{\nabla}_f\mathcal{L}\,\delta f\,dt \quad + \int_{t_0}^{T} \boldsymbol{\nabla}_y\mathcal{L}\,\delta y\,dt.
\end{aligned}
\tag{9.16}
$$

9.3.2 Solving Problem

The final form of $\delta\mathcal{L}$ obtained in (9.16) provides an analytical formula for the adjoint-based gradient derived in L_2 functional space

$$\nabla_f\mathcal{L} = -\psi(t) \tag{9.17}$$

based on the solution of the adjoint ODE problem

$$
\begin{aligned}
(a_1\psi)'' - (a_2\psi)' + a_3\psi &= y - \tilde{y}, \\
\psi(t_0) = 0, \quad \psi(T) &= 0
\end{aligned}
\tag{9.18}
$$

to be solved to find adjoint state $\psi(t)$. Now we could start solving the optimization problem of Example 9.1 numerically after performing discretization for

- time t:

$$t_0, \ t_1 = t_0 + \Delta t, \ t_2 = t_0 + 2\Delta t, \ \ldots, \ t_n = T, \quad \Delta t = \frac{T - t_0}{n}, \tag{9.19}$$

- states $y(t) \to \mathbf{y} \in \mathbb{R}^{n+1}$, controls $f(t) \to \mathbf{f} \in \mathbb{R}^{n+1}$, and adjoints $\psi(t) \to \boldsymbol{\psi} \in \mathbb{R}^{n+1}$:

$$
\mathbf{y} = \begin{bmatrix} y_0 \\ y_1 \\ \cdots \\ y_n \end{bmatrix} = \begin{bmatrix} y(t_0) \\ y(t_1) \\ \cdots \\ y(T) \end{bmatrix}, \quad
\mathbf{f} = \begin{bmatrix} f_0 \\ f_1 \\ \cdots \\ f_n \end{bmatrix}, \quad
\boldsymbol{\psi} = \begin{bmatrix} \psi_0 \\ \psi_1 \\ \cdots \\ \psi_n \end{bmatrix}, \tag{9.20}
$$

- objective function $\mathcal{J}(y(t); f(t))$, e.g.,

$$\mathcal{J}(\mathbf{f}) = \frac{1}{2}\sum_{i=1}^{n-1} \left(y_i(\mathbf{f}) - \tilde{y}_i\right)^2 \Delta t. \tag{9.21}$$

Here, we have to note that the discretization of the objective in (9.21) supports the continuous structure of available measurements $\tilde{y}(t)$ by using the rectangular or midpoint rule for numerical integration.

? *In the case measurements $\{\tilde{y}_j\}_{j=1}^{M}$ are pointwise (not continuous, $M \neq n+1$), how does this affect the derivation of the gradient?*

! *Think about a more general way to discretize objectives, e.g., as suggested in (8.63), i.e.,*

$$\mathcal{J}(\mathbf{f}) = \frac{1}{2}\sum_{j=1}^{M} \left(y_j(\mathbf{f}) - \tilde{y}_j\right)^2. \tag{9.22}$$

We now summarize all steps in the practical implementation of the "optimize–then–discretize" approach to perform numerical computations for solving the optimization problem of Example 9.1. Algorithm 9.1 provides the entire adjoint-based procedure.

Algorithm 9.1 (Example 9.1 by "Optimize–then–Discretize")

1. *Discretize time t by (9.19) and initialize vectors for states, controls, and adjoints by (9.20)*

2. *Obtain and store measurement data* $(\tilde{t}_i, \tilde{y}_i)$

3. *Set $k = 0$ and choose initial guess* $\mathbf{f}^0$ *for control*

4. *Solve numerically (discretized) forward problem (9.3) to find* $\mathbf{y}^k$

5. *Evaluate objective in (9.21)*

6. *For $k = 1, 2, \ldots$ check optimality of* $\mathbf{f}^k$*; if optimal* $\Rightarrow$ **TERMINATE**

7. *Solve numerically (discretized) adjoint problem (9.18) to find* $\boldsymbol{\psi}^k$

8. *Evaluate gradient* $\nabla_{\mathbf{f}}\mathcal{L}(\mathbf{y}^k; \mathbf{f}^k, \boldsymbol{\psi}^k)$ *by (9.17)*

9. *Improve solution by finding optimal step size* α^k

$$\mathbf{f}^{k+1} = \mathbf{f}^k - \alpha^k \nabla_{\mathbf{f}}\mathcal{L}(\mathbf{y}^k; \mathbf{f}^k, \boldsymbol{\psi}^k) \qquad (9.23)$$

10. *Set $k \leftarrow k + 1$ and go to Step 4*

As a final comment, we note that Step 9 could be practically implemented by using any gradient-based strategies and methods for optimal step size α discussed in Chapters 5–7. We also advise the reader to compare the structure of this procedure with the one we derived for adjoint-gradient equality-based optimization in Chapter 8; refer to Algorithm 8.1.

9.4   Discretize–then–Optimize

As we complete our discussion on the "optimize–then–discretize" approach, we turn to its alternative and provide the reader with a brief review of the computational algorithm to follow the "discretize–then–optimize" scheme. Its practical implementation first requires the discretization of the main objects ("discretize" phase): such as time t, e.g., by (9.19), state, control, and adjoint variables by (9.20), and objective function by (9.21). A continuous formulation of the forward problem given in (9.3) should be also discretized by applying, e.g., any finite difference scheme. To provide a reader with a big picture, we define the discretized form of the forward problem by

$$\mathbf{g}(y_1, y_2, \ldots, y_{n-1}; f_0, f_1, \ldots, f_n) = \mathbf{0}, \tag{9.24}$$

described in general as a large system of linear equations. Finally, we have to create the (discretized) Lagrangian based on the objective function (9.21) augmented by the set of adjoint variables (Lagrange multipliers) $\psi_1, \psi_2, \ldots, \psi_{n-1}$, i.e.,

$$\mathcal{L}(y_1, y_2, \ldots, y_{n-1}; f_0, f_1, \ldots, f_n, \psi_1, \psi_2, \ldots, \psi_{n-1})$$
$$= \frac{1}{2} \sum_{i=1}^{n-1} (y_i - \tilde{y}_i)^2 \Delta t + \sum_{i=1}^{n-1} \psi_i \cdot g_i(y_1, y_2, \ldots, y_{n-1}; f_0, f_1, \ldots, f_n). \tag{9.25}$$

To derive gradients, we also refer to the KKT optimality conditions discussed in Chapter 8:

$$\frac{\partial \mathcal{L}}{\partial \psi_i} = 0, \quad \frac{\partial \mathcal{L}}{\partial y_i} = 0, \quad i = 1, 2, \ldots, n-1,$$
$$\frac{\partial \mathcal{L}}{\partial f_j} = 0, \quad j = 0, 1, \ldots, n. \tag{9.26}$$

Similarly to seen before, these conditions define

- equations of states (linear system of $n-1$ equations)

$$\frac{\partial \mathcal{L}}{\partial \psi_i} = g_i(y_1, y_2, \ldots, y_{n-1}; f_0, f_1, \ldots, f_n) = 0, \tag{9.27}$$

- adjoint equations (linear system of $n-1$ equations)

$$\frac{\partial \mathcal{L}}{\partial y_i} = (y_i - \tilde{y}_i)\Delta t$$
$$+ \sum_{k=1}^{n-1} \psi_k \frac{\partial g_k}{\partial y_i}(y_1, y_2, \ldots, y_{n-1}; f_0, f_1, \ldots, f_n) = 0, \tag{9.28}$$

- and $n+1$ components of gradient $\nabla_{\mathbf{f}}\mathcal{L}$

$$\frac{\partial \mathcal{L}}{\partial f_j} = \sum_{i=1}^{n-1} \psi_i \frac{\partial g_i}{\partial f_j}(y_1, y_2, \ldots, y_{n-1}; f_0, f_1, \ldots, f_n) \tag{9.29}$$

to be applied during iterative optimization ("optimize" phase).

As the reader may notice, the main complication in implementing "discretize–then–optimize" in computations is that every gradient evaluation requires computing $(n-1)^2$ derivatives $\frac{\partial g_k}{\partial y_i}$ in (9.28) and $(n-1)(n+1)$ derivatives $\frac{\partial g_i}{\partial f_j}$ in (9.29). If discretization parameter n changes frequently and is large, this approach is suitable if paired with any automatic tools for computing derivatives (e.g., automatic differentiation techniques). In Algorithm 9.2, we summarize all steps in the practical implementation of the "discretize–then–optimize" approach to perform numerical computations for solving the optimization problem of Example 9.1.

Algorithm 9.2 (Example 9.1 by "Discretize–then–Optimize")

1. *Discretize time t by (9.19) and initialize vectors for states, controls, and adjoints by (9.20)*

2. *Obtain and store measurement data $(\tilde{t}_i, \tilde{y}_i)$*

3. *Set $k = 0$ and choose initial guess $\mathbf{f}^0$ for control*

4. *Solve numerically (discretized) forward problem (9.24) to find $\mathbf{y}^k$*

5. *Evaluate objective in (9.21)*

6. *For $k = 1, 2, \ldots$ check optimality of $\mathbf{f}^k$; if optimal $\Rightarrow$ **TERMINATE***

7. *Evaluate derivatives $\frac{\partial g_k}{\partial y_i}$ and solve numerically (discretized) adjoint problem*

$$\sum_{k=1}^{n-1} \psi_k \frac{\partial g_k}{\partial y_i} = -(y_i - \tilde{y}_i)\Delta t, \quad i = 1, 2, \ldots, n-1$$

based on (9.28) to find ψ^k

8. *Evaluate derivatives $\frac{\partial g_i}{\partial f_j}$ and compute $(n+1)$ components of gradient $\nabla_{\mathbf{f}} \mathcal{L}(\mathbf{y}^k; \mathbf{f}^k, \psi^k)$ by (9.29)*

9. *Improve solution by finding optimal step size α^k*

$$\mathbf{f}^{k+1} = \mathbf{f}^k - \alpha^k \nabla_{\mathbf{f}} \mathcal{L}(\mathbf{y}^k; \mathbf{f}^k, \psi^k) \tag{9.30}$$

10. *Set $k \leftarrow k + 1$ and go to Step 4*

As for Algorithm 9.1, designed to implement the "optimize–then–discretize" approach, we note that Step 9 could be practically implemented using any gradient-based strategies and methods for optimal step size α discussed in Chapters 5–7. We also advise the reader to compare both algorithms to conclude on the similarity of their general (adjoint-gradient-based) structure and details related to the practical implementation.

? *Do you think both methods ("optimize–then–discretize" and "discretize– then–optimize") may converge to the same solution? Under what condition?*

! *You may think about setting $n \to \infty$ for both methods. However, proving the equivalence of both methods under this condition is still an open problem.*

9.5 Numerical ODE Solvers

In Sections 9.3 and 9.4, we created Algorithms 9.1 and 9.2 for solving ODE-based optimization problems by constructing associated adjoint-based gradients. Both algorithms feature two common steps: solving numerically (discretized) forward (Step 4) and adjoint (Step 7) problems. These two steps refer to f- and d-evaluators and are integral parts of the entire optimization framework, as discussed in Chapter 3. The reader may also refer to Figure 3.2 for refreshing the general concept of this framework structure. Here, we discuss some approaches for solving (integrating) numerically ODEs or ODE systems to supply Steps 4 and 7 with required solutions.

9.5.1 Runge–Kutta Integration Methods

To start, we approach the problem of solving the system of ODEs by generalizing this problem in the following context: solve the system of N coupled 1-order ODEs for functions $y_i(x)$, i.e.,

$$\frac{dy_i(x)}{dx} = f_i(x, y_1, y_2, \ldots, y_N), \quad i = 1, 2, \ldots, N. \tag{9.31}$$

One of the easiest numerical solutions is available by the *Euler's method*: for each $x_{n+1} = x_n + h$ we recover (iteratively) solution

$$\mathbf{y}_{n+1} = \mathbf{y}_n + h \cdot \mathbf{f}(x_n, \mathbf{y}_n) + \mathcal{O}(h^2),$$

where h is the step size in discretizing x-interval. This method has known drawbacks:

- depending on f_i, it might not be stable, and

- locally it is $\mathcal{O}(h^2)$ method (2-order); however, globally, it is only 1-order accurate, i.e., $\mathcal{O}(h)$.

For better performance, finite difference solutions use more accurate *Runge–Kutta (RK) nth order methods.* For example, one could use the 2-order RK2 (or midpoint method) with a "trial" step to the midpoint of the new interval $[x_n, x_{n+1}]$, i.e., for each $x_{n+1} = x_n + h$ the solution is

$$\mathbf{y}_{n+1} = \mathbf{y}_n + \mathbf{k}_2 + \mathcal{O}(h^3), \qquad (9.32)$$

where

$$\begin{aligned}
\mathbf{k}_1 &= h \cdot \mathbf{f}(x_n, \mathbf{y}_n), \\
\mathbf{k}_2 &= h \cdot \mathbf{f}\left(x_n + \frac{h}{2}, \mathbf{y}_n + \frac{\mathbf{k}_1}{2}\right).
\end{aligned} \qquad (9.33)$$

Among methods with even higher accuracy, the 4-order Runge–Kutta (RK4) method is probably the most popular: for each $x_{n+1} = x_n + h$, it provides the solution

$$\mathbf{y}_{n+1} = \mathbf{y}_n + \frac{\mathbf{k}_1}{6} + \frac{\mathbf{k}_2}{3} + \frac{\mathbf{k}_3}{3} + \frac{\mathbf{k}_4}{6} + \mathcal{O}(h^5), \qquad (9.34)$$

based on 4-step computations

$$\begin{aligned}
\mathbf{k}_1 &= h \cdot \mathbf{f}(x_n, \mathbf{y}_n), \\
\mathbf{k}_2 &= h \cdot \mathbf{f}\left(x_n + \frac{h}{2}, \mathbf{y}_n + \frac{\mathbf{k}_1}{2}\right), \\
\mathbf{k}_3 &= h \cdot \mathbf{f}\left(x_n + \frac{h}{2}, \mathbf{y}_n + \frac{\mathbf{k}_2}{2}\right), \\
\mathbf{k}_4 &= h \cdot \mathbf{f}\left(x_n + h, \mathbf{y}_n + \mathbf{k}_3\right).
\end{aligned} \qquad (9.35)$$

The reader, however, should be careful as higher-order does **not always mean** higher accuracy. Also,

- method's stability depends on the behavior of individual functions f_i in $\mathbf{f}(x, \mathbf{y})$, and

- may require adaptive step size h control.

In general, RK4 is **superior** to the midpoint (RK2) method but requires **four** evaluations $\mathbf{f}(x, \mathbf{y})$ per one step h.

9.5.2   Solving ODEs by MATLAB

Alternative to implementing the RK (or other) numerical schemes for solving ODEs numerically, the reader may consider software packages that provide ready-to-use utilities with no need to check the correctness of their implementation (just considering debugging for tuning their performance). In MATLAB, for example, there are several functions available for numerical solutions of ODEs. We focus on two of them: `ode23` and `ode45`. They solve nonstiff ODEs employing 2-/3-order and 4-/5-order, respectively, Runge–Kutta formulas for **medium** (`ode23`) and **higher** (`ode45`) accuracy; refer to the syntax below.

MATLAB: `ode23` and `ode45` syntax (mandatory/*optional*)

> [t, y, *te, ye, ie*] = ode23(odefun, tspan, y0, *options*)
> [t, y, *te, ye, ie*] = ode45(odefun, tspan, y0, *options*)

- **t, y**: each row in solution array y corresponds to a value returned in column vector t

- **odefun**: function handle that defines the functions to be integrated

- **tspan**: integration interval specified as a vector $[t_0\, t_1\, t_2\, \ldots\, t_f]$; at least two elements t_0 and t_f should be provided

- **y0**: initial conditions specified as a vector; it must be the same length as **odefun**

- *options*: options structure; look at MathWorks [23]

- *te, ye, ie*: optionally find where (event) functions of (t, y) are zero

---

9.6 Dynamics of Biological Systems by Lotka–Volterra Equations

In this section, we consider an example of an ODE-based problem to model dynamical processes in various biological systems described by the *Lotka-Volterra (LV) predator–pray model*. It is represented by a pair of 1-order nonlinear ODEs

$$\dot{x}_1 = (\alpha - \beta\, x_2)\, x_1,$$
$$\dot{x}_2 = (-\gamma + \delta\, x_1)\, x_2 \tag{9.36}$$

to describe changes in two populations, namely predator and prey. In (9.36), we apply $\dot{x}$ for the time derivative commonly used in dynamical systems. Other notations used in the LV system (9.36) are

- *number of prey* $x_1(t)$, e.g., rabbits, hares, etc.,

- *number of predators* $x_2(t)$, e.g., foxes, wolves, etc.,

- instantaneous *growth rates* of two populations $\dot{x}_1 = \frac{dx_1}{dt}$ and $\dot{x}_2 = \frac{dx_2}{dt}$ with respect to time t, and

- *parameters* $\alpha, \beta, \gamma, \delta > 0$ describing interaction between two species.

The reader may also find useful the physical meaning and some assumptions usually made in modeling when employing the LV model (9.36).

(1) The prey population finds ample food at all times.

(2) The food supply of the predator population depends entirely on the size of the prey population.

(3) The rate of change in populations is proportional to their size.

(4) Within the modeled time frame, the environment does not change in favor of one species.

(5) Predators have a limitless appetite.

A keen reader may go further to come up with more comprehensive models. For instance, *Kolmogorov's model* creates a general framework that models the dynamics of ecological systems with predator-prey interactions, including competition, disease, and mutualism.

Before using the LV model as a constraint in any optimization problem, we first experiment with numerical integration in MATLAB applied to the system (9.36) to check the performance discussed at length in Chapter 3. All components for this experiment (files Chapter_9_LV_example.m and fn_LV.m) are placed in the separate folder example. We start with the following parameters: $\alpha = 1$, $\beta = 0.01$, $\gamma = 1$, $\delta = 0.02$, $t \in [0, 15]$, $\mathbf{x}_0 = [20\ 20]^T$ (the readers may use their own sets to explore more the solution structure and its sensitivity to any changes applied to parameters).

MATLAB: Chapter_9_LV_example.m

Parameters for solving the LV model:

- *time frame:* ti = 0; tf = 15; $t \in [0, 15]$

- *initial conditions:* x0 = [20 20]'; $\mathbf{x}^0 = [20\ 20]^T$

- *interaction parameters:* params = [1 0.01 1 0.02]; $\alpha = 1$, $\beta = 0.01$, $\gamma = 1$, $\delta = 0.02$

MATLAB: fn_LV.m

```
function xp = fn_LV(params,t,x)                              1
                                                             2
    xp = diag([params(1)  -  params(2)*x(2),  ...           3
                    -params(3)  + params(4)*x(1)])  *  x;    4
                                                             5
return                                                       6
```

We first run `ode23` with its default relative accuracy (10^{-3}) for time interval $[0, 15]$ (two full cycles). Figure 9.2 shows the *time history* and *phase plane plots* for solutions $x_1(t)$ (prey) and $x_2(t)$ (predators). We also run it with different initial conditions $\mathbf{x}_0$ (e.g., $[25\ 25]$, $[30\ 30]$, etc.) to generalize the structure of the solution shown by multiple phase plane plots in Figure 9.3. It also features two *equilibrium* (also, fixed or stationary) points: at $(0, 0)$ and $(\gamma/\delta, \alpha/\beta) = (50, 100)$.

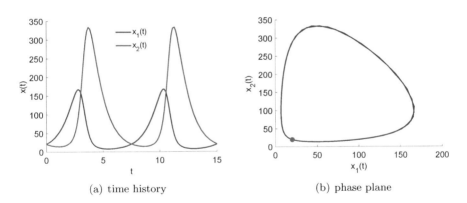

| (a) time history | (b) phase plane |

FIGURE 9.2
(a) Time history plot for solutions $x_1(t)$ and $x_2(t)$ for time interval $[0, 15]$, and (b) phase plane plot $(x_1(t), x_2(t))$ obtained by `MATLAB`'s `ode23`.

Now we run `ode45` to compare the results. The system 9.36 consists of nonlinear ODEs, so the analytical solution may not be available to check the results objectively. However, as seen in Figure 9.4, the solution curve provided by `ode45` is smooth. By default, it uses a continuous extension formula to produce its output at four equally spaced time points in the span of each step taken. The number of points is also adjustable by playing with the optional settings of the solver. The final note on the performance is related to the computational time: `ode45` tends to make larger steps, but computations are longer for each step (refer to Section 9.5.1 for explanations). We also advise

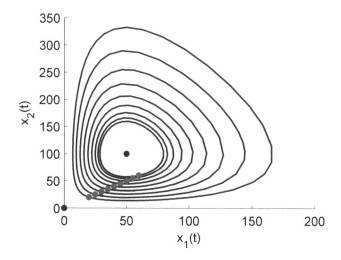

FIGURE 9.3
Phase plane plots for different initial conditions (red dots) with two equilibrium points (blue dots).

the reader to refer to the description of other properties of both `ode23` and `ode45` in [23].

9.7 Optimization Problem Constrained by LV Model

9.7.1 Statement of Problem

At this point, we are ready to derive a new optimization problem to practice the following:

- adjusting the optimization framework used in Chapter 8 for solving the problem of Example 8.1 (with a simple linear equality constraint) to enable solutions for the ODE-based optimization,

- deriving adjoint-based gradients for such optimization, and

- solving it numerically using discretized settings for forward and adjoint problems, objective function, and gradients.

For example, we could consider reconstructing the exact values of the interaction (constant) parameters α, β, γ, and δ in the LV model (9.36). However, we would like to include more challenges and instead consider adding

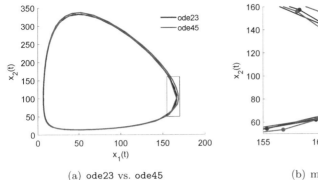 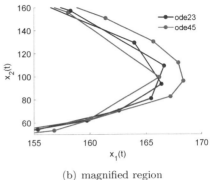

(a) ode23 vs. ode45   (b) magnified region

FIGURE 9.4
(a) Phase plane plot comparison for `ode23` and `ode45`. Magnified image for the region inside the rectangular box is shown in (b).

a new control, the function $u(t)$. Its physical meaning is to represent some environmental factors influencing the change in the populations of both prey and predators. The following problem formulates the identification of such function.

Example 9.2 Parameter Identification in the LV Model
Identify parameter $u(t)$ in the nonlinear IVP-1 system defined by the LV model by fitting (continuous) data $\tilde{x}_1(t)$ and $\tilde{x}_2(t)$, $t \in (0, T]$

$$\min_{u(t)} \beta_1 \int_0^T (x_1 - \tilde{x}_1)^2 \, dt + \beta_2 \int_0^T (x_2 - \tilde{x}_2)^2 \, dt, \qquad (9.37)$$

(a) subject to the following (modified) Lotka–Volterra predator–prey model

$$\begin{aligned}
\dot{x}_1 &= (\alpha - \beta x_2) x_1 - u(t) x_1, \\
\dot{x}_2 &= -(\gamma - \delta x_1) x_2 - u(t) x_2,
\end{aligned} \qquad (9.38)$$

$$x_1(0) = x_1^0, \qquad x_2(0) = x_2^0, \qquad t \in [0, T], \qquad (9.39)$$

(b) when data $\tilde{x}_1(t)$ and $\tilde{x}_2(t)$ are available continuously over interval $(0, T]$, and

(c) constants $\alpha, \beta, \gamma, \delta, \beta_1, \beta_2 \geq 0$ are given.

In the rest of this chapter, we will solve this problem using the *adjoint-based gradient* (*Lagrange multiplier*) method by deriving gradients using the "*optimize–then–discretize*" approach discussed in Section 9.3.

9.7.2   Deriving Gradients

We remind the reader that the "optimize–then–discretize" approach consists of applying optimality conditions and deriving the optimization algorithm using the *continuous formulation* of the problem meaning the form provided in Example 9.2. Before proceeding with the derivation of the adjoint-based gradients, we identify the objective function

$$\mathcal{J}(x_1(t), x_2(t); u(t)) = \beta_1 \int_0^T (x_1 - \tilde{x}_1)^2 \, dt + \beta_2 \int_0^T (x_2 - \tilde{x}_2)^2 \, dt \qquad (9.40)$$

defined over the *control space*

$$\mathcal{U} = \{u(t) \in L_2[0, T] : 0 \leq u(t) \leq 1, \ t \in [0, T]\}. \qquad (9.41)$$

Briefly, (9.41) means that we will obtain the solution function (control variable) $u(t)$ and derived gradients $\nabla_u \mathcal{J}$ in the $\chi = L_2$ functional space as described by (9.5) and exemplified in Section 9.3.1. We also assume that this function is bounded; however, we will discuss implementing this requirement later in Chapter 10.

Further analysis of the Example 9.2 statement provides a forward problem (governing equation or equation of state) given by the modified LV model (9.38)–(9.39) with state variables $x_1(t)$ and $x_2(t)$. Finally, we define two adjoint variables (Lagrange multipliers) $\psi_1(t)$ and $\psi_2(t)$, as we have two equations in (9.38).

Following the procedure previously discussed in Section 9.3.1, we start to derive the formula for adjoint-based gradients by defining the Lagrangian (augmented objective function) in the L_2 functional space, i.e.,

$$\begin{aligned}
\mathcal{L}(x_1, x_2; u, \psi_1, \psi_2) = \ & \beta_1 \int_0^T (x_1 - \tilde{x}_1)^2 \, dt + \beta_2 \int_0^T (x_2 - \tilde{x}_2)^2 \, dt \\
& + \int_0^T [\dot{x}_1 - (\alpha - \beta \, x_2)x_1 + ux_1] \, \psi_1 \, dt \qquad (9.42) \\
& + \int_0^T [\dot{x}_2 + (\gamma - \delta \cdot x_1)x_2 + ux_2] \, \psi_2 \, dt.
\end{aligned}$$

The 1-order *total variation* for this Lagrangian

$$\delta\mathcal{L}(x_1, x_2; u, \psi_1, \psi_2)$$

$$= 2\beta_1 \int_0^T (x_1 - \tilde{x}_1)\delta x_1 \, dt + 2\beta_2 \int_0^T (x_2 - \tilde{x}_2)\delta x_2 \, dt$$

$$+ \int_0^T [\delta\dot{x}_1 + \beta x_1 \,\delta x_2 - (\alpha - \beta\, x_2)\delta x_1 + x_1\delta u + u\delta x_1]\,\psi_1 \, dt$$

$$+ \int_0^T [\dot{x}_1 - (\alpha - \beta\, x_2)x_1 + ux_1]\,\delta\psi_1 \, dt \tag{9.43}$$

$$+ \int_0^T [\dot{x}_2 + (\gamma - \delta \cdot x_1)x_2 + ux_2]\,\delta\psi_2 \, dt$$

$$+ \int_0^T [\delta\dot{x}_2 - \delta \cdot x_2 \,\delta x_1 + (\gamma - \delta \cdot x_1)\delta x_2 + x_2\delta u + u\delta x_2]\,\psi_2 \, dt$$

should be consistent (by its shape) with the structure provided by the Riesz Representation Theorem in (9.11), i.e.,

$$\delta\mathcal{L} = \int_0^T \boldsymbol{\nabla}_u\mathcal{L}\,\delta u \, dt + \int_0^T \boldsymbol{\nabla}_{\mathbf{x}}\mathcal{L}\,\delta\mathbf{x} \, dt + \int_0^T \boldsymbol{\nabla}_{\boldsymbol{\psi}}\mathcal{L}\,\delta\boldsymbol{\psi} \, dt \tag{9.44}$$

and set to 0 as required by the KKT optimality conditions discussed in Chapter 8.

Due to the forward problem (9.38), the fourth and fifth terms on the right side of (9.43) are zeros:

$$\int_0^T [\dot{x}_1 - (\alpha - \beta\, x_2)x_1 + ux_1]\,\delta\psi_1 \, dt$$

$$+ \int_0^T [\dot{x}_2 + (\gamma - \delta \cdot x_1)x_2 + ux_2]\,\delta\psi_2 \, dt = 0. \tag{9.45}$$

In addition, parts $\psi_1\delta\dot{x}_1$ and $\psi_2\delta\dot{x}_2$ (the third and seventh terms, respectively) are not consistent with the structure of (9.44): we use the integration by parts for them to get rid of $\delta\dot{x}_1$ and $\delta\dot{x}_2$, i.e.,

$$\int_0^T \psi_1\delta\dot{x}_1 \, dt = [\psi_1\delta x_1]_0^T - \int_0^T \dot{\psi}_1\delta x_1 \, dt,$$

$$\int_0^T \psi_2\delta\dot{x}_2 \, dt = [\psi_2\delta x_2]_0^T - \int_0^T \dot{\psi}_2\delta x_2 \, dt. \tag{9.46}$$

Finally, boundary terms $\psi_1\delta x_1|_0$ and $\psi_2\delta x_2|_0$ in (9.46) are zeros (for $t = 0$ only) due to the perturbation system (more specifically, its homogeneous boundary conditions)

$$\delta\dot{x}_1 = -\beta x_1\delta x_2 + (\alpha - \beta\, x_2)\,\delta x_1 - x_1\delta u - u\,\delta x_1,$$

$$\delta\dot{x}_2 = \delta \cdot x_2\delta x_1 - (\gamma - \delta \cdot x_1)\,\delta x_2 - x_2\delta u - u\,\delta x_2, \tag{9.47}$$

$$\delta x_1(0) = 0, \quad \delta x_2(0) = 0, \quad t \in [0, T],$$

obtained by perturbing forward problem (9.38). And now we have (9.43) in the fully "Riesz-consistent" form

$$\delta\mathcal{L} = 2\beta_1 \int_0^T (x_1 - \tilde{x}_1)\delta x_1 \, dt + 2\beta_2 \int_0^T (x_2 - \tilde{x}_2)\delta x_2 \, dt$$

$$+ \psi_1(T)\delta x_1 - \int_0^T \dot{\psi}_1 \delta x_1 \, dt + \psi_2(T)\delta x_2 - \int_0^T \dot{\psi}_2 \delta x_2 \, dt$$

$$+ \int_0^T [\beta x_1 \, \delta x_2 - (\alpha - \beta \, x_2)\delta x_1 + x_1\delta u + u\delta x_1] \, \psi_1 \, dt$$

$$+ \int_0^T [-\delta \cdot x_2 \, \delta x_1 + (\gamma - \delta \cdot x_1)\delta x_2 + x_2\delta u + u\delta x_2] \, \psi_2 \, dt$$

(9.48)

that is confirmed after grouping all terms in (9.48) by factoring δu, δx_1, and δx_2:

$$\delta\mathcal{L} = \int_0^T [x_1\psi_1 + x_2\psi_2] \, \delta u \, dt + \psi_1(T)\delta x_1 + \psi_2(T)\delta x_2$$

$$+ \int_0^T \left[-\dot{\psi}_1 + 2\beta_1(x_1 - \tilde{x}_1) - (\alpha - \beta \, x_2)\psi_1 + u\psi_1 - \delta \cdot x_2\psi_2 \right] \delta x_1 \, dt \quad (9.49)$$

$$+ \int_0^T \left[-\dot{\psi}_2 + 2\beta_2(x_2 - \tilde{x}_2) + \beta x_1\psi_1 + (\gamma - \delta \cdot x_1)\psi_2 + u\psi_2 \right] \delta x_2 \, dt.$$

9.7.3   Optimization Algorithm

The final form of $\delta\mathcal{L}$ obtained in (9.49) provides an analytical formula for the adjoint-based gradient derived in the L_2 functional space

$$\nabla_u\mathcal{L} = x_1(t)\psi_1(t) + x_2(t)\psi_2(t) \tag{9.50}$$

based on the solution of the adjoint ODE problem

$$\dot{\psi}_1 = 2\beta_1(x_1 - \tilde{x}_1) - (\alpha - \beta \, x_2)\psi_1 + u\psi_1 - \delta x_2\psi_2,$$
$$\dot{\psi}_2 = 2\beta_2(x_2 - \tilde{x}_2) + \beta x_1\psi_1 + (\gamma - \delta x_1)\psi_2 + u\psi_2, \tag{9.51}$$
$$\psi_1(T) = 0, \quad \psi_2(T) = 0, \quad t \in [0, T],$$

to be solved (backward in time) to find adjoint states $\psi_1(t)$ and $\psi_2(t)$ with *terminal conditions* $\psi_1(T)$ and $\psi_2(T)$. Now we could start solving the optimization problem of Example 9.2 numerically after performing discretization for

- time t: $\mathbf{t} = [t_0 \ t_1 \ \ldots \ t_n]^T$, where

$$t_0 = 0, \ t_1 = t_0 + \Delta t, \ t_2 = t_0 + 2\Delta t, \ \ldots, \ t_n = T, \quad \Delta t = \frac{T - t_0}{n}, \tag{9.52}$$

- states $x_i(t) \to \mathbf{x}_i \in \mathbb{R}^{n+1}$, controls $u(t) \to \mathbf{u} \in \mathbb{R}^{n+1}$, and adjoints $\psi_i(t) \to \psi_i \in \mathbb{R}^{n+1}$, $i = 1, 2$:

$$
\mathbf{x} = [\mathbf{x}_1^k \; \mathbf{x}_2^k] = \begin{bmatrix} x_1^0 & x_2^0 \\ x_1^1 & x_2^1 \\ \cdots & \cdots \\ x_1^n & x_2^n \end{bmatrix}, \quad \mathbf{u} = \begin{bmatrix} u_0 \\ u_1 \\ \cdots \\ u_n \end{bmatrix},
$$
$$
\psi = [\psi_1^k \; \psi_2^k] = \begin{bmatrix} \psi_1^0 & \psi_2^0 \\ \psi_1^1 & \psi_2^1 \\ \cdots & \cdots \\ \psi_1^n & \psi_2^n \end{bmatrix},
\tag{9.53}
$$

- objective function $\mathcal{J}(x_1(t), x_2(t); u(t))$ by

$$
\mathcal{J}(\mathbf{x}; \mathbf{u}) = \beta_1 \sum_{i=1}^{n} \left(x_1^i(\mathbf{u}) - \tilde{x}_1^i \right)^2 \Delta t + \beta_2 \sum_{i=1}^{n} \left(x_2^i(\mathbf{u}) - \tilde{x}_2^i \right)^2 \Delta t.
\tag{9.54}
$$

We have to make a comment similar to that made for the discretized objective (9.21). The discretization method in (9.54) also supports the continuous structure of available measurements $\tilde{x}_i(t)$, $i = 1, 2$. It employs the rectangular (or midpoint) rule of numerical integration used to approximate the objective (9.40). It is important to mention that making **continuous and discretized formulations consistent** is critical for applying correctly kappa-tests while checking the quality of the discretized gradients.

We now summarize all steps in the practical implementation of the "optimize–then–discretize" approach to perform numerical computations for solving the optimization problem of Example 9.2 in Algorithm 9.3 below.

Algorithm 9.3 ("Optimize–then–Discretize" for LV Model)

1. *Discretize time t by (9.52) and initialize vectors for states, controls, and adjoints by (9.53)*

2. *Obtain and store (analytic/synthetic) data $(\tilde{t}_i, \tilde{x}_1^i)$ and $(\tilde{t}_i, \tilde{x}_2^i)$*

3. *Set $k = 0$ and choose initial guess $\mathbf{u}^0$ for control*

4. *Solve numerically (discretized) forward problem (9.38)–(9.39) to find state $\mathbf{x}^k = [\mathbf{x}_1^k \; \mathbf{x}_2^k]$*

5. *Evaluate objective $\mathcal{J}(\mathbf{x}^k; \mathbf{u}^k)$ by (9.54)*

6. *For $k = 1, 2, \ldots$ check optimality of $\mathbf{u}^k$; if optimal $\Rightarrow$ **TERMINATE***

7. *Solve numerically (discretized) adjoint problem (9.51) to find adjoint state*
   $\boldsymbol{\psi}^k = [\boldsymbol{\psi}_1^k \ \boldsymbol{\psi}_2^k]$

8. *Evaluate gradient by*

$$\nabla_{\mathbf{u}}\mathcal{L}(\mathbf{x}^k; \mathbf{u}^k, \boldsymbol{\psi}^k) = \mathbf{x}_1^k \circ \boldsymbol{\psi}_1^k + \mathbf{x}_2^k \circ \boldsymbol{\psi}_2^k, \qquad (9.55)$$

   where $\circ$ denotes Hadamard (element-wise, entrywise, or Schur) product

9. *Improve solution by finding optimal step size α^k*

$$\mathbf{u}^{k+1} = \mathbf{u}^k - \alpha^k \nabla_{\mathbf{u}}\mathcal{L}(\mathbf{x}^k; \mathbf{u}^k, \boldsymbol{\psi}^k) \qquad (9.56)$$

10. *Set $k \leftarrow k + 1$ and go to Step 4*

As before, we note that Step 9 could be practically implemented by using any gradient-based strategies and methods for optimal step size α discussed in Chapters 5–7. Finally, the reader should not be confused by using gradient $\nabla_{\mathbf{u}}\mathcal{L}$ obtained for Lagrangian $\mathcal{L}$ while evaluating the original objective $\mathcal{J}$. The adjoint analysis used for deriving the gradients here sets $\nabla_{\psi}\mathcal{L}$ to zero and, as such, makes the Lagrangian in (9.42) equivalent to the original objective in (9.40) by assuming that the numerical solution of the forward problem (9.38)–(9.39) is fairly accurate.

9.8   Optimization Framework in MATLAB

After discussing solutions for ODEs given by MATLAB (Section 9.6) and the "optimize–then–discretize" framework for identifying parameters in the LV model (Section 9.7), we are ready to experiment with the computational part of solving the ODE-based optimization of Example 9.2. Back to Chapter 7, our computational framework completed its modular structure by establishing proper communication between its different components ("logical level"), being constructed from multiple files organized within several folders ("physical level"). Next, in Chapter 8, we checked its functionality while adding the ability to solve problems supplied with constraints. Currently, we have another opportunity to challenge the ergonomics of the created structure in application to a more complicated problem that has

- a new objective constructed on the solutions of the system of ODEs (our LV model as a forward problem),

- variable size of the control vector that depends on the chosen discretization, and

- gradients that require solving adjoint equations (another system of ODEs).

9.8.1 Benchmark Models for Controls

Before proceeding to the computational part and the analysis of the obtained results, we have to discuss a choice of the *benchmark models* for the established controls (i.e., the function $u(t)$, in our case). The main goal of playing with different models is to check the computational performance of our optimization framework, while reconstructing control functions $u(t)$ of various levels of complexity (refer, e.g., to the robustness and applicability tests mentioned in Section 3.5).

To follow this idea, we created three models represented by a constant $u(t) = 0.3$ (model #1), smooth periodic $u(t) = \frac{1}{5}\cos\left(\frac{\pi}{2}t\right) + 0.5$ (model #2), and discontinuous (step or signum) function $u(t) = \frac{1}{5}\text{sign}\left[\cos\left(\frac{\pi}{2}t\right)\right] + 0.5$ (model #3); refer to Figure 9.5 for their images paired with associated solutions $x_1(t)$ and $x_2(t)$. The reader could examine file `models_ctrl.m` added to folder `algorithms` in the code modified for this chapter. We discuss more details about these modifications to implement the ODE-based constraints in Section 9.8.3. The reader may also experiment freely with other functions added to the framework by expanding MATLAB's `switch` block in `models_ctrl.m` with one or more `case` statements.

MATLAB: <code>models_ctrl.m</code>

```
switch modelU                                                    1
  case 1                                                         2
    u_ex = 0.3*ones(size(tt));                                   3
  case 2                                                         4
    u_ex = cos(pi/2*tt)/5+0.5;                                   5
  case 3                                                         6
    u_ex = sign(cos(pi/2*tt))/5+0.5;                             7
  otherwise                                                      8
    disp(['error: Unknown model ' num2str(modelU) ' is chosen!']); 9
  return;                                                       10
end                                                            11
```

9.8.2 Analytic vs. Synthetic Measurements

Another important issue we have to discuss before starting framework modifications is how we intend to obtain measured data (or, simply, measurements)

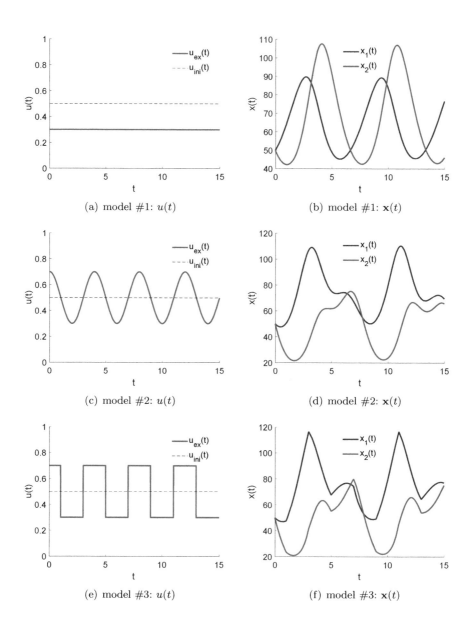

FIGURE 9.5
(a,c,e) Benchmark models for control $u(t)$ paired with (b,d,f) solution functions $x_1(t)$ and $x_2(t)$ obtained by solving forward problem (9.38)–(9.39).

$\tilde{x}_1$ and $\tilde{x}_2$ appeared in the objective (9.40). For many problems, especially those based on real-life models, there exist data obtained directly or indirectly from actual experiments or observations. It is good, but it may not be available for every model. In addition, it may not help validate the accuracy and performance of new computational frameworks (e.g., if these data are very noisy). For doing that, people first use analytic or synthetic measurements during the debugging and validation stages to remove any concerns about the consistency between the model and associated data.

The *analytic measurements (data)* could be produced if a forward problem is solvable analytically. A simple Algorithm 9.4 illustrates the procedure for getting such data.

Algorithm 9.4 (Analytic Measurements)

1. *Find analytic functions (exact solutions) $x_{1,ex}(t)$, $x_{2,ex}(t)$, and $u_{ex}(t)$ that satisfy forward problem, e.g., (9.38)–(9.39)*

2. *Discretize time, e.g., by using (9.52)*

3. *Evaluate $x_{1,ex}(\mathbf{t})$, $x_{2,ex}(\mathbf{t})$ to obtain two vectors $\tilde{\mathbf{x}}_1$ and $\tilde{\mathbf{x}}_2$*

4. *Use analytic measurements $(\tilde{t}_i, \tilde{x}_1^i)$ and $(\tilde{t}_i, \tilde{x}_2^i)$ to substitute real data while running optimization*

Here, we notice that it is hard to implement as usually complex models do not possess analytical solutions. However, if they are available, in addition to checking the optimization performance (convergence from initial guess $\mathbf{u}^0$ to known optimal solution $\mathbf{u}^* = u_{ex}(t)$), they may also be used to check the accuracy of d- (in our case $\mathcal{J}$-) evaluators; refer to Figure 9.6 for notation review.

Opposite to analytic measurements, the *synthetic data* do not require a solution for a forward problem, i.e., state variables defined by analytically given functions. Algorithm 9.5 shows the process for obtaining such data.

Algorithm 9.5 (Synthetic Measurements)

1. *Derive analytic function (exact solution) $u_{ex}(t)$*

2. *Discretize time, e.g., by using (9.52)*

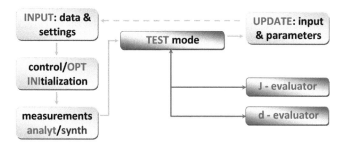

FIGURE 9.6
Generalized optimization framework: parts included in the TEST mode (updated from Figure 3.11 to reflect notation changes in the new framework structure).

> 3. *Solve numerically forward problem* (9.38)–(9.39) *to obtain solutions for* $x_1(t)$ *and* $x_2(t)$ *in terms of two vectors* $\tilde{\mathbf{x}}_1$ *and* $\tilde{\mathbf{x}}_2$
>
> 4. *Use synthetic measurements* $(\tilde{t}_i, \tilde{x}_1^i)$ *and* $(\tilde{t}_i, \tilde{x}_2^i)$ *to substitute real data while running optimization*

As seen from this process, getting synthetic measurements is simple to implement. However, it requires interpolation if t-discretization changes over the optimization process or in the case of pointwise (or scattered) data used. Practically, obtaining synthetic data requires running $\mathcal{J}$-evaluator once with the control $u(t)$ set to its exact value $u_{ex}(t)$; refer to file `measurements.m` added to folder `algorithms` in the code modified for this chapter.

MATLAB: `measurements.m`

```
% preallocation to structure data matrix                         1
data = [tt zeros(Nt+1,1) zeros(Nt+1,1)];                          2
                                                                  3
% calling J-evaluator                                            4
[Cf,x1meas,x2meas] = f(0.0,u_ex,zeros(size(u_ex)),params,data);   5
                                                                  6
% updating data matrix                                           7
data(:,2:3) = [x1meas x2meas];                                    8
```

Finally, we remark on synthetic measurements used during the validation stage and beyond. Without a doubt, it is a powerful tool that we suggest using actively during the debugging phase and making analysis for the entire computational framework performance. However, an opinion exists that, in some circumstances (e.g., when the adaptive mesh is used), synthetic data can "cheat" the optimization by giving the cue for the optimal solution used to generate such data. It requires further analysis to recognize and prove such cases; e.g., the "cheating" claim is debatable under the assumption of noise added inevitably to any numerical (and, as such, approximate) solution of a forward problem. We may also suggest using noise at a reasonable level added to synthetic data or using different spatial discretizations (meshes) while generating data and running optimization.

9.8.3 Adjusting Framework and Choosing Parameters

In Chapter 8, we left our computational framework with an added functionality to solve an optimization problem with a simple constraint. As we did before, now we advise the readers to either perform the transition of this framework to the state of supporting ODE-based constraints by themselves or compare the modified framework (Chapter_9_opt_lotka_volterra.m) with its version used previously in Chapter 8; refer to Table 9.1 for a summary of all modifications.

element	implementation	software
main **OPT**	**minimally adjusted to support ODE-based constraints**	MATLAB
$\mathcal{J}$-evaluator	**m-function using ODE solver** ode23/45 **for** (9.38)–(9.39)	MATLAB
d-evaluator	**m-function using ODE solver** ode23/45 **for** (9.51) **& analytically defined** $\nabla_u \mathcal{J}$ **for SD, CG, and BFGS**	MATLAB
1D α search	**m-code for CONST, GS, and BB only**	MATLAB
visualizer	**m-code with updated subplots**	MATLAB

TABLE 9.1
Computational elements of Chapter_9_opt_lotka_volterra.m modified from Chapter_8_opt_constrained.m to solve constrained optimization problem of Example 9.2.

To help the reader trace all incorporated modifications, we start by noticing the **main changes** in the code.

(a) Control vector is changed from **a** to **u**. While we could still use vector **a**, we perform this change anticipating the need for further experiments with ODE- and PDE-based optimization problems.

(b) Functions for evaluating objective $\mathcal{J}(\mathbf{x}; \mathbf{u})$ and computing gradient $\nabla_{\mathbf{u}}\mathcal{J}$ undergo significant changes as now they include calls for MATLAB function ode23 solving forward (9.38)–(9.39) and adjoint (9.51) problems. To emphasize the significance of these changes, we created new functions named fn_eval_obj.m and fn_eval_grad.m.

(c) Finally, the visualization part (m-code visualize.m) has been upgraded from a 4-window to a 5-window interface; the 6th window is reserved for the convergence analysis performed, as usual, after the optimization ends.

The list of other (minor) changes includes the following.

- To keep the focus on implementing ODE-based constraints (and as commented in the main file Chapter_9_opt_lotka_volterra.m), we will use only the steepest descent SD, conjugate gradient CG, and quasi-Newton BFGS gradient-based methods. For the same reason, we will keep using only the constant value CONST, golden section GS search, and bracketing-Brent BB approaches for step size α search. However, we invite the reader to "activate" all other methods considered in Chapters 5–7 or add new approaches to check their performance and practice in expanding the framework with new functionalities. From here (after completing the initialization phase), we also call the new m-code measurements.m to start creating synthetic measurements (steps 3 and 4 in Algorithm 9.5).

- As before, the entire structure (including variable names) of the parameter file params.m is preserved to allow the reader to experiment with new methods. We added an extra part to accommodate parameters specific to the LV model: e.g., t-domain discretization, initial conditions and parameters for the forward problem, and the choice of the function for control $u(t)$. Another addition (section OPT part) includes setting a constant as the initial guess $\mathbf{u}^0$ and defining coefficients α, β, γ, and δ in objective $\mathcal{J}(\mathbf{x}; \mathbf{u})$. Finally, we added restart parameter n_r for BFGS as discussed in Section 5.3.5.

- In folder functions, we added files fn_Lotka_Volterra_fwd.m and fn_Lotka_Volterra_adj.m containing the description of ODEs (9.38)–(9.39) and (9.51). Files fn_convergence_sol_norm.m and FUNC.m are minimally modified to be consistent with other changes in the code.

- In folder algorithms, all other files not mentioned before are also modified to reflect the changes discussed above.

Now, we could check the functionality of the modified computational framework to test the quality of the discretized gradients and solve

ODE-based optimization by running `Chapter_9_opt_lotka_volterra.m` with the updated termination conditions to address the relative decrease in the objective and the solution norm, given respectively by

$$\left| \frac{\mathcal{J}(\mathbf{u}^{k+1}) - \mathcal{J}(\mathbf{u}^k)}{\mathcal{J}(\mathbf{u}^k)} \right| < \epsilon_1 \tag{9.57}$$

and

$$\frac{\|\mathbf{u}^{k+1} - \mathbf{u}^k\|_2}{\|\mathbf{u}^k\|_2} < \epsilon_2, \tag{9.58}$$

and the following settings (we list only important ones, those related closely to the implemented changes and the description of the LV model).

MATLAB: <code>params.m</code>

Settings for solving ODE-based optimization of Example 9.2:

- *discretized t-domain:* `ti = 0; tf = 15; Nt = 100;` $t \in [0, 15]$, $n = 100$

- *initial conditions:* `x1i = 50.0; x2i = 50.0;` $\mathbf{x}^0 = [50\ 50]^T$

- *interaction parameters:* `paramsODE = [1.0 0.01 1.0 0.02];` $\alpha = 1$, $\beta = 0.01$, $\gamma = 1$, $\delta = 0.02$

- *control model:* `modelU = 2;` #1, #2, or #3 as shown in Figure 9.5

- *objective:* `objWeight = [1.0 1.0];` $\beta_1 = \beta_2 = 1.0$

- *initial guess:* `u_ini = 0.5;` $u_{ini}(t) = 0.5$

- *BFGS restarts:* `BFGSrst = 5;` $n_r = 5$

- *termination #1:* `epsilonJ = 1e-9;` for ϵ_1, refer to (9.57)

- *termination #2:* `epsilonU = 1e-9;` for ϵ_2, refer to (9.58)

- *termination #3:* `kMax = 200;`

- *BB-search:* `AXini = 0; BXini = 1e-6; MAXITER = 20; GLIMIT = 100.0; TOL = 1e-9; ITMAX = 2;`

9.8.4 Checking and Improving Quality of Discretized Gradients

We advise the reader to review the procedure (*kappa-test*) we used before to check the correctness of gradient computations, as described in Section 3.6.

Here, we consider the same test extended for use in our current $(n + 1)$-dimensional (as $\mathbf{u} \in \mathbb{R}^{n+1}$) case, i.e.,

$$\kappa(\epsilon) = \frac{\mathcal{J}(\mathbf{u} + \epsilon\,\delta\mathbf{u}) - \mathcal{J}(\mathbf{u})}{\epsilon\,\langle \boldsymbol{\nabla}_{\mathbf{u}}\mathcal{J}(\mathbf{u}), \delta\mathbf{u}\rangle}, \quad \epsilon \to 0, \tag{9.59}$$

where $\delta\mathbf{u} = [\Delta u_0\ \Delta u_1\ \dots\ \Delta u_n]^T$ is the *perturbation* applied to current control $\mathbf{u}$. We note that in (9.59), $\boldsymbol{\nabla}_{\mathbf{u}}\mathcal{J}(\mathbf{u})$ is the gradient of objective $\mathcal{J}(\mathbf{u}) = \mathcal{J}(\mathbf{x}; \mathbf{u})$ defined analytically with respect to control $\mathbf{u}$ and evaluated for the current values of state $\mathbf{x}$ and control $\mathbf{u}$ using the formula (9.55), i.e., $\boldsymbol{\nabla}_{\mathbf{u}}\mathcal{J}(\mathbf{u}) = \boldsymbol{\nabla}_{\mathbf{u}}\mathcal{L}(\mathbf{x}^k; \mathbf{u}^k, \psi^k)$. We also use the notation $\langle \cdot, \cdot \rangle$ for inner products in the L_2 space as defined in (9.5). As before, we will apply the kappa-test of two types.

(a) *Cheap test.* It requires **two** $\mathcal{J}$-evaluations: for fixed $\delta\mathbf{u}$, e.g., $\delta\mathbf{u} = \mathbf{u}$, we compute $\kappa(\epsilon)$ for a range of ϵ, e.g., $\epsilon = 10^{-15} \div 10^0$.

(b) *Expensive test.* It requires $n + 1$ $\mathcal{J}$-evaluations in n steps: for fixed ϵ, e.g., $\epsilon = 10^{-6}$, we perform kappa-test to check sensitivity for every component of $\mathbf{u}$ by changing $\delta\mathbf{u}$, i.e.,
    step #1: $\delta\mathbf{u} = [u_0\ 0\ 0\ \dots\ 0]^T$,
    step #2: $\delta\mathbf{u} = [0\ u_1\ 0\ \dots\ 0]^T$,
    ...
    step #n: $\delta\mathbf{u} = [0\ 0\ 0\ \dots\ u_{n-1}\ 0]^T$.

> **?** *Why is the last gradient component related to control u_n removed from the expensive kappa-test?*

> **!** *We will discuss the answer after exploring the results of our numerical experiments with all three models in Section 9.8.5. An eager reader may also explain it now after reviewing the structure of the solution for the adjoint problem (9.51).*

Our previous statement about the meaningfulness of choosing $\delta\mathbf{u} = \mathbf{u}$ is still valid here as the magnitudes of components in vector $\mathbf{u}$ may vary significantly, sometimes by order. As such, the suggested strategy helps keep uniform the applications of the kappa-test to all components of the control vector.

Now we could check the correctness of changes implemented in our algorithm related to the $\mathcal{J}$- (objective function) and d- (gradient) evaluators facilities by setting `mode = 'TEST'` in `params.m` and running the main m-file `Chapter_9_opt_lotka_volterra.m` in the TEST mode for model #2; review Figure 9.5(c). Although model #1 ($u(t) = const$) looks the least sophisticated out of all three models, this simplicity (as we will see it from our experiments) is deceptive. We also choose constant initial guess $u_{ini}(t) = 0.5$ for all our numerical experiments with all models for the current problem. Figure 9.7 provides the results of running a cheap version of our kappa-test. This time, its plateau region forms more distant from 1 in comparison with the ones observed in Figures 3.12(a) and 8.6(a) obtained, respectively, for low dimensional problems of Examples 1.3 (3D) and 8.1 (4D). It is seen very often when the problem's complexity increases. We assume, however, that the gradient is computed **correctly**, as this distance is not too big and the parameter ϵ spans a large range (8-11 orders of magnitude) of those values where κ is close to 1. As explained before, Figure 9.7(b) supplements extra information added to this analysis: quantity $\log_{10}|\kappa(\epsilon) - 1|$ shows that the gradient evaluation captures between one and two significant digits of accuracy. We also remind the reader about the well-known effects of the very small and very large ϵ-values for which $\kappa(\epsilon)$ deviates from the unity (to the left and right sides of the plateau regions) due to, respectively, subtractive cancelation (roundoff) and truncation errors.

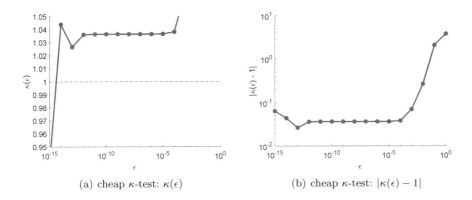

(a) cheap κ-test: $\kappa(\epsilon)$ (b) cheap κ-test: $|\kappa(\epsilon) - 1|$

FIGURE 9.7
Results of cheap kappa-test.

The results of the expensive version of the kappa-test are shown in Figure 9.8. To perform computations for this component-wise sensitivity (accuracy) analysis, we use $\epsilon = 10^{-6}$ taken arbitrarily from the plateau region ϵ-interval $[10^{-12}, 10^{-4}]$. It confirms the correctness of almost all 100 gradient components related to u_i, $i = 0, 1, \ldots, 99$, except for two outliers, which is a regular issue for problems of increased complexity. We refer the reader to the discussion initiated by the question posed on p. 73, dealing with the process of identifying problems and controlling errors throughout the entire optimization. Finally, we advise the reader to experiment a bit more by, e.g., allowing any programming "errors" in describing the procedures for computing objectives and/or gradients to see the immediate reflection of the associated miscomputations to the results of both cheap and expensive kappa-tests.

Before we discuss the results of the new numerical experiments, we have to address an important issue of "quality *controllability*."

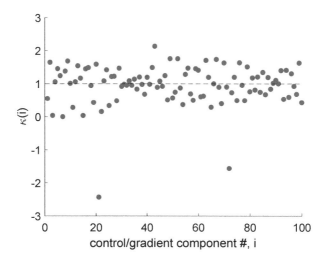

FIGURE 9.8
Results of expensive kappa-test.

> **?** *That is: what can we do to improve the quality of gradients, and thus, to expect a better performance of optimization?*

For example, for our current problem, we may suggest the following.

(1) Refine the **time discretization**: e.g., repeat kappa-tests when $n = 50$, 100 (used in current t-discretization), $500, 1000$, etc.

(2) Use **higher-order** ODE solvers: e.g., replace MATLAB's `ode23` (current solver) with `ode45`.

(3) Experiment with **parameters** for the ODE solvers.

We encourage the reader to experiment with options (1) and (2) as homework and proceed with the last one here as an example. First, we may use MATLAB's keyword `odeset` to check (in the command line) the default settings for the ODE solver (`ode23`) currently used to support our $\mathcal{J}$- and d-evaluators.

MATLAB: odeset

Checking default settings for ODE solvers:

```
>> odeset
        AbsTol:   [ positive scalar or vector {1e-6} ]
        RelTol:   [ positive scalar {1e-3} ]
   NormControl:   [ on | {off} ]
   NonNegative:   [ vector of integers ]
         ...   ...
```

Two settings (tolerances), `RelTol = 1e-3` and `AbsTol = 1e-6`, relate directly to the solution accuracy; look for more details at [23]. Roughly, if we decrease these tolerances, the accuracy of the result and, correspondingly, the runtime will be increased. Therefore, we modify files `fn_eval_obj.m` and `fn_eval_grad.m` by adding optional parameters given by `odeset`; see the example of the modified $\mathcal{J}$-evaluator (`fn_eval_obj.m`) below.

MATLAB: `fn_eval_obj.m`

```
function [obj,x1,x2] = fn_eval_obj(step, u, d, params, data)      1
                                                                  2
   % updating solution                                            3
   u = u + step*d;                                                4
                                                                  5
   % solving numerically LV ODE using ode23                       6
   options = odeset('RelTol', 1e-9, 'AbsTol', 1e-12);             7
   [T X] = ode23(@(t,x) fn_Lotka_Volterra_fwd(t,x,params,...      8
       data,u), data(:,1), params(5:6), options);                 9
                                                                  10
   % computing objective                                         11
   x1 = X(:,1); x2 = X(:,2);                                     12
   obj = params(7) * trapz(data(:,1), (x1 - data(:,2)).^2) + ... 13
       params(8) * trapz(data(:,1), (x2 - data(:,3)).^2);        14
                                                                  15
return                                                           16
```

Next, we run the code in the TEST mode again with `RelTol = 1e-9` and `AbsTol = 1e-12`; Figure 9.9 shows the results for cheap and expensive kappa-tests. These results reveal improvement in the gradient quality: namely, accuracy and consistency with its analytical formula (9.50). It is what we could refer to as the "*controllable*" quality. The plateau part of the cheap test moves closer to 1, and the "cloud" of individual κ-values in the expensive test increases its density in notable proximity to the line $\kappa = 1$. Although this test also detects several outliers, their proximity to 1 shows improvement compared to the results provided with the default tolerances.

The last thing to mention is the runtime: about 1s for default settings versus 72s after the update. Here, we leave the decision on the trade-off between

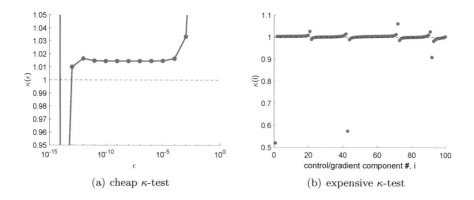

(a) cheap κ-test   (b) expensive κ-test

FIGURE 9.9
Results of cheap and expensive kappa-tests with modified settings (`RelTol = 1e-9` and `AbsTol = 1e-12`) for ODE solver `ode23`.

accuracy and computational time to the reader: the solution may vary significantly depending on the problem size, type/speed of a particular computer, and even the user's preferences. The readers may also consider adding various `odeset` parameters to the `params.m` file in case they want to allow more flexibility, while changing them to find a proper balance between accuracy and feasible runtime. We will use the default settings for all our experiments in the rest of this chapter.

9.8.5 Optimization Results

As we have already confirmed the correctness of the gradient and objective function computations, we are ready to start our experiments with optimization itself (`mode = 'OPT'`). We first run the main m-file `Chapter_9_opt_lotka_volterra.m` with the steepest descent method (`method = 'SD'`) and other `params.m` settings as given on p. 261 for model #2 (`modelU = 2`), where the true (exact) control function $u(t) = u_{ex}(t)$ is smooth and periodic; refer back to Figure 9.5(c). We remind the reader that we use the analytical formula for this function only to generate synthetic measurements as described in Section 9.8.2 and for visualization purposes to compare the exact function $u_{ex}(t)$ with the obtained solution $\mathbf{u}(t)$. This analytical formula is unknown to the optimization started with the initial guess $u_{ini}(t) = 0.5$ (dashed line) chosen as a constant to make optimization "fair" to all models. We invite the reader to run the code and look at the dynamics of the entire optimization process at different angles now provided by five `MATLAB` windows: objective $\mathcal{J}(\mathbf{u})$, control $\mathbf{u}(t)$, ODE solution $\mathbf{x}(t)$, gradient $\nabla_{\mathbf{u}}\mathcal{J}(\mathbf{u})$, and step size α. To allow more idle time between iterations (for capturing changes) or setting it to 0 (for running faster and evaluating real runtime), we advise changing the parameter pauseTime (default value is 2s) in params.m.

Figure 9.10 shows the results of this optimization run terminated with condition #3 at $k = k_{max} = 200$ by six MATLAB subplots (the order is different in the actual MATLAB window) updated dynamically except the last one for the computational convergence analysis created right before the termination. Our analysis of graphs (a) and (e) suggests a conclusion on a satisfactory performance – the decrease of the objective function spans about four orders of magnitude. However, analysis of the step size α history predicts a possibility of better results if we allow optimization to run for longer. The values of taken steps span several orders of magnitude and confirm the performance of the bracketing-Brent approach (methodAlpha = 'BB'). However, this history lacks the signature of the (small-scale) convergence to the optimal solution – the "tail" created in the proximity of the local optimum (we will see it later in other runs). Still, the overall quality of the solution is acceptable according to graphs (b) and (c). The shape of the reconstructed control $\mathbf{u}(t)$ is close to that of the exact solution $u_{ex}(t)$, with slight deviations at peak points. Both solutions $\mathbf{x}_1(t)$ and $\mathbf{x}_2(t)$ (circles) match the data (dots) rather accurately. We also note that in graph (c), we use dashed lines to represent solutions $\mathbf{x}_1^0(t)$ and $\mathbf{x}_2^0(t)$ obtained with the initial state of control $\mathbf{u}^0(t)$. We defer the analysis of the gradient structure in graph (d) to the time of available comparison with the results of other experiments. Finally, graph (f) shows linear convergence ($r = 1.013$), which is consistent with our previous observations for the SD performance.

In the next iteration, we replace the 1-order SD with the 2-order BFGS (method = 'BFGS') to check the performance of this quasi-Newton's approach. We keep the rest parameters in params.m the same and run m-file Chapter_9_opt_lotka_volterra.m again for model #2. Figure 9.11 shows the results of this optimization run terminated with condition #1 (inability to minimize further the objective) at $k = 56$. Similar to the previous case, we start our analysis with graphs (a) and (e), showing this time much better performance. For example, the magnitude of the objective function has decreased by about six orders. In addition, the step size α history confirms the convergence to the local optimum: the reader now could observe the salient feature, the "tail" created by the four last iterations. According to graphs (b) and (c), the solution shows the superior quality: the shape of the reconstructed control $\mathbf{u}(t)$ is just the same as that of the exact solution $u_{ex}(t)$, and both solutions $\mathbf{x}_1(t)$ and $\mathbf{x}_2(t)$ match the data with no visible deviations. We, again, skip the analysis of the gradient structure in the (d) graph and endorse the superlinear convergence ($r = 1.2329$ seen in the (f) graph) that is typical for BFGS applied to many complicated problems.

For the next step, we keep using the quasi-Newton's BFGS method with its confirmed performance (paired with the BB approach for choosing optimal step size) in application to a model that might be mistakenly considered a simpler one, model #1 (modelU = 1). Figure 9.12 shows the results of this

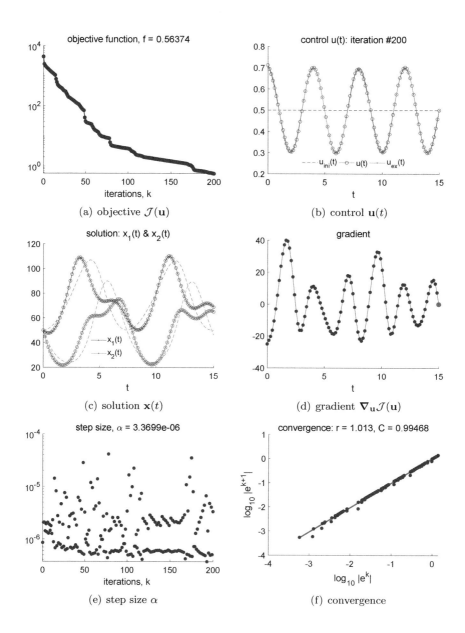

FIGURE 9.10
MATLAB window with six plots updated dynamically (finalized with SD at
$k = 200$), while solving Example 9.2 for control $\mathbf{u}(t)$ with model #2.

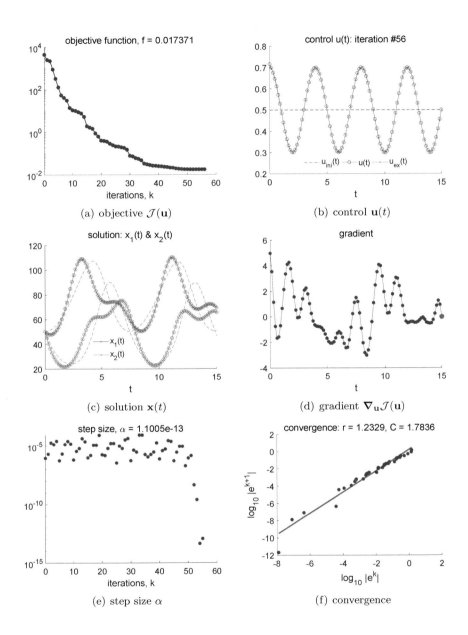

(a) objective $\mathcal{J}(\mathbf{u})$

(b) control $\mathbf{u}(t)$

(c) solution $\mathbf{x}(t)$

(d) gradient $\nabla_{\mathbf{u}}\mathcal{J}(\mathbf{u})$

(e) step size α

(f) convergence

FIGURE 9.11

MATLAB window with six plots updated dynamically (finalized with BFGS at $k = 56$), while solving Example 9.2 for control $\mathbf{u}(t)$ with model #2.

optimization run (with the same rest parameters in `params.m`) terminated
with condition #2 (inability to update further the control) at $k = 69$. A similar
analysis applied to this solution returns the following. As seen in graph (e),
the "tail" is present, meaning that computationally the solution qualifies to
be a local optimal one. However, the objective decrease spans only four orders
of magnitude in graph (a), and the convergence analysis in graph (f) returns
the linear rate ($r = 1.0916$). The data match is also satisfactory except for the
solution at the final time, i.e., $\mathbf{x}_1(T)$ and $\mathbf{x}_2(T)$, where we could visibly identify
some deviations. Now it is a turn for the control $\mathbf{u}(t)$, where its shape for the
major part of the interval $[0, T]$ is, more a less, consistent with the $u_{ex}(t)$ that
is a straight line. However, a crimp (small oscillation) at the left side makes the
solution imperfect within the time interval roughly $[0, 1]$. In general, we do not
need to worry about that. Although we cannot remove this "defect" entirely,
we may try to make the oscillation smaller by setting both tolerances ϵ_1 and
ϵ_2 to smaller values and allowing the optimization to run a bit longer. We
may find a simple explanation in the structure of the gradient computed from
the solutions of the forward and adjoint problems. Due to the continuity of
these solutions (in the continuous settings), constructing a particular gradient
component in the discretized settings utilizes the information obtained from
its left and right "neighbors." Obviously, several components on the left side
lack this information as there is no solution for $t < 0$. As such, the associated
elements of the solution are "late" compared with its other parts. To illustrate
this phenomenon, we refer the reader to Figure 9.13, showing intermediate
solutions $\mathbf{u}^k(t)$ recorded at $k = 10$ and $k = 20$.

The reader now could suggest the same explanation for the problem seen
on the right side of $\mathbf{u}(t)$ that is valid only in part. As Figure 9.12(b) shows, this
problem is aggravated by the fact that the last component of the control $\mathbf{u}(t)$,
i.e., u_{100}, does not move for the entire course of optimization. We suggest
the reader watch carefully for the associated component in the gradient –
it is 0 at every iteration. That is why we did not check its quality by the
expensive kappa-test. To complete the answer to the question posed on p. 262,
we advise the reader to review the formulation of the adjoint problem in
(9.51). As its terminal conditions are $\psi_1(T) = 0$ and $\psi_2(T) = 0$, the gradient
component for time $t = T$ is also 0. Therefore, it is not a coding problem
but rather a problem of the mathematical approach chosen for finding the
solution. To add more, many other time-dependent optimization problems
solved with backward-in-time adjoints face similar outcomes. We will discuss
some approaches to mitigate the consequences of such issues, including the
insufficient gradient sensitivity on the left side discussed before, in Chapter 10.

Finally, we move to our last model #3 (`modelU = 3`), where control $u(t)$
is a continuous but nonsmooth periodic step-function; refer, again, to Figure 9.5(e) for review. As we did for model #1, here we keep using the paired
quasi-Newton's BFGS and BB methods to run optimization with the same
parameters in `params.m`. Figure 9.14 shows the results of this numerical experiment also terminated with condition #2 (inability to update further the

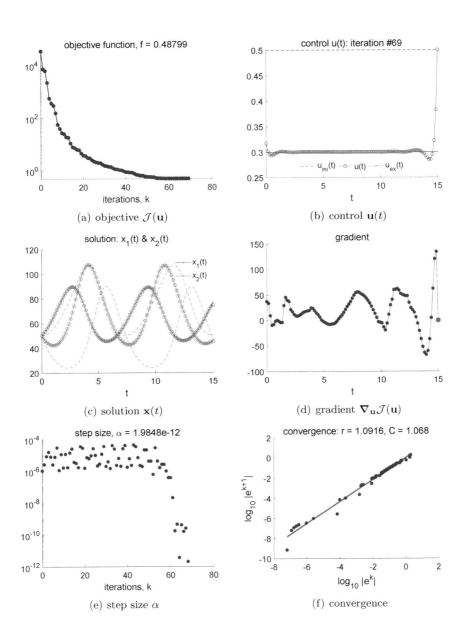

(a) objective $\mathcal{J}(\mathbf{u})$

(b) control $\mathbf{u}(t)$

(c) solution $\mathbf{x}(t)$

(d) gradient $\boldsymbol{\nabla}_{\mathbf{u}}\mathcal{J}(\mathbf{u})$

(e) step size α

(f) convergence

FIGURE 9.12

MATLAB window with six plots updated dynamically (finalized with BFGS at $k = 69$), while solving Example 9.2 for control $\mathbf{u}(t)$ with model #1.

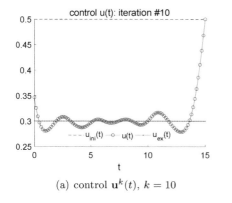
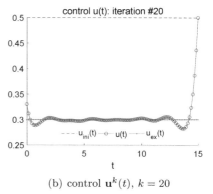

(a) control $\mathbf{u}^k(t)$, $k = 10$ (b) control $\mathbf{u}^k(t)$, $k = 20$

FIGURE 9.13
Intermediate solutions $\mathbf{u}^k(t)$ for model #1 recorded at $k = 10$ and $k = 20$.

control) at $k = 46$ in the proximity of the local optimum (find the "tail" in the (e) graph). Although the objective function decreases by only three orders of the magnitude seen in the (a) graph, the optimization performs with the superlinear rate of convergence ($r = 1.2321$, according to the (f) graph reading). Even if the data match is satisfactory, the solution $\mathbf{u}(t)$ quality is not as good as in the two previous models; refer to graphs (c) and (b), respectively. Based (partially) on the analysis of the previous models we may suggest the following.

(1) The gradient suffers from the lack of sensitivity on the left and right sides of the interval $[0, T]$, as discussed for model #1.

(2) It also exhibits the missing information for its component u_{100} (similar to model #1).

(3) As observed in the (d) graph, the gradient shape is not smooth. It is due to the nonsmoothness of the true control function $u_{ex}(t)$ used to generate measurements, meaning the data is also "nonsmooth."

As the final remark, we invite the reader to play more with all three models and experiment with various settings to see possible improvements in the solution quality and the overall performance of the entire optimization. For instance, we could change the termination tolerances ϵ_1 and ϵ_2 and tolerances used to control ODE solvers `ode23` and `ode45`, as discussed before, to deal with the issue (1) above. We also recommend finer t-discretization as a commonly used remedy for nonsmooth (or underresolved) problems, similar to that mentioned in (3). In the next chapter, we will discuss some regularization techniques that might suggest a resolution in addition to those mentioned here.

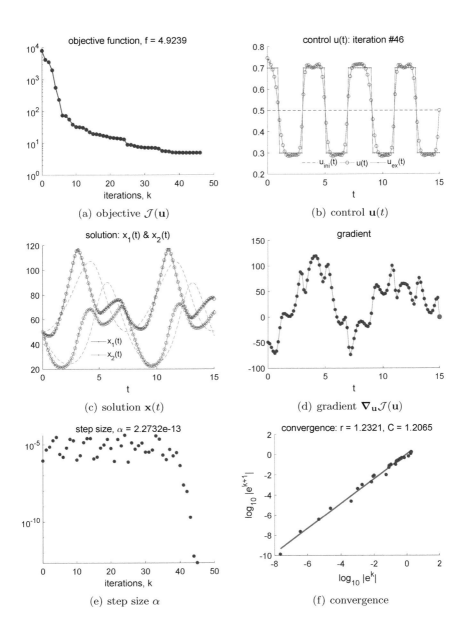

(a) objective $\mathcal{J}(\mathbf{u})$

(b) control $\mathbf{u}(t)$

(c) solution $\mathbf{x}(t)$

(d) gradient $\boldsymbol{\nabla}_{\mathbf{u}}\mathcal{J}(\mathbf{u})$

(e) step size α

(f) convergence

FIGURE 9.14

`MATLAB` window with six plots updated dynamically (finalized with BFGS at $k = 46$), while solving Example 9.2 for control $\mathbf{u}(t)$ with model #3.

9.9   Homework Problems

1. Derive the analytical formula for gradients in Example 8.4 of Chapter 8.

2. Derive the analytical formula for gradients in Example 8.5 of Chapter 8.

3. Solve problems in Examples 8.4 and 8.5 using "discretize–then–optimize" approach for very coarse discretization in t, e.g., when $n = 3$.

4. Use MATLAB code `Chapter_9_opt_lotka_volterra.m` for the LV model of Example 9.2 to compare the quality of discretized gradients by:

   (a) refining time discretization using $n = 50$, 100, 500, and 1000,

   (b) using different ODE solvers, e.g., `ode23` and `ode45`, and

   (c) changing parameters (e.g., absolute `AbsTol` and relative `RelTol` tolerances) for current ODE solver `ode23`. Record CPU elapsed time (total runtime) for every change and derive your strategy in choosing optimal parameters for the ODE solver. Observe the results, and conclude.

5. Perform optimization using different gradient-based (SD, CG, and BFGS) and step size search (CONST, GS, and BB) methods changing their settings. Conclude on the performance in applications to all three models.

6. Modify MATLAB code `Chapter_9_opt_lotka_volterra.m` by adding other gradient-based and derivative-free approaches and methods for the optimal step size search. Apply all available techniques to the problem of Example 9.2. Compare the performance in terms of accuracy of the obtained solutions, computational time, and rate of convergence, and conclude.

7. Modify MATLAB code `Chapter_9_opt_lotka_volterra.m` by expanding its functionality to enable reconstruction of the interaction parameters: α, β, γ, and δ.

⎡READ⎤   Where to Read More

Gockenbach (2011), [14]
   Chapter 4 (Essential Ordinary Differential Equations)

Higham (2005), [18]

MathWorks, [23]

Press (2007), [26]
   Chapter 17 (Integration of Ordinary Differential Equations)

Tarantola (2005), [28]

Chapter 5 (Functional Inverse Problems)

Vogel (2002), [29]

Chapter 2 (Analytical Tools), Chapter 3 (Numerical Optimization Tools), Chapter 6 (Parameter Identification)

RUN   MATLAB **Codes for Chapter 9**

- **root** folder:

  - Chapter_9_opt_lotka_volterra.m

  - params.m

- **example** folder:

  - Chapter_9_LV_example.m

  - fn_LV.m

- folder algorithms:

  - direction_search.m

  - file_output.m

  - golden_section_search.m

  - initialize.m

  - kappa_test.m

  - measurements.m

  - mode_OPT.m

  - mode_TEST.m

  - models_ctrl.m

  - stepsize_search.m

  - visualize.m

- folder functions:

  - fn_brent.m

  - fn_convergence_sol_norm.m

  - fn_eval_grad.m

  - fn_eval_obj.m

  - fn_Lotka_Volterra_adj.m

  - fn_Lotka_Volterra_fwd.m

  - fn_min_brack.m

– `FUNC.m`

- folder `output`

- folder `temp`

10

Implementing Regularization Techniques

We devote this entire chapter to discussing the regularity of the obtained solutions in terms of reaching an agreement between their properties and imposed requirements. The motivating examples of problems where regularization is needed include mathematical issues that might naturally appear, while deriving the solution approaches, special requirements for the solution's boundness, and appropriate functional spaces. We briefly review several techniques commonly used in various optimization problems: namely, quadratic penalty function, barrier functions, and slack variable methods, also Tikhonov-type regularization, gradient preconditioning, and imposing bounds via simple projections. Due to the simplicity of their computational implementation, we added the last three methods to our existing framework for solving the optimization problem based on the LV model developed in the previous chapter. We also discuss the obtained results and create other examples for additional practice to help the reader feel comfortable with regularization, while solving problems where its use might benefit.

10.1 Motivation for Regularization

In a very broad context, *regularization* means making any model, process, or obtained results regular. The meaning of *regularity*, in its turn, is to satisfy some prescribed properties, e.g., continuity, smoothness, boundness, etc. These properties may come together with the description of the optimization problem, or they could come into light after analyzing the obtained solutions and deciding on their appropriateness for use in the model description. In this chapter, we discuss some regularization techniques and methods for their practical implementation applied within our computational framework modified in the previous chapter to solve the ODE-based optimization problem.

As a motivation example, we will consider multiple issues discovered for the solutions in reconstructing parameters for the LV optimization model of Example 9.2 of Chapter 9. We advise the reader to review the formulation of this problem given by the objective function $\mathcal{J}(u)$ defined over the control space $\mathcal{U}$ subject to the ODE-based constraint (forward problem) described, respectively, by (9.40), (9.41), and (9.38)–(9.39). As the reasons to consider

DOI: 10.1201/9781003275169-10

regularization of any kind may be very different and vary from problem to problem, here we mention just a few of them.

(1) In general, the overall **solution quality** may be questionable due to various reasons, e.g.,

  (a) issues of **mathematical nature** affecting derivation or computations of objectives, gradients, solutions for forward or adjoint problems, etc. We refer the reader to solutions for models #1 and #3 (Example 9.2) shown in Section 9.8.5. The chosen mathematical approach for deriving the analytical gradients $\nabla_{\mathbf{u}}\mathcal{J}(\mathbf{u})$ requires them to take 0 values when evaluated at the final time, $t = T$, due to the established terminal conditions $\psi_1(T) = 0$ and $\psi_2(T) = 0$; look back at Figures 9.12(d) and 9.14(d). It affects not only the solution at the particular point $t = 15$. The problem propagates to several neighbors (to the left) by creating an unnatural oscillation (irregularity) at the right end of the solution interval $[0, 15]$; refer now to Figures 9.12(b) and 9.14(b).

  (b) presence of **multiple local optimizers** imposing the dependence of the solution on the chosen initial guess, etc.

(2) The solution may have a requirement to be in a particular **functional space**: e.g., L_2 (square-integrable) functions, *Sobolev space* H^1 smooth (continuous 1-order derivative) functions, etc. Here, we note that the requirement for the solution in Example 9.2 is $u(t) \in L_2[0, T]$, which we assume is satisfied for all three models. However, if, for some reason, the oscillations appearing on the right side start amplifying, this requirement will no longer be fulfilled. We may also require the solution to be, to some extent, smooth, e.g., $u(t) \in H^1[0, T]$.

(3) Also, many solutions related to real-world application models require **boundness** to keep consistent with the underlying physical laws. In Example 9.2, we expect the control $u(t)$ to honor both lower and upper bounds, i.e., $0 \leq u(t) \leq 1$. As seen in outcomes for all three models (Figures 9.10 through 9.14), the obtained solutions do not violate these constraints. However, this might be the case if the functions $u_{ex}(t)$ used to generate data for models #2 and #3 span larger intervals of allowable values.

(4) We will discuss other reasons that might depend on the structure of problems and the approaches used for solutions.

Before proceeding to a brief review of some regularization techniques, we create a new model (#4)

$$u(t) = \frac{1}{2}\text{sign}\left[\cos\left(\frac{\pi}{2}t\right)\right] + 0.5 \tag{10.1}$$

based on the existing model #3 by magnifying the amplitude of its steps to allow variation within the interval $[0, 1]$. To do that, we expand MATLAB's switch block in models_ctrl.m with an extra case statement (case 4) and run the computational framework (m-file Chapter_9_opt_lotka_volterra.m) created in Chapter 9 with model #4; see Figure 10.1 for the main results. The (b) graph shows now multiple points (filled circles) that violate conditions established for lower ($u(t) \geq 0$) and upper ($u(t) \leq 1$) bounds. A closer comparison of the gradients in Figures 9.14(d) and 10.1(d) reveals a rougher structure for our new model. It explains the small-scale oscillations in the (b) graph when the solution is around 0 and 1. Therefore, this model will serve as a good experimental exemplar to apply various regularization strategies, while dealing with issues (1a), (2), and (3) mentioned above.

MATLAB: models_ctrl.m (case 4 for model #4)

```
case 4                                                                    1
    u_ex = sign(cos(pi/2*tt))/2+0.5;                                      2
```

10.2 Some Regularization Theory

In this section, we select several approaches commonly used for regularization due to the simplicity of their computational implementation while solving various optimization problems. Before proceeding, we note that this brief overview is far from comprehensive as we pursue the idea of studying how to merge these techniques with the solution stream. Hence, we invite the interested readers to research and read more about other methods that may better benefit the solutions to their particular problems.

10.2.1 Quadratic Penalty Function Method

We start our review by mentioning two approaches we considered before. In Section 8.6, we compared the augmented Lagrangian and penalty function methods (used separately!) by discussing their application and computational results obtained, while solving the constrained optimization problem of Example 8.1. We often see enforcing constraints as adding regularity to the solutions. Hence, here we consider the two approaches mentioned above paired together in the form of the *quadratic penalty function method* as an example of regularization.

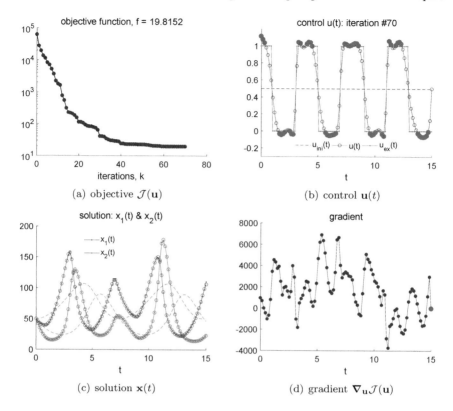

(a) objective $\mathcal{J}(\mathbf{u})$ (b) control $\mathbf{u}(t)$

(c) solution $\mathbf{x}(t)$ (d) gradient $\nabla_{\mathbf{u}}\mathcal{J}(\mathbf{u})$

FIGURE 10.1
MATLAB window with selected (four) plots updated dynamically (finalized with
BFGS at $k = 70$), while solving Example 9.2 for control $\mathbf{u}(t)$ with model #4.

We refer the reader back to the equality constrained optimization problem
(8.27) written below in the generalized form

$$\min_{\mathbf{x} \in \mathbb{X}} f(\mathbf{x})$$
$$\text{s.t. } \mathbf{h}(\mathbf{x}) = 0 \tag{10.2}$$

and assume that $\mathbb{X} \subset \mathbb{R}^n$, $f(\mathbf{x}) : \mathbb{R}^n \to \mathbb{R}$ and $\mathbf{h}(\mathbf{x}) : \mathbb{R}^n \to \mathbb{R}^m$ are all
continuous and **real-valued** (nonlinear) functions. We also note that in-
equalities may be treated similarly. We combine the use of (8.28) with added
penalization term $\|\mathbf{h}(\mathbf{x})\|^2$, in the same fashion as done in (8.47), to create a
new augmented *Lagrangian function* $\mathcal{L}_{c^k}(\mathbf{x}, \boldsymbol{\lambda}^k) : \mathbb{R}^n \times \mathbb{R}^m \to \mathbb{R}$

$$\mathcal{L}_{c^k}(\mathbf{x}, \boldsymbol{\lambda}^k) = f(\mathbf{x}) + (\boldsymbol{\lambda}^k)^T \mathbf{h}(\mathbf{x}) + c^k \|\mathbf{h}(\mathbf{x})\|^2, \tag{10.3}$$

where c^k is a *penalty parameter*. Now we have to solve the equality constrained problem (10.2) in the modified form as a **sequence** of unconstrained optimization problems

$$\min_{\mathbf{x} \in \mathbb{X}} \mathcal{L}_{c^k}(\mathbf{x}, \boldsymbol{\lambda}^k) \tag{10.4}$$

defined by the penalty parameter sequence $\{c^k\}$.

Among different approaches for solving (10.4), we would mention the simplest one introduced in the early 1960s by setting $\boldsymbol{\lambda}^k = \mathbf{0}$ for every k. The reader may notice the only way this approach differs from the plain penalty function method used in Section 8.6: it simply allows penalization weight to change when moving within the sequence of problems (10.4), i.e., $\beta = c^k$. However, the modified approaches commonly used nowadays are different. For example, the *method of Lagrange multiplier estimates* allows $\boldsymbol{\lambda}^k$ to change in the course of the optimization algorithm while increasing c^k ($c^k \to \infty$ as $k \to \infty$). The convergence of the quadratic penalty function method is proven by the following theorem.

Theorem 10.1 Convergence of the Quadratic Penalty Function Method

Assume that

- *functions $f(\mathbf{x})$ and $\mathbf{h}(\mathbf{x})$ are continuous, feasible set $\mathbb{X}$ is closed, and set $\{\mathbf{x} \in \mathbb{X} : \mathbf{h}(\mathbf{x}) = \mathbf{0}\}$ is nonempty; and*

- *for $k = 0, 1, \ldots,$ let $\mathbf{x}^k$ be a global minimizer for modified problem (10.4), where $\{\boldsymbol{\lambda}^k\}$ is bounded, $0 < c^k < c^{k+1}$ for every k, and $c^k \to \infty$.*

Then every limit point of $\{\mathbf{x}^k\}$ is a global minimizer of the original problem (10.2).

We comment on the statement provided in Theorem 10.1 by analyzing three possible outcomes seen in practice while solving problems (10.2) and (10.4) computationally.

1. This method **breaks down** after the inability to find solution $\mathbf{x}^k$ at iteration k with preselected termination tolerance ϵ^k, i.e., termination condition $\|\nabla_{\mathbf{x}} \mathcal{L}_{c^k}\| < \epsilon^k$ cannot be reached.

2. Sequence $\{\mathbf{x}^k : \forall k \, \|\nabla_{\mathbf{x}} \mathcal{L}_{c^k}\| < \epsilon^k\}$ is obtained, but

   (a) it has **no limit points**, or

   (b) for each limit point $\mathbf{x}^*$, matrix $\nabla \mathbf{h}(\mathbf{x}^*)$ has **linearly dependent columns**, i.e., rank$[\nabla \mathbf{h}(\mathbf{x}^*)] < m$.

3. **Success** in finding the solution for problem (10.4): $\mathbf{x}^*$ and $\boldsymbol{\lambda}^*$ satisfy 1-order necessary optimality conditions, i.e.,

$$\nabla_{\mathbf{x}} \mathcal{L}_{c^k}(\mathbf{x}^*, \boldsymbol{\lambda}^*) = \mathbf{0}. \tag{10.5}$$

Finally, we have to mention a well-known *ill-conditioning problem*: the *condition number* of Hessian $\nabla^2_{\mathbf{xx}}\mathcal{L}_{c^k}(\mathbf{x}^k, \boldsymbol{\lambda}^k)$ tends to increase with $c^k \to \infty$. Possible resolutions for this problem might include, e.g.,

- using any **Newton-like method** paired with increased accuracy for all computational steps (by using, e.g., **double precision**),

- using a **good starting point** (initial guess), and

- increasing c^k at a **moderate rate**.

The reader now has an excellent opportunity to experiment with the quadratic penalty function method practically. We advise reviewing the computational framework created in Chapter 8 (with `Chapter_8_opt_constrained.m` as the main m-file) and solving the constrained optimization problem of Example 8.1 by the discussed approach utilizing the existing facilities for the Lagrange multiplier and penalization methods minimally modified to

- allow them to work together, and

- change penalty parameter c^k.

? *Could you suggest any approaches for automatic adjustments made to penalty parameter c^k while solving the sequence of optimization problems (10.4)?*

! *The reader may consider any parameterized formula for c^k to ensure $0 < c^k < c^{k+1}$ (for every k) and $c^k \to \infty$, and any ways to tune up this parameterization.*

10.2.2   Barrier Functions Method

We proceed now with reviewing an inequality constrained optimization problem (8.66) by reconsidering it, in a very general sense, as a nonlinear-inequality-constrained problem

$$\min_{\mathbf{x} \in \mathbb{X}} f(\mathbf{x})$$
$$\text{s.t. } g_j(\mathbf{x}) \leq 0, \quad j = 1, \ldots, r \tag{10.6}$$

In (10.6), we assume that $\mathbb{X} \subset \mathbb{R}^n$ and $f(\mathbf{x})$, $g_j(\mathbf{x}) : \mathbb{X} \to \mathbb{R}$ are all **continuous** and **real-valued** functions.

As discussed in general in Section 8.4, we may solve the constrained problem (10.6) by using different approaches. However, a feasible direction is suitable if constraints are linear. Application of the necessary optimality conditions (Lagrange Multipliers Theory) may guarantee only a local convergence depending on the choice of the initial guess. Although we advise the reader to review all other methods mentioned in Section 8.4, we stop at the group of penalization techniques and introduce the *barrier function method*.

We could set any function, say $B(\mathbf{x})$, to be the *barrier function* if it satisfies two basic properties:

- $B(\mathbf{x})$ is **continuous (!)** and

- $B(\mathbf{x}) \to \infty$ if any $g_j(\mathbf{x}) \to 0_-$.

While there are many possibilities to define $B(\mathbf{x})$, we provide two examples of the barrier function used widely as they both are **convex** whenever $g_j(\mathbf{x})$ are convex:

- *logarithmic*

$$B(\mathbf{x}) = -\sum_{j=1}^{r} \ln\left[-g_j(\mathbf{x})\right], \qquad (10.7)$$

- *inverse*

$$B(\mathbf{x}) = -\sum_{j=1}^{r} \frac{1}{g_j(\mathbf{x})}. \qquad (10.8)$$

To solve the nonlinear-inequality-constrained optimization (10.6), we have to augment its objective and consider a new problem

$$\mathbf{x}^k = \operatorname*{argmin}_{\mathbf{x} \in \mathbb{S}} \left[f(\mathbf{x}) + \epsilon^k B(\mathbf{x})\right], \quad k = 0, 1, \ldots, \qquad (10.9)$$

where ϵ^k are weighting coefficients changed at every iteration. Understanding the concept of the *barrier function method* is easy if combined with the notation of the *feasibility* discussed in Sections 1.2 and 1.4.1, i.e.,

$$\mathbb{S} = \{\mathbf{x} \in \mathbb{X} : \ g_j(\mathbf{x}) < 0, \ j = 1, \ldots, r\}, \qquad (10.10)$$

as the method defines the interior of the *feasible set* $\mathbb{S}$. A successful implementation requires enforcing two conditions:

(1) sequence $\{\epsilon^k\}$ satisfies

$$0 < \epsilon^{k+1} < \epsilon^k, \quad \epsilon^k \to 0 \qquad (10.11)$$

to ensure finding a solution whenever it is **close to** or **at a boundary**, and

(2) iterates $\mathbf{x}^k$ must be **interior points**;

refer to Figure 10.2 to explicate both requirements. When implemented by honoring both conditions (1) and (2) above, the barrier functions method exhibits a nice property given by the following theorem.

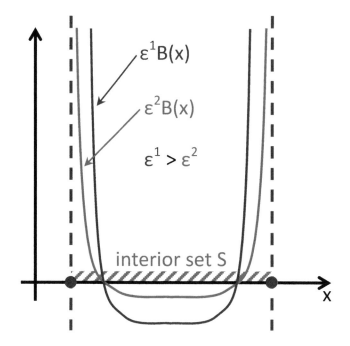

FIGURE 10.2
Schematic showing the concept of the barrier functions method.

Theorem 10.2 Global Minima in Barrier Problems
Every limit point of sequence $\{\mathbf{x}^k\}$ generated by the barrier functions method is the **global minimum** *of the problem* (10.6).

The proof may be easily found, e.g., in [4]. In addition to easy implementation, this approach has numerous advantages. Here, we list just the major ones, e.g.,

- it avoids *combinatorial issues* associated with *active constraints* (refer to Section 8.8 for more details), and

- shows *convergence* under **mild** conditions; also,

- barrier functions are **convex** if the problem (10.6) is convex, and

- barrier problems are typically solved by *Newton's method* with **fast convergence**.

In addition to all these advantages, we also need to mention one complication. The barrier functions method requires some caution, while defining sequence $\{\epsilon^k\}$ by (10.11) as both fast and slow convergence ($\epsilon^k \to 0$) may cause a problem with finding a solution if located close to or near the boundary of feasible set $\mathbb{S}$ defined in (10.10).

? *Could you suggest any approaches for automatic adjustments made to weights ϵ^k through optimization iterations?*

! *Think about strategies already discussed in Section 10.2.1 for changing penalty parameter c^k. For example, we could use a formula for ϵ^k parameterized to ensure (10.11) and control the performance.*

It is the right time now to experiment with the barrier functions method practically. To allow the reader to apply this method both analytically and computationally, we modified Example 8.1 of Chapter 8 by changing the objective to a linear function, introducing a nonlinear inequality constraint, and keeping the same number (four) of variables (controls).

Example 10.1 Linear Objective + Nonlinear (Quadratic) Inequality Constraint
Solve minimization problem

$$\min_{\mathbf{x}\in\mathbb{R}^4} \quad f(\mathbf{x}) = x_1 + x_2 + x_3 + x_4$$
$$\text{s.t.} \quad x_1^2 + x_2^2 + x_3^2 + x_4^2 \le 4 \tag{10.12}$$

First, we solve the optimization problem of Example 10.1 analytically by employing the barrier functions method discussed in this section. We construct a *logarithmic barrier function*

$$B(\mathbf{x}) = -\ln(x_1^2 + x_2^2 + x_3^2 + x_4^2 - 4) \tag{10.13}$$

and solve a new optimization problem to minimize the augmented objective $\tilde{f}(\mathbf{x}) = f(\mathbf{x}) + \epsilon^k B(\mathbf{x})$, i.e.,

$$\min_{\mathbf{x}\in\mathbb{R}^4} \tilde{f}(\mathbf{x}) = x_1 + x_2 + x_3 + x_4 - \epsilon^k \ln(x_1^2 + x_2^2 + x_3^2 + x_4^2 - 4), \tag{10.14}$$

where ϵ^k are small positive constants. Then the 1-order necessary condition (in Theorems 5.1 and 5.2) $\nabla\tilde{f}(\mathbf{x}^*) = \mathbf{0}$ provides the following system of four(nonlinear equations

$$1 - \frac{2\epsilon^k x_1}{x_1^2 + x_2^2 + x_3^2 + x_4^2 - 4} = 0,$$

$$1 - \frac{2\epsilon^k x_2}{x_1^2 + x_2^2 + x_3^2 + x_4^2 - 4} = 0,$$

$$1 - \frac{2\epsilon^k x_3}{x_1^2 + x_2^2 + x_3^2 + x_4^2 - 4} = 0,$$ (10.15)

$$1 - \frac{2\epsilon^k x_4}{x_1^2 + x_2^2 + x_3^2 + x_4^2 - 4} = 0.$$

By solving (10.15), we could find a *stationary point* $\mathbf{x}^k = [x_1^k \ x_2^k \ x_3^k \ x_4^k]^T$, where

$$x_i^k = \frac{1}{4}\epsilon^k \pm \sqrt{\left(\frac{\epsilon^k}{4}\right)^2 + 1}, \quad i = 1, 2, 3, 4.$$ (10.16)

Finally, we keep only "$-$" in (10.16) to deal with a local minimum and conclude that if $k \to \infty$, then $\epsilon^k \to 0$ making $\mathbf{x}^* = [-1 \ -1 \ -1 \ -1]^T$ a unique solution to the problem (10.12).

? *Could we conclude on $\mathbf{x}^*$ as a strict local and global minimizer of $\tilde{f}(\mathbf{x})$?*

Here, we complete the analytical solution by answering this question similarly as we did for Example 8.1 and make two conclusions.

(1) Assuming that we allow ϵ^k and $((x_1^k)^2 + (x_2^k)^2 + (x_3^k)^2 + (x_4^k)^2 - 4)^2$ to converge to 0_+ at the same rate, we may evaluate Hessian as $\nabla^2 \tilde{f}(\mathbf{x}^*) = \alpha\mathcal{E} \succeq 0$, where $\mathcal{E}$ is a unity matrix, and α is some constant (we leave proving this fact and formula (10.16) to the reader as homework). Unfortunately, it is the 2-order necessary condition (in Theorem 5.1) only, not sufficient one required by Theorem 5.2 to state that optimal solution $\mathbf{x}^*$ is a **strict local minimizer** of $\tilde{f}(\mathbf{x})$.

(2) However, functions $\tilde{f}(\mathbf{x})$ and $g(\mathbf{x}) = x_1^2 + x_2^2 + x_3^2 + x_4^2 - 4$ are both convex. We refer to Theorem 5.3 (Convex Cost Function) to confirm that $\mathbf{x}^*$ is a **global minimizer** of the augmented objective $\tilde{f}(\mathbf{x})$ and also the original one $f(\mathbf{x})$ (as $\epsilon^k \to 0$), which is consistent with the statement of Theorem 10.2.

Now the reader may check the obtained solution computationally. The easiest way is to modify the computational framework we created in Chapter 8 (with `Chapter_8_opt_constrained.m` as the main m-file) to solve the constrained optimization problem of Example 8.1 by the Lagrange multiplier and penalization methods. Implementing the latter is similar to the barrier functions approach, so the readers may enjoy adding it at minimal costs after reviewing the modification process and computational results discussed in Section 8.6.

10.2.3 Tikhonov-type Regularization

Generally speaking, both methods discussed so far (quadratic penalty function and barrier functions) apply to almost any constraints imposed on states and controls (respectively, **x** and **u**, in our established notations). Due to that, in Sections 10.2.1 and 10.2.2, we used vector **x** to denote both. Now we turn to the methods commonly used to regularize control variables. For many problems, we could find functions used as the *reference solutions* with the requirement for the optimal solution **not to deviate too much** from these references. Implementing such requirements is relatively straightforward and is broadly known as the *Tikhonov-type regularization*. It consists of augmenting the original objective function $\mathcal{J}(u)$ with a new term that penalizes (in some functional spaces) any deviations of control u from the provided reference solution $\bar{u}$, i.e.,

$$\mathcal{J}_\gamma(u) = \mathcal{J}(u) + \gamma \|u - \bar{u}\|_{\mathcal{X}}^2,$$

where $\gamma \in \mathbb{R}_+$ is adjustable (constant) regularization parameter and $\|\cdot\|_{\mathcal{X}}$ is the functional (Hilbert) space norm, e.g., $\mathcal{X} = L_2$, in which the deviation $(u - \bar{u})$ is measured.

We exemplify this type of regularization by applying it to the LV model designed for solving the optimization problem of Example 9.2 in Chapter 9. Implementing is straightforward, as noted before, and includes modifying the objective and tracing associated changes in the gradient scheme. First, we augment the original objective in (9.40) as discussed above and solve a new optimization problem

$$\min_{u(t)} \mathcal{J}_\gamma(x_1(t), x_2(t); u(t)), \tag{10.17}$$

where

$$\mathcal{J}_\gamma(x_1(t), x_2(t); u(t)) = \beta_1 \int_0^T (x_1 - \tilde{x}_1)^2 \, dt + \beta_2 \int_0^T (x_2 - \tilde{x}_2)^2 \, dt$$
$$+ \gamma \int_0^T (u - \bar{u})^2 \, dt, \tag{10.18}$$

with some assumptions made for the reference solution $\bar{u}(t)$. We invite the reader one more time to go through the complete process of deriving the gradient-based scheme discussed in Section 9.7 to determine necessary modifications; we leave it as homework. Briefly, incorporated changes in the objective **do not affect** the adjoint system (9.51). We would remind that the gradient $\nabla_u \mathcal{L}(u)$ evaluates the sensitivity of the objective $\mathcal{J}(u)$ (in our case, Lagrangian $\mathcal{L}(u)$) to any changes in the control u. As u is now present **explicitly!** in $\mathcal{J}$ (in the added regularization part), another modification is required only in the structure of the gradient (9.50), i.e.,

$$\nabla_u \mathcal{L}_\gamma = \nabla_u \mathcal{L} + 2\gamma (u - \bar{u})$$
$$= x_1(t)\psi_1(t) + x_2(t)\psi_2(t) + 2\gamma (u - \bar{u}). \tag{10.19}$$

We will discuss the technical aspects of implementing the Tikhonov regularization numerically in Section 10.3.

10.2.4 Gradient Preconditioning

The Tikhonov-type regularization discussed in the previous section may improve the regularity of the solution (e.g., its continuity and smoothness) by penalizing the deviation between this solution and the chosen "template" (reference function $\bar{u}$) based on the selected functional space norm. Amplifying the regularization effect requires enforcing penalization by playing with the constant parameter γ. This process is quite delicate: small and large γ values will make all incorporated changes almost imperceptible or lead to the obvious "copying" effect $u = \bar{u}$, respectively. Also, the choice of $\bar{u}$ may be a question. Finally, the penalization format itself allows the solution to violate the regularization requirement.

Here, we consider a possible alternative to the Tikhonov-type regularization known as *preconditioning*. It projects the obtained gradients (and, as such, the associated solutions) from their original functional space onto another one. We refer again to the *Riesz Representation Theorem* (9.11). It implies the equivalence of inner products in various functional (Hilbert) spaces $\mathcal{X}$, e.g., in L_2 and H^1, i.e.,

$$\delta \mathcal{J}(u; \delta u) = \langle \boldsymbol{\nabla}_u^{L_2} \mathcal{J}, \delta u \rangle_{L_2} = \langle \boldsymbol{\nabla}_u^{H^1} \mathcal{J}, \delta u \rangle_{H^1}, \qquad (10.20)$$

where $\boldsymbol{\nabla}_u^{L_2} \mathcal{J}$ and $\boldsymbol{\nabla}_u^{H^1} \mathcal{J}$ are gradients obtained in the respected spaces. In (10.20), the *Sobolev space* H^1 is endowed with the inner product, e.g., in 1D

$$\langle f_1, f_2 \rangle_{H^1} = \int_a^b \left[f_1 f_2 + \ell^2 \frac{df_1}{dx} \frac{df_2}{dx} \right] dx, \quad f_1, f_2 \in H^1, \qquad (10.21)$$

where $\ell \in \mathbb{R}$ is a length–scale *smoothing parameter*.

We note that (10.20) has a strict mathematical meaning of projecting gradients from the L_2 space of square-integrable functions onto the Sobolev H^1 space of smooth (continuous 1-order derivative) functions. However, (10.21) is meaningful only in practical computations as its parameter ℓ serves as a measure of the "completeness" of such projection. For example, $\ell = 0$ returns (10.21) to the original definition of the L_2 inner product; see, e.g., (9.5). Altered ℓ changes the smoothness requirements: e.g., for keeping the term $\ell^2 \dfrac{df_1}{dx} \dfrac{df_2}{dx}$ integrable, the increase in ℓ^2 will be compensated by the corresponding decrease in $\dfrac{df_1}{dx} \dfrac{df_2}{dx}$, leading to a projection represented by a **smoother** function. Finally, if $\ell \to \infty$, the projection inevitably becomes a constant function.

? *How to perform this projection computationally?*

! *We allow the interested readers to answer this question by exercising their mathematical skills: combine (10.21) where ℓ is an arbitrary constant and $\ell = 0$ with (10.20) to show the equivalence*

$$\int_a^b \nabla_u^{L_2} \mathcal{J} \delta u \, dx = \int_a^b \left[\nabla_u^{H^1} \mathcal{J} \delta u + \ell^2 \frac{d \nabla_u^{H^1} \mathcal{J}}{dx} \frac{d \delta u}{dx} \right] dx \qquad (10.22)$$

and obtain the result shown below.

As we may easily prove, the *Sobolev* (smoothed) *gradients* can be determined as a solution of the following elliptic BVP

$$\nabla_u^{H^1} \mathcal{J} - \ell^2 \frac{d^2}{dx^2} \nabla_u^{H^1} \mathcal{J} = \nabla_u^{L_2} \mathcal{J}, \qquad x \in (a, b)$$

$$\frac{d}{dx} \nabla_u^{H^1} \mathcal{J} = 0, \qquad x = a, b \qquad (10.23)$$

solved numerically in Section 10.3 by using built-in `MATLAB` solvers.

10.2.5 Bounds by Simple Projections and Slack Variables

In this section, we address some issues related to the necessity for establishing boundness for the obtained solutions. We refer again to the description (9.38)–(9.41) of the optimization problem created for the LV model of Example 9.2. In particular, we need to consider the lower and upper bounds, $0 \le u(t) \le 1$.

An apparent and easy-to-implement solution would be applying the barrier functions method as discussed in Section 10.2.2. However, it might be possible for some problems to find even simpler approaches, such as a *simple projection*. We describe it in application to the general boundness condition

$$u_\ell \le u_i^k \le u_u \qquad (10.24)$$

via simple **componentwise** projection performed for every iteration k, i.e.,

- if $u_i^k > u_u$, then $u_i^k = u_u$, or
- if $u_i^k < u_\ell$, then $u_i^k = u_\ell$,

where u_i^k is the ith component of the control vector $\mathbf{u}^k$; u_u and u_ℓ are the upper and lower bounds, respectively. The simplicity of both math and practical implementation is evident. However, this componentwise procedure may

- cause a problem of **losing regularity** of obtained solutions,
- **increase objective** at small-scale iterations, and
- **reduce the rate of convergence**.

The abrupt changes in the solution may create an inconsistency with the information provided by the gradients. We will discuss these side effects while observing them in practice later in this chapter.

Our last but not least method to control bounds is introducing *slack variables*. It is a simplistic and elegant approach for incorporating the *positiveness* (non-negativeness) *condition*

$$u > 0 \qquad \text{or} \qquad u \geq 0$$

via slack variable $v(t)$. For example, we may consider the change of variables

$$u = v^2 + \epsilon_u, \tag{10.25}$$

where $\epsilon_u \geq 0$ is a small constant.

We illustrate the method's implementation in application to our Example 9.2. By using the change of variables (10.25), we have to solve the modified optimization problem

$$\min_{v(t)} \mathcal{J}(x_1(t), x_2(t); v(t)) = \beta_1 \int_0^T (x_1 - \tilde{x}_1)^2 \, dt + \beta_2 \int_0^T (x_2 - \tilde{x}_2)^2 \, dt \tag{10.26}$$

defined over the control space

$$\mathcal{U} = \{v(t) \in L_2[0, T] : \epsilon_u \leq v(t) \leq 1, \ t \in [0, T]\} \tag{10.27}$$

subject to the updated ODE-based constraint (forward problem)

$$\begin{aligned}
\dot{x}_1 &= (\alpha - \beta \, x_2) \, x_1 - (v^2 + \epsilon_u) \, x_1, \\
\dot{x}_2 &= -(\gamma - \delta \, x_1) \, x_2 - (v^2 + \epsilon_u) \, x_2, \\
x_1(0) &= x_1^0, \quad x_2(0) = x_2^0, \quad t \in [0, T].
\end{aligned} \tag{10.28}$$

We leave adjusting the gradient-based scheme for solving a new optimization problem (10.26)–(10.28) to the reader; refer to Section 9.7 for review.

10.3   Examples of Numerical Implementation

In Section 10.1, we discussed some factors that might motivate us to consider using regularization to improve the obtained solutions. Although we devote Section 10.2 to briefly reviewing just several approaches, here we examine the practical implementation only for three of them: namely, preconditioning (to improve the solution's smoothness), Tikhonov-type regularization (to overcome the gradient issue of mathematical nature mentioned in (1a) of Section 10.1), and the simple projection (to incorporate the required bounds).

10.3.1 From Theory to Practice

Adding the simple projection to our existing framework is pretty straightforward, so we proceed to the numerical implementation of preconditioning using MATLAB. In a nutshell, we use the formula in (9.50) for computing L_2 gradients $\nabla_u^{L_2}\mathcal{J}$ that will serve as the right-hand side of the elliptic BVP (10.23), while we consider the new Sobolev H^1 gradients $\nabla_u^{H^1}\mathcal{J}$ as solutions of this equation. The reader may consider various software packages that provide ready-to-use utilities with no need to check the correctness of their implementation; we have to learn how to include them in our framework and tune their performance. Here, we review two options for solving BVP (10.23) numerically.

Option #1. Similar to solving IVPs, MATLAB has several functions available for numerical solutions of BVPs, e.g., bvp4c and bvp5c; refer to the syntax below.

MATLAB: bvp4c and bvp5c syntax (**mandatory**/*optional*)

---

  sol = bvp4c(**odefun**, **bcfun**, **solinit**, *options*)
  sol = bvp5c(**odefun**, **bcfun**, **solinit**, *options*)

- **sol**: solution structure; look at MathWorks [23]

- **odefun**: function handle that defines the functions to be integrated

- **bcfun**: function handle that defines the boundary conditions

- **solinit**: initial guess for the solution, also specified as a structure; look at MathWorks [23]

- *options*: options structure; look at MathWorks [23]

If the reader is willing to explore this option in more detail, we note that both bvp4c and bvp5c use the description of the boundary value problem of the nth order (ODE-n) as a system of n ODE-1 equations. As (10.23) is BVP-2, we must first rearrange it to have two ODE-1 equations; refer to [23] for more details. In our updated computational framework (with the main m-file Chapter_10_opt_lotka_volterra_enhanced.m), m-function fn_precond_bvp4c.m returns an updated (preconditioned) gradient (search direction) after applying the preconditioning procedure with parameter ℓ.

MATLAB: fn_precond_bvp4c.m

```
function grad_precond = fn_precond_bvp4c(grad_old,tt,ell)      1
   % initial guess for solution                                2
   solinit = bvpinit(tt,[mean(grad_old),0]);                   3
   % solving BVP-2 using bvp4c                                 4
   sol = bvp4c(@(t,y) deriv(t,y,ell,grad_old,tt),@bcs,solinit); 5
   yint = deval(sol,tt);                                       6
   grad_precond = (yint(1,:))';                                7
return                                                         8
                                                               9
% deriving RHS of ODE system                                  10
function dydx = deriv(t,y,ell,grad_old,tt)                     11
   dydx = [y(2); (y(1)-interp1(tt,grad_old,t))/ell^2];        12
                                                               13
% boundary conditions y'(a)=0, y'(b)=0                        14
function res = bcs(ya,yb)                                      15
   res = [ya(2)-0; yb(2)-0];                                   16
```

Option #2. We may also solve BVP (10.23) by discretizing it with any FD scheme for solving ODEs using linear algebra. For example, we may use the time discretization with step size $h = \Delta t$ as in (9.52) with, e.g., an implicit scheme utilizing the 2-order central FD formulas for all derivatives. The matrix equation created for BVP (10.23) using such discretization is

$$\mathcal{P}\,\mathcal{G}_{new} = \mathcal{G}_{old}, \qquad (10.29)$$

where vectors $\mathcal{G}_{old}$ and $\mathcal{G}_{new}$ represent, respectively, discretized gradients $\nabla_u^{L_2}\mathcal{J}$ and $\nabla_u^{H^1}\mathcal{J}$, and matrix $\mathcal{P}$ is the preconditioning operator defining the discretized form of BVP with already added boundary conditions as shown below.

$$\mathcal{P} = \begin{bmatrix} 1+\frac{2\ell^2}{3h^2} & -\frac{2\ell^2}{3h^2} & 0 & \cdots & 0 & 0 & 0 \\ -\frac{\ell^2}{h^2} & 1+\frac{2\ell^2}{h^2} & -\frac{\ell^2}{h^2} & \cdots & 0 & 0 & 0 \\ 0 & -\frac{\ell^2}{h^2} & 1+\frac{2\ell^2}{h^2} & \ddots & 0 & 0 & 0 \\ \cdots & \cdots & & \ddots & \ddots & \cdots & \cdots \\ 0 & 0 & 0 & \ddots & 1+\frac{2\ell^2}{h^2} & -\frac{\ell^2}{h^2} & 0 \\ 0 & 0 & 0 & \cdots & -\frac{\ell^2}{h^2} & 1+\frac{2\ell^2}{h^2} & -\frac{\ell^2}{h^2} \\ 0 & 0 & 0 & \cdots & 0 & -\frac{2\ell^2}{3h^2} & 1+\frac{2\ell^2}{3h^2} \end{bmatrix}.$$

We assume that matrix $\mathcal{P}$ is invertible and solve (10.29) by

$$\mathcal{G}_{new} = \mathcal{P}^{-1}\mathcal{G}_{old}. \qquad (10.30)$$

In the updated computational framework, we created another m-function, fn_precond_fd.m, to solve the preconditioning problem (10.23) numerically using this option.

MATLAB: fn_precond_fd.m

```
function grad_precond = fn_precond_fd(grad_old,ht,ell)          1
  Nt = length(grad_old) - 2; % # of grid points               2
  E = ones(Nt,1);                                             3
                                                              4
  % matrix P for preconditioning operator                     5
  P = spdiags([(-ell^2/(ht^2))*E (1.0+2.0*ell^2/(ht^2))*E ...  6
     (-ell^2/(ht^2))*E], -1:1, Nt, Nt);                        7
  % boundary conditions                                       8
  P(1,1)  = 1 + 2*ell^2/(3*ht^2);                             9
  P(1,2)  = -2*ell^2/(3*ht^2);                                10
  P(Nt,Nt-1) = -2*ell^2/(3*ht^2);                            11
  P(Nt,Nt) = 1 + 2*ell^2/(3*ht^2);                           12
                                                              13
  % gradient update                                           14
  grad_new = P\grad_old(2:Nt+1);                             15
  grad_precond = [(4*grad_new(1)-grad_new(2))/3; grad_new; ... 16
     (4*grad_new(Nt)-grad_new(Nt-1))/3];                      17
return                                                        18
```

We invite the readers to examine algorithms for implementing options #1 and #2 in detail or use them as black-box tools for future experiments. Both options have the same input structure asking for L_2 gradients $\nabla_{\mathbf{u}}\mathcal{J}(\mathbf{u})$ (grad_old) and smoothing parameter ℓ (ell) except for time discretization information. bvp4c requires discretized vector $\mathbf{t}$ (tt), while the FD approach needs just the step size Δt (ht) in this discretization.

Finally, we elaborate more on some practical approaches to deal with the last component of the gradient $\nabla_{\mathbf{u}}\mathcal{J}(\mathbf{u})$ related to its inability to improve solution $\mathbf{u}(t)$ at $t = T$ as seen in models #1, #3, and #4; refer to graphs (b) and (d) in Figures 9.12, 9.14, and 10.1, respectively. As mentioned in (1a) of Section 10.1, the nature of this issue is mathematical and associated with the chosen method to compute the adjoint-based gradients: the adjoint system (9.51) "suppresses" the sensitivity input for $\mathbf{u}(T)$ by having terminal conditions $\psi_1(T) = 0$ and $\psi_2(T) = 0$. Therefore, we must surpass this restraint by adding this missing information to the gradient "manually." Again, we suggest two methods for this addition by considering different techniques involved and, as such, two methodologies of implementing the remedy.

Method #1. We apply regularization in the **space of the state variables** $x_1(t)$ and $x_2(t)$ by adding two explicit constraints

$$x_1(T) = \tilde{x}_1(T), \quad x_2(T) = \tilde{x}_2(T) \tag{10.31}$$

to the augmented objective (9.40) by two **penalization** terms with their own weights β_3 and β_4, e.g.,

$$
\begin{aligned}
\mathcal{J}(x_1(t), x_2(t); u(t)) = {}& \beta_1 \int_0^T (x_1 - \tilde{x}_1)^2 \, dt + \beta_2 \int_0^T (x_2 - \tilde{x}_2)^2 \, dt \\
& + \beta_3 \left[x_1(T) - \tilde{x}_1(T) \right]^2 + \beta_4 \left[x_2(T) - \tilde{x}_2(T) \right]^2 .
\end{aligned}
\tag{10.32}
$$

Speaking mathematically, the third and fourth terms on the right-hand side of (10.32) are already present in the first and second terms, respectively. However, we want to move them through the adjoint procedure on their own and check any associated modifications. We also advise the reader to do that (look for details in Section 9.7) and prove that these modifications affect (only!) the terminal conditions in the adjoint system (9.51):

$$
\begin{aligned}
\psi_1(T) &= -2\beta_3 \left[x_1(T) - \tilde{x}_1(T) \right], \\
\psi_2(T) &= -2\beta_4 \left[x_2(T) - \tilde{x}_2(T) \right].
\end{aligned}
\tag{10.33}
$$

We note the elegance of this method as it corrects the issue on the mathematical side and requires neither modifications in the problem statement nor additional data.

Method #2. We could also apply regularization directly in the **control space** by adding the Tikhonov term to the augmented objective (9.40) in the following manner.

$$
\begin{aligned}
\mathcal{J}(x_1(t), x_2(t); u(t)) = {}& \beta_1 \int_0^T (x_1 - \tilde{x}_1)^2 \, dt + \beta_2 \int_0^T (x_2 - \tilde{x}_2)^2 \, dt \\
& + \beta_5 \left[u(T) - \bar{u}(T) \right]^2 .
\end{aligned}
\tag{10.34}
$$

However, we have to assume that observations of the control $u(t)$ at terminal time T may become available as a part of measured data. We advise the reader to review Section 10.2.3 for the general concept of regularizing controls and help come up with the ideas on the required changes to the adjoint-gradient structure of the solution. We refer to formulas (10.18) and (10.19) (consider $\gamma = \beta_5$) to arrive at the modified form of the gradient

$$
\boldsymbol{\nabla}_u \mathcal{J} = x_1(t)\psi_1(t) + x_2(t)\psi_2(t) + 2\beta_5 \left[u(T) - \bar{u}(T) \right]
\tag{10.35}
$$

as the only requirement for necessary changes.

Although both methods are lightweight for implementation into our existing computational framework, the availability of data (e.g., in case of method #2) may add some constraints to their practical use.

10.3.2   Adjusting Framework

In Chapter 9, we left our computational framework modified for solving the ODE-based optimization problem based on the LV model of Example 9.2. It requires further modifications to incorporate the selected regularization techniques discussed in Section 10.3.1. However, these modifications are so minimal that it is worth considering only the "affected" parts of the framework "pointwisely" for adding new functionalities. We see it as an outstanding experience to exercise all these changes by the readers themselves. Therefore, we advise you to transition this framework to the state of supporting the selected types of regularization or careful comparison of the modified framework (`Chapter_10_opt_lotka_volterra_enhanced.m`) with its previous version used in Chapter 9; refer to Table 10.1 for a summary of all modifications.

element	implementation	software
main **OPT**	**minimally adjusted to support new parameters & bounds**	MATLAB
$\mathcal{J}$-**evaluator**	**minimally adjusted to support added regularization**	MATLAB
d-**evaluator**	**minimally adjusted to support added regularization**	MATLAB
1D α search	m-code for CONST, GS, and BB only	MATLAB
visualizer	**m-code with updated subplot for precond**	MATLAB

TABLE 10.1
Computational elements of `Chapter_10_opt_lotka_volterra_enhanced.m` modified from `Chapter_9_opt_lotka_volterra.m` to solve the ODE-based optimization problem of Example 9.2 with added regularization functionalities.

To help the reader with this assignment, we note all changes in the code grouped below by a particular implemented technique.

(a) The changes related to all methods require adding new parameters to file `params.m`: e.g., lower u_ℓ and upper u_u bounds in (10.24) (variables `u_l` and `u_u`, respectively), preconditioning (or smoothing) parameter ℓ in (10.21) (`ell`), and the method of application, MATLAB's bvp4c (option #1, `precondMethod = 'MATLAB'`) or FD with linear algebra (option #2, `precondMethod = 'FD'`). To further minimize all changes, we may add new weights β_3, β_4, and β_5 of the augmented objectives in (10.32) and (10.34) at the end of the existing vector `objWeight`. The rest structure of `params.m` is preserved with the main settings kept the same as used in our computations for Chapter 9; refer to p. 261. We also remind you that we added the description of the new model #4 to the existing file `models_ctrl.m` (folder `algorithms`), as described in Section 10.1.

(b) To implement both options for preconditioning, we add two m-functions, `fn_precond_bvp4c.m` (option #1) and `fn_precond_fd.m` (option #2), to folder `functions` as discussed in Section 10.3.1. Both functions are called from m-file `direction_search.m` according to the method specified in precondMethod. We may also want to update the way of visualizing gradients (m-file `visualize.m`) to show their shapes before and after applying preconditioning.

(c) For the simple projection, we have to add a small part to m-file `mode_OPT.m` to describe the algorithm as provided in Section 10.2.5; see the example below.

(d) To incorporate methods #1 and #2 related to the augmented objective, we have to store one piece of our synthetic data $\bar{u}(T) = u_{ex}(T)$. Again, to minimize the changes, we could do it directly in the `measurements.m` file by adding this value to the end of vector params (i.e., `params = [params u_ex(end)];`). We also need to modify both $\mathcal{J}$- and d-evaluators in our m-functions `fn_eval_obj.m` and `fn_eval_obj.m` as exemplified below.

(e) And we make a final comment on the vector params. It currently stores 12 different parameters to share between various functionalities: namely, interaction (forward ODE) parameters α, β, γ, and δ, initial conditions x_1^0 and x_1^0, objective function weights from β_1 through β_5, and extra data $\bar{u}(T)$ (in the order of their location).

MATLAB: `mode_OPT.m` **(implementing simple projection)**

```
for i = 1:length(u)                                        1
  if u(i) > u_u                                            2
    u(i) = u_u;                                            3
  end                                                      4
  if u(i) < u_l                                            5
    u(i) = u_l;                                            6
  end                                                      7
end                                                        8
```

MATLAB: **modified** `fn_eval_obj.m`

```
function [obj,x1,x2] = fn_eval_obj(step, u, d, params, data)    1
                                                                2
% updating solution                                             3
u = u + step*d;                                                 4
                                                                5
% solving numerically LV ODE using ode23                        6
[T X] = ode23(@(t,x) fn_Lotka_Volterra_fwd(t,x,params,data,u),...7
       data(:,1), params(5:6));                                8
```

```
% computing objective                                                  9
                                                                      10
x1 = X(:,1); x2 = X(:,2);                                             11
obj = params(7) * trapz(data(:,1), (x1 - data(:,2)).^2) + ...        12
      params(8) * trapz(data(:,1), (x2 - data(:,3)).^2);             13
                                                                      14
% adding regularization parts                                        15
obj = obj + params(9) * (x1(end) - data(end,2))^2 + ...             16
      params(10) * (x2(end) - data(end,3))^2 + ...                   17
      params(11) * (u(end) - params(12))^2;                         18
                                                                      19
return                                                                20
```

MATLAB: modified `fn_eval_grad.m`

```
function grad = fn_eval_grad(u, x, params, data)                      1
                                                                      2
  [T PSI] = ode23(@(t,psi) fn_Lotka_Volterra_adj(t,psi,params, ...   3
      data,u,x), data(end:-1:1,1), ...                               4
      [-2*params(9)*(x(end,1)-data(end,2)) ...                       5
       -2*params(10)*(x(end,2)-data(end,3))]);                       6
                                                                      7
  % reverting time as adjoint problem is solved backward in time     8
  grad = x(:,1).*PSI(end:-1:1,1) + x(:,2).*PSI(end:-1:1,2);          9
                                                                      10
  % updating regularization part                                     11
  grad(end) = grad(end) + 2 * params(11) * (u(end)-params(12));     12
                                                                      13
return                                                                14
```

10.3.3 Results and Food for Thought

Now, when the theory is discussed, and the framework is ready for the new portion of numerical experiments, we could start running optimization with added regularization methods. To get the best performance on the side of the gradient-based methodology, we keep using quasi-Newton's BFGS paired with the bracketing-Brent step size search (`method = 'BFGS'` and `methodAlpha = 'BB'`). Unless stated otherwise, we fix the maximum number of optimization iterations to 200 (`kMax = 200`) and option #2 (FD with linear algebra; `precondMethod = 'FD'`) for applying preconditioning to the obtained gradients.

First, let us experiment with preconditioning. To exclude overlapping effects by applying different regularization methods, we start with model #3. Although our new facility for checking bounds $0 \leq u_i^k \leq 1$ is active, the model will not require any adjustments to control $\mathbf{u}(t)$; refer to the results shown in

Figure 9.14(b). We run optimization with different values of the smoothing parameter ℓ, say $\ell = 0.5$ and $\ell = 1.0$, and compare the results from multiple perspectives. As seen in Figure 10.3, the effect of preconditioning is noticeable – we iron out the gradients, and the control becomes smoother in the same manner. The good news is that the preconditioned gradients make control $\mathbf{u}(t)$ at $t = T$ move. Thus, the solutions on the right side now are pretty good; see Figures 10.3(c) and (e). However, the drawback is also apparent: smoothed solutions cannot reconstruct the desired shape (periodic steps). As discussed before, by increasing ℓ, we project the gradients onto the smaller subspaces of H^1 (Sobolev space of smooth functions). We have to note that currently, there are no proven methods for choosing optimal ℓ, and the use of preconditioning is very dependent on a problem. The general recommendation, though, is to use this powerful technique with caution: removing undesired oscillations improves the quality of gradients but makes them **less informative!** The reader may check this statement by running optimization with a large ℓ, e.g., $\ell = 10^5$, to examine the gradient shape (almost straight line) and its inability to improve the solution.

To complete the discussion on the preconditioning, we have to check the performance. Figure 10.4(a) shows the objective function changes for all three cases, $\ell = 0$, 0.5, and 1.0. It confirms the issues discussed above – smoother solution does not necessarily mean a better solution, as we define the latter by better fitting the available data. Also, making the solution space (a subspace of H^1) smaller makes the problem of finding an optimizer harder. Figures 10.5(a,c,e) confirm this statement by the data obtained after performing convergence analysis: it moves from being initially superlinear ($r = 1.2321$ for $\ell = 0$), which is natural for BFGS, down to linear ($r = 1.0093$ for $\ell = 1.0$). We invite the reader to experiment more with preconditioning to get more practical knowledge of this technique and some experience with choosing the appropriate degree of smoothness by varying its parameter ℓ.

We may now get even more insight into the effects of preconditioning as we move to our new model #4. This model amplifies the magnitude of the periodic steps of model #3 and, as such, starts controlling the provided bounds by applying the simple projection method. As we see in Figure 10.6(a), the boundness goal is reached. We also add some preconditioning (the same $\ell = 0.5$ and 1.0 as for model #3) to check if the result is improvable. Figures 10.6(c-f) illustrate the concept of choosing the "right" value for the smoothing parameter ℓ. Value 0.5 slightly modifies the gradient and does not have enough "power" to complete the desired shape on the right side. We may choose a higher value of 1.0, and the result is here: Figure 10.6(e) confirms the "proper" shape of the solution. We may also notice an example of a good job done by the preconditioning: Figure 10.3(f) shows how the 7-point oscillation on the left side has been smoothed. But is this solution really good? We are back to Figure 10.5 to check the convergence by comparing plots (b), (d), and (f) – the difference is not too big (between linear and superlinear for all three cases). However, Figure 10.4(b) now tells us more. In fact, $\ell = 0.5$ provides better

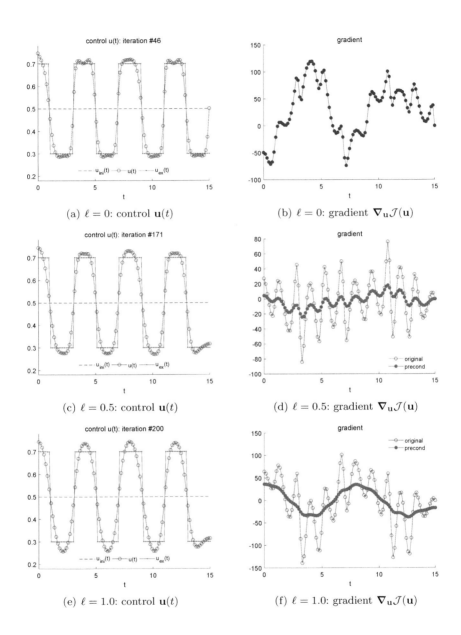

(a) $\ell = 0$: control $\mathbf{u}(t)$

(b) $\ell = 0$: gradient $\boldsymbol{\nabla}_{\mathbf{u}}\mathcal{J}(\mathbf{u})$

(c) $\ell = 0.5$: control $\mathbf{u}(t)$

(d) $\ell = 0.5$: gradient $\boldsymbol{\nabla}_{\mathbf{u}}\mathcal{J}(\mathbf{u})$

(e) $\ell = 1.0$: control $\mathbf{u}(t)$

(f) $\ell = 1.0$: gradient $\boldsymbol{\nabla}_{\mathbf{u}}\mathcal{J}(\mathbf{u})$

FIGURE 10.3

Results of applying preconditioning to model #3 with different smoothing parameter ℓ: (a,b) $\ell = 0$, (c,d) $\ell = 0.5$, and (e,f) $\ell = 1.0$.

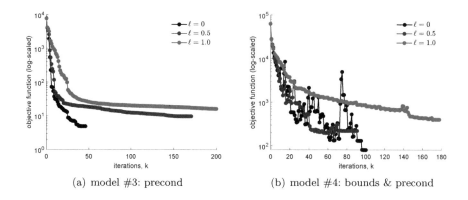

(a) model #3: precond   (b) model #4: bounds & precond

FIGURE 10.4
Performance of (a) preconditioning applied to model #3 and (b) preconditioning paired with the simple projection to control bounds in model #4.

results: it is two times faster and about half-order more accurate in fitting
the data. To convince the readers, we advise them to return to Figure 10.6
and compare plots (c) and (e) carefully to notice a better overall fit (not just
points where the control takes 0 and 1 values).

Finally, we have more to add about the objective function behavior in
Figure 10.4(b). The simple projection controls the bounds by correcting them
manually, not through the information "sent" to the solution in the gradient.
Therefore, it cannot guarantee monotonicity in the objective decrease observed
so far in our previous experiments. In addition, other side effects may include,
e.g., numerical instabilities while solving forward and adjoint problems that
appear after forced changes in the control. The readers may notice unusually
higher computational time for these iterations. If these changes break (even
locally) the control smoothness, ODE solvers will need a longer time to get corresponding solutions for the state and adjoint variables. To help the gradient
scheme "recover" faster after such alterations, we may also advise disallowing
the BB method using the search intervals "inherited" from previous iterations by commenting command `BXini = alphaOpt` in `stepsize_search.m`.
The better advice, however, would be to consider more advanced mechanisms
to control boundness, e.g., the Lagrange multipliers or barrier functions discussed earlier in this chapter.

In the next set of our experiments, we would play more with other available methods to help the adjoint gradients improve solution $\mathbf{u}(t)$ at the final
time, $t = T$. As we checked previously, applying preconditioned gradients re-
mediates this situation, however, at the expense of the entire solution being
"over-smoothed." We now tend to use two regularization approaches, methods #1 and #2, discussed in Section 10.3.1, to deal with this problem explicitly. In the first approach, we apply regularization in the **state space** by

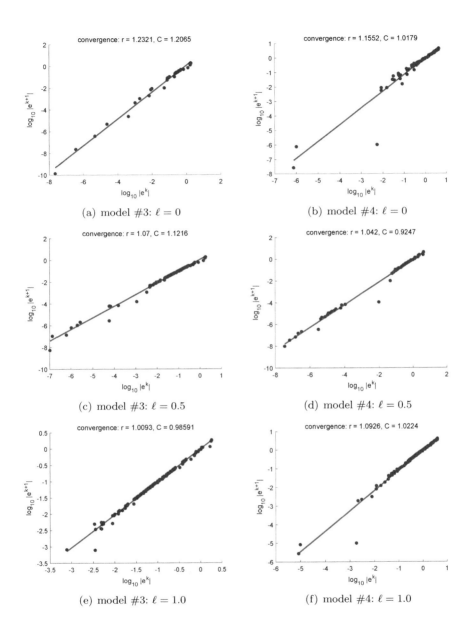

(a) model #3: $\ell = 0$

(b) model #4: $\ell = 0$

(c) model #3: $\ell = 0.5$

(d) model #4: $\ell = 0.5$

(e) model #3: $\ell = 1.0$

(f) model #4: $\ell = 1.0$

FIGURE 10.5
The analysis of the computational convergence evaluated for (a,c,e) model #3 and (b,d,f) model #4 with the use of preconditioning: (a,b) $\ell = 0$, (c,d) $\ell = 0.5$, and (e,f) $\ell = 1.0$.

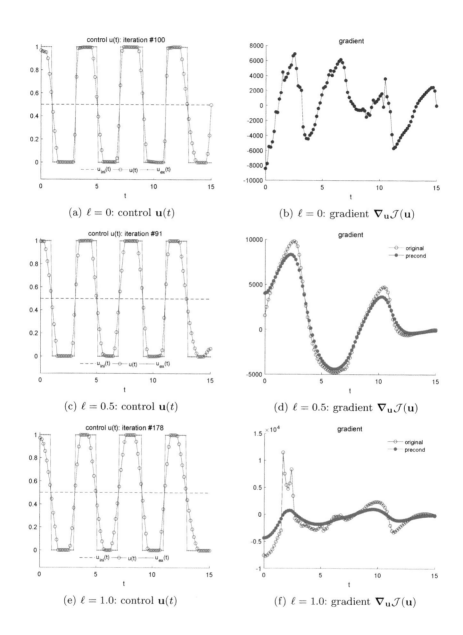

(a) $\ell = 0$: control $\mathbf{u}(t)$

(b) $\ell = 0$: gradient $\boldsymbol{\nabla}_{\mathbf{u}} \mathcal{J}(\mathbf{u})$

(c) $\ell = 0.5$: control $\mathbf{u}(t)$

(d) $\ell = 0.5$: gradient $\boldsymbol{\nabla}_{\mathbf{u}} \mathcal{J}(\mathbf{u})$

(e) $\ell = 1.0$: control $\mathbf{u}(t)$

(f) $\ell = 1.0$: gradient $\boldsymbol{\nabla}_{\mathbf{u}} \mathcal{J}(\mathbf{u})$

FIGURE 10.6

Results of applying preconditioning to model #4 with different smoothing parameter ℓ: (a,b) $\ell = 0$, (c,d) $\ell = 0.5$, and (e,f) $\ell = 1.0$.

augmenting the objective and modifying correspondingly terminal conditions in the adjoint system; refer to (10.32)–(10.33) for review. Here, we are back to our model #3 to "deactivate" simple projection and focus only on the results of the considered regularization. We also need to play with regularization weights β_3 and β_4 to balance the contributions of all parts in the objective: for simplicity, we keep $\beta_3 = \beta_4$.

Figure 10.7 shows the results of three runs with the weights set to 0.1, 1.0, and 10. A regular practice to explore the regularizing effect is to span the coefficients within a broad range and find limiting values experimentally. We will see later that these values may differ significantly (sometimes by several orders of magnitude) for different types of regularization. We explore the solutions on plots (a), (c), and (e), having a particular focus on their quality on the right side. It appears that the value of 0.1 is too small. However, 1.0 and 10 provide much better and about similar quality results. The convergence analysis on plots (b), (d), and (f) confirms the superior performance as it moves from linear ($r = 1.0602$ for a value of 0.1) to superlinear ($r = 1.232$ and $r = 1.2896$ for values of 1.0 and 10, respectively) convergence. If the solution quality satisfies our needs, what to choose? One of the possible recommendations, again, might be checking accuracy vs. the computational time. As seen in Figure 10.8(a), the last two options (1.0 and 10 values) fit the data at the same level of precision; however, optimization with $\beta_3 = \beta_4 = 1.0$ terminates about 20 iterations faster.

Our next step is to experiment with the second approach, where we apply regularization in the **control space** by augmenting the objective and directly modifying the gradient formula; refer to (10.34)–(10.35) for review. The reader may try the same span of values (0.1, 1.0, and 10) applied to the regularization coefficient β_5 to check that now it has almost no effect. As seen in Figures 10.9(a,c,e), the regularization starts at a value of 100 and improves the solution noticeably when β_5 is 1000. In this case, the analysis of the overall optimization performance based on plain numbers only may be deceptive. The convergence results from Figures 10.9(d,f) "vote" toward the last choice: we observe the superlinear ($r = 1.3936$) against linear ($r = 1.0774$) rates. The objectives in Figure 10.8(b) are also supportive: optimization with $\beta_5 = 1000$ terminates faster. Still, we advise looking at Figure 10.9(e) one more time and checking the solution shape. On the right side, it creates a kink as the gradient modification makes the $\mathbf{u}(T)$ part move faster than the rest of the solution in its neighborhood. Here, we refer to this nice situation exemplifying the necessity to check the quality of the final solution (or the intermediate solutions dynamically) versus relying blindly on the numbers.

? *Anyway, how to improve the quality?*

To answer this question, the reader could use all the discussions applied to the results of our prior experiments. For example, we could experiment more with the weight β_5 to find a better value (say, between 100 and 1000). Or,

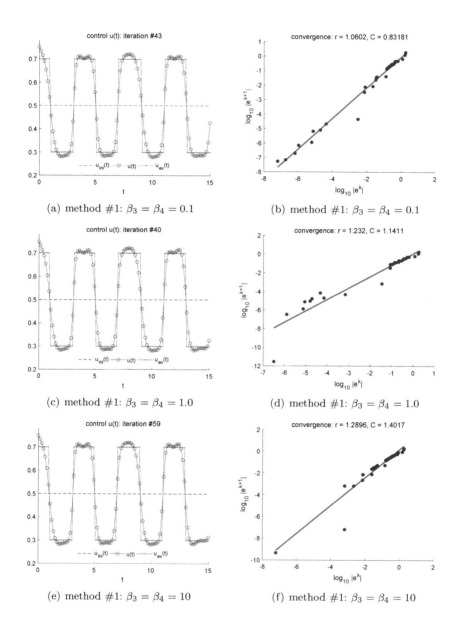

(a) method #1: $\beta_3 = \beta_4 = 0.1$

(b) method #1: $\beta_3 = \beta_4 = 0.1$

(c) method #1: $\beta_3 = \beta_4 = 1.0$

(d) method #1: $\beta_3 = \beta_4 = 1.0$

(e) method #1: $\beta_3 = \beta_4 = 10$

(f) method #1: $\beta_3 = \beta_4 = 10$

FIGURE 10.7

Results of applying regularization in the space of state variables (method #1) for model #3 for various values of $\beta_3 = \beta_4$: (a,b) 0.1, (c,d) 1.0, and (e,f) 10.

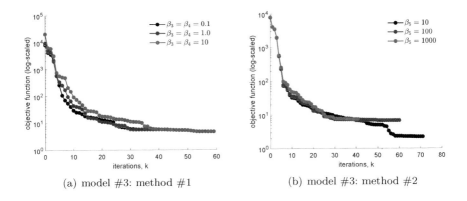

(a) model #3: method #1　　　　　(b) model #3: method #2

FIGURE 10.8
Performance of regularization applied in (a) the state space (method #1) and
(b) the control space (method #2) in optimization for model #3.

one could try to couple this regularization with preconditioning to eliminate,
or smoothen, the kink. However, we should notice that pairing these two very
(mathematically) different approaches is not an easy task – an "optimal"
tandem of found parameters β_5 and ℓ may break apart easily once other
settings in the problem are changed. One of the best suggestions may be to
reconsider the approach and use the regularization of method #1 instead.

? *How about the quality of the gradient itself? May we suspect that adding
regularization on top of the existing adjoint-based procedure spoils the accuracy of computing gradients?*

It is a good point – we will check it now as it is never late to do it. For
example, we could compare two "extreme" cases of applying methods #1 and
#2 to model #3 at the same time with the weak ($\beta_3 = \beta_4 = 0.1$, $\beta_5 = 10$)
and strong ($\beta_3 = \beta_4 = 10$, $\beta_5 = 1000$) regularizing effects applied. The cheap
kappa-test in Figure 10.10(a) confirms the correctness of implementing both
regularization methods with more consistency toward smaller weights as they
ensure lesser "invasion" into the original structure of the adjoint procedure.
However, comparing the behavior of the individual controls in the expensive
test of Figure 10.10(b) does not reveal any difference. It confirms the safe
usability of these types of regularization. Nevertheless, as discussed many
times before, performing gradient testing periodically or after applying any
changes to its computing procedures is a very good habit for protecting the
framework by keeping it error-free.

Our final numerical experiment will check the performance of the discussed regularization methods in application to "simple" model #1. We choose
state-space regularization only (method #1) with $\beta_3 = \beta_4 = 10$ paired with

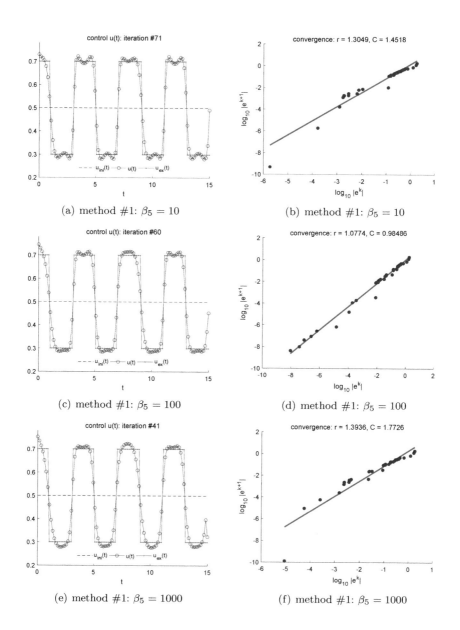

FIGURE 10.9
Results of applying regularization in the space of control variables
(method #2) for model #3 for various values of β_5: (a,b) 10, (c,d) 100, and
(e,f) 1000.

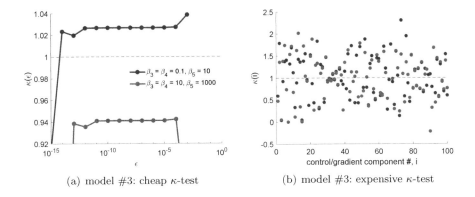

(a) model #3: cheap κ-test (b) model #3: expensive κ-test

FIGURE 10.10
Results of (a) cheap and (b) expensive kappa-tests applied to model #3 with both state- and control-space regularization activated.

preconditioning ($\ell = 1$). This time we run it for longer by allowing the maximum number of iterations to be 500 (kMax = 500) and put all six plots recorded from the full MATLAB window after termination (at $k = 381$) in Figure 10.11. We advise the reader to practice analyzing all available information to see how it is consistent with the previously obtained results and the associated discussions. We would allow just a general comment that these results are good.

? *But are they good enough to satisfy the accuracy requirements set by a potential user while solving a particular problem?*

We also leave this discussion to the readers, advising first to look into the zoomed-out image of the obtained solution in Figure 10.12 and answer the following questions. Why do you need regularization, and what do you expect to satisfy your needs? Are you ready to sacrifice the simplicity of the solutions and, as such, the computational speed for getting more accurate results? As applying and tuning regularization is not an easy problem, your time invested here may really matter.

10.4 Homework Problems

1. Modify MATLAB code Chapter_8_opt_constrained.m by adding the quadratic penalty function method to solve the constrained optimization

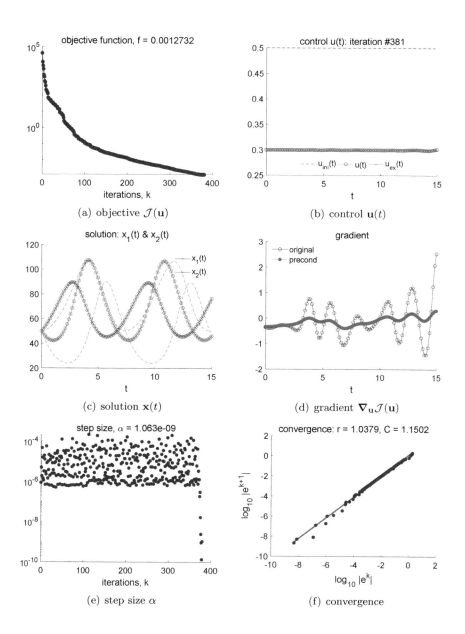

FIGURE 10.11
MATLAB window with six plots updated dynamically (finalized at $k = 381$), while solving Example 9.2 for model #1 using preconditioning ($\ell = 1.0$) and state-space regularization (method #1) with $\beta_3 = \beta_4 = 10$.

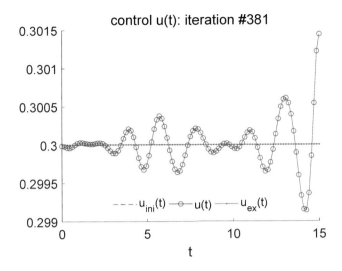

FIGURE 10.12
The zoomed-out image of control $\mathbf{u}(t)$ from Figure 10.11(b).

problem of Example 8.1. Check the results and conclude on the performance.

2. Find all stationary points for the optimization problem of Example 10.1 and check if they are strict local and global minimizers.

3. Modify MATLAB code Chapter_8_opt_constrained.m by adding the barrier functions method to solve the constrained optimization problem of Example 10.1. Check the results and conclude on the performance.

4. Review the derivation of the gradient-based scheme for solving the optimization problem of Example 9.2 to determine necessary modifications. Prove the updated gradient formula (10.19).

5. Prove that projecting gradients from L_2 to Sobolev H^1 space could be done by solving the elliptic BVP (10.23).

6. Derive the gradient-based scheme for solving the optimization problem of Example 9.2 by introducing the slack variable $v(t)$ to incorporate the non-negativeness condition for control $u(t)$.

7. Use MATLAB code Chapter_10_opt_lotka_volterra_enhanced.m for the LV model of Example 9.2 to combine all discussed regularization approaches (preconditioning, simple projection, state- and control-space regularization) by varying the values of their parameters in applications to different models. Conclude on your results (performance, applicability, etc.).

8. Modify MATLAB code `Chapter_10_opt_lotka_volterra_enhanced.m` by implementing boundness conditions $u_\ell \leq u(t) \leq u_u$ using

- slack variables,
- barrier functions method,
- Lagrange multiplier approach, or
- others methods of your choice.

Conclude on implementation, performance, and applicability of each method.

⎡READ⎤ Where to Read More

Bertsekas (2016), [4]
Chapter 5 (Lagrange Multiplier Algorithms)

Boyd (2004), [6]
Chapter 11 (Interior-Point Methods)

Griva (2009), [15]
Chapter 16 (Penalty and Barrier Methods)

Nocedal (2006), [25]
Chapter 17 (Penalty and Augmented Lagrangian Methods)

⎡RUN⎤ MATLAB Codes for Chapter 10

- **root** folder:
    - Chapter_10_opt_lotka_volterra_enhanced.m
    - params.m

- folder `algorithms`:
    - direction_search.m
    - file_output.m
    - golden_section_search.m
    - initialize.m
    - kappa_test.m
    - measurements.m
    - mode_OPT.m
    - mode_TEST.m
    - models_ctrl.m

- – stepsize_search.m
- – visualize.m

- folder functions:

  - – fn_brent.m
  - – fn_convergence_sol_norm.m
  - – fn_eval_grad.m
  - – fn_eval_obj.m
  - – fn_Lotka_Volterra_adj.m
  - – fn_Lotka_Volterra_fwd.m
  - – fn_min_brack.m
  - – fn_precond_bvp4c.m
  - – fn_precond_fd.m
  - – FUNC.m

- folder output

- folder temp

10.5   Lab Assignment #4: Review Chapters 8–10

For all three problems, consider the optimization problem based on the LV model of Example 9.2 in Chapter 9. Assume that synthetic data $\tilde{x}_1(t)$ and $\tilde{x}_2(t)$ are available continuously over the interval $[0, 15]$, and all constants are given as follows: $\alpha = 1.0, \beta = 0.01, \gamma = 1.0, \delta = 0.02$, and $x_1^0 = x_2^0 = 50$.

Problem 1: Modify MATLAB code `Chapter_10_opt_lotka_volterra_enhanced.m` to implement boundness conditions $0 \leq u(t) \leq 1$ using the barrier functions method discussed in Section 10.2.2. Present the results of optimization applied to model #4 before and after implementing these constraints (graphs for objective, optimal solutions for control and state variables, and gradient structure). Compare the results with those obtained previously with the simple projection method. Approximate convergence parameter r in each case. Compare the performance and conclude on both positive and negative aspects of implementing constraints by both approaches.

Keep the boundness conditions implemented in Problem 1 while solving Problems 2 and 3.

Problem 2: Show how regularization for both state (method #1) and control (method #2) spaces used to improve the solution $\mathbf{u}(t)$ at $t = T$ changes the adjoint-based algorithm for solving the LV model optimization problem. Use MATLAB code updated in Problem 1 to run optimization using the following regularization weights:

- for state-space: $\beta_3 = \beta_4 = 10^{-2}, 10^{-1}, 10^0, 10^1, 10^2, 10^3$,

- for control-space $\beta_5 = 10^0, 10^1, 10^2, 10^3, 10^4, 10^5$.

Present the results of optimization (graphs for objective, optimal solutions for control and state variables, and gradient structure) in applications to models #4 and #1 for every case. Conclude on both positive and negative aspects of implemented regularization techniques on the quality of the obtained solutions and the overall performance of the optimization process.

Do not use in Problem 3 any regularization added while solving Problem 2.

Problem 3: Use MATLAB code updated in Problem 1 to run optimization for model #4 and compare the methods for projecting gradients from

L_2 to Sobolev H^1 space by solving an elliptic BVP (10.23) using built-in MATLAB function bvp4c and the FD linear algebra scheme discussed in Section 10.3.1. For both methods, perform optimization with different values of parameter $\ell = 0.1, 0.5, 1.0, 5.0$. Present the results of optimization (graphs for objective, optimal solutions for control and state variables, and gradient structure) for all cases. Conclude on both positive and negative aspects of the implemented preconditioning techniques. Discuss the quality of the obtained solutions and the overall performance of the optimization process.

For all computations, use the following:

- number of discrete intervals in t-discretization $n = 100$,

- initial guess $\mathbf{u}^0 = 0.5$ and objective weights $\beta_1 = \beta_2 = 1.0$,

- termination parameters for optimization: $\epsilon_1 = \epsilon_2 = 10^{-9}$ and $k_{max} = 200$,

- BFGS and BB methods for finding, respectively, search directions and optimal steps with the following parameters: (BFGS) restarts every 5 iterations; (BB) initial interval $[0, 10^{-6}]$; bracketing: MAXITER = 20, GLIMIT = 100.0; Brent: TOL = 1e-9, ITMAX = 2.

11

Moving to PDE-based Optimization

This chapter starts a smooth transition from ODE-based to PDE-based optimization. First, we generalize the problem of identifying multiple parameters in the differential equations by reconsidering their solutions as functions of several variables based on the example of the parabolic PDE. The reader will have enough space for practicing both theoretically and computationally, while moving forward and exploring the process of reconfiguring the existing computational framework to enable it to solve PDE-based optimization problems. Second, we create a simplified practice example using an elliptic PDE as a forward problem to evaluate the scope of the required changes. Here, we review the "optimize–then–discretize" concept by deriving the adjoint-based gradient formula and constructing the gradient-based algorithm for this example. At this point, we naturally come up with the idea of employing external software, namely `FreeFEM`, to support our solutions for PDEs in higher dimensions. A brief introduction to the Finite Elements Method is included in this chapter to familiarize the readers with its principles which support discretization used for numerical PDE solutions. Finally, we concisely discuss the `FreeFEM` functionalities and coding technicalities required for solving a single PDE problem and also for including `FreeFEM` into the existing computational framework as a PDE solver anticipated in the next chapter.

11.1   Generalized Problem of Fitting Data

In Chapters 9 and 10, we discussed solution methods applied to optimization problems constrained by (the systems of) ODEs. We also modified our existing computational framework to enable such solutions with some requirements for their properties. So far, we can utilize `MATLAB`'s built-in functionalities, while constructing all components of our framework, including ones for solving ODEs as forward and adjoint problems. Now it is time to move forward up to a level of the PDE-based optimization by analyzing the required changes and examining practically how smoothly the created computational structure allows the new portion of these modifications. We assume that, at this point, the reader is comfortable enough with the *continuous formulation* of the DE-based optimization problems and has collected sufficient experience with solving them in discrete settings. Therefore, we consider the following generalization of the (multiple) *parameter identification* problems in parabolic PDEs by fitting available (probably, continuous) data.

Example 11.1 Identifying Multiple Parameters in Parabolic PDEs
Identify multiple parameters, e.g., functions $a(x,t), b(x,t), c(x,t)$, and $f(x,t)$ in a parabolic PDE by fitting (continuous) data $\tilde{u}(x,t)$

$$\min_{\mathbf{v}(x,t)} \ \mathcal{J}(u(x,t); \mathbf{v}(x,t)) = \frac{1}{2} \int_0^T \int_\Omega (u(x,t) - \tilde{u})^2 \, dx \, dt. \tag{11.1}$$

(a) In (11.1), data $\tilde{u}(x,t)$ is available continuously at each (x,t), and

(b) the general form of the equation of state (governing PDE) is given by

$$\left(a(x,t)\,u_x\right)_x + b(x,t)\,u_x + c(x,t)\,u - u_t = f(x,t), \quad x \in \Omega, \ t \in [0,T] \tag{11.2}$$

subject to the properly chosen boundary and initial conditions.

(c) We define $\Omega \subseteq \mathbb{R}^n$ to be the domain of problem (11.2), where $u(x,t)$ and $a(x,t), b(x,t), c(x,t), f(x,t)$ are the state and control variables, respectively; i.e., $u, a, b, c, f : \Omega \times [0,T] \to \mathbb{R}$.

(d) Finally, we determine the general form of the control vector (set) as

$$\mathbf{v}(x,t) = \{a(x,t), b(x,t), c(x,t), f(x,t), \ldots\} \tag{11.3}$$

by assuming that other functions (or parameters) from the problem (11.2) or its boundary and initial conditions may be added to the set (11.3) as additional controls.

Bearing in mind the current state of our computational framework, we are interested in solving the optimization problem of Example 11.1 using the adjoint-based gradient method by the employed Lagrange multiplier approach discussed in detail in Chapters 8 and 9. To diversify the experience gained from applying this approach previously, we will derive gradients using the same *"optimize–then–discretize"* concept, where optimality conditions and the optimization algorithm are both based on the continuous form of the problem.

11.2   Practice Example

11.2.1   Elliptic Equation as Governing PDE

Although the comparison of Examples 11.1 and 9.2 in terms of the increased level of complexity may be a bit scary, it should not discourage the reader in any way. We could start and experiment both theoretically and numerically with a simplified problem that keeps the constraint in (11.1) in the PDE format. By doing that, we remove the time dependency and set all controls to zero except function $a(x)$. Whenever the readers feel comfortable to proceed to the next level of complexity, they may do it themselves easily by adding several more layers of sophistication (i.e., time and other removed controls), reverting the problem to its original definition as stated in Example 11.1. Therefore, we proceed with the following example we will use onward.

Example 11.2 Identifying Parameters in Elliptic PDEs
By minimizing the following objective function

$$J(u(x); a(x)) = \frac{1}{2} \int_\Omega (u(x) - \tilde{u})^2 \, dx, \qquad (11.4)$$

identify the space-dependent parameter, i.e., function $a(x)$ in the following elliptic PDE (forward problem)

$$\begin{aligned} \nabla \cdot [a(x)\nabla u(x)] &= f(x), & x \in \Omega \\ u(x) &= \phi(x), & x \in \partial\Omega \end{aligned} \qquad (11.5)$$

by fitting available (continuous) data $\tilde{u}(x)$ and assuming $\Omega \subseteq \mathbb{R}^n$, $n = 2, 3$, $u, a, f : \Omega \to \mathbb{R}$, and $\phi : \partial\Omega \to \mathbb{R}$. We also assume functions $f(x)$ and $\phi(x)$ to be given.

We notice the consistency of this problem with the problem of Example 11.1. They both have the same state variable, $u(x)$. However, the former has function $a(x)$ as the only control variable with the associated *control space*

$$\mathcal{V} = \{a(x) \in L_2 : a(x) > 0, \ x \in \Omega\}. \qquad (11.6)$$

11.2.2   Deriving Gradient by Optimize–then–Discretize

In this section, we will follow the same methodology to derive an analytical formula for the adjoint-based gradients consistent with the "optimize–then–discretize" concept. Before we start, we advise the reader to review the examples of such derivation in Chapter 9 to compare its application to the PDE-based problems and notice the similarity in the structure.

As before, we define the *Lagrangian (augmented objective function)*

$$
\begin{aligned}
\mathcal{L}(u(x); a(x), \psi(x)) &= \mathcal{J}(u(x); a(x)) \\
&\quad + \langle \boldsymbol{\nabla} \cdot [a(x)\boldsymbol{\nabla} u(x)] - f(x),\ \psi(x) \rangle_\chi,
\end{aligned}
\tag{11.7}
$$

where $\psi(x)$ is the *adjoint variable (Lagrange multiplier)*, and we use the notation $\langle \cdot,\ \cdot \rangle_\chi$ to define the inner product in χ-space. For simplicity, we consider $\chi = L_2$; review also (9.5) for the corresponding definition used to determine the exact form of the Lagrangian (11.7) in the L_2 functional space, i.e.,

$$
\mathcal{L}(u; a, \psi) = \frac{1}{2} \int_\Omega (u - \tilde{u})^2\, dx + \int_\Omega [\boldsymbol{\nabla} \cdot (a\boldsymbol{\nabla} u) - f]\, \psi\, dx.
\tag{11.8}
$$

The 1-order *total variation* for this Lagrangian

$$
\begin{aligned}
\delta\mathcal{L}(u; a, \psi) &= \int_\Omega (u - \tilde{u})\, \delta u\, dx + \int_\Omega [\boldsymbol{\nabla} \cdot (\delta a \boldsymbol{\nabla} u)]\, \psi\, dx \\
&\quad + \int_\Omega [\boldsymbol{\nabla} \cdot (a\boldsymbol{\nabla} \delta u)]\, \psi\, dx + \int_\Omega [\boldsymbol{\nabla} \cdot (a\boldsymbol{\nabla} u) - f]\, \delta\psi\, dx
\end{aligned}
\tag{11.9}
$$

should be consistent with the structure provided by the Riesz Representation Theorem in (9.11), i.e.,

$$
\delta\mathcal{L}(u; a, \psi) = \int_\Omega \boldsymbol{\nabla}_a \mathcal{L}\, \delta a\, dx + \int_\Omega \boldsymbol{\nabla}_u \mathcal{L}\, \delta u\, dx + \int_\Omega \boldsymbol{\nabla}_\psi \mathcal{L}\, \delta\psi\, dx
\tag{11.10}
$$

and set to 0 as required by the KKT optimality conditions discussed in Chapter 8.

Due to the forward problem (11.5), the fourth term on the right side of (11.9) is zero:

$$
\int_\Omega [\boldsymbol{\nabla} \cdot (a\boldsymbol{\nabla} u) - f]\, \delta\psi\, dx = 0.
\tag{11.11}
$$

In addition, parts $\boldsymbol{\nabla} \cdot (\delta a \boldsymbol{\nabla} u)$ and $\boldsymbol{\nabla} \cdot (a\boldsymbol{\nabla} \delta u)$ (in the second and third terms of (11.9), respectively) are not consistent with the structure of (11.10). We use the integration by parts one time for the second term, i.e.,

$$
\int_\Omega [\boldsymbol{\nabla} \cdot (\delta a \boldsymbol{\nabla} u)]\, \psi\, dx = \int_{\partial\Omega} \psi \frac{\partial u}{\partial n}\, \delta a\, ds - \int_\Omega \boldsymbol{\nabla} u \cdot \boldsymbol{\nabla} \psi\, \delta a\, dx,
\tag{11.12}
$$

and two times for the third one, i.e.,

$$
\begin{aligned}
\int_\Omega [\boldsymbol{\nabla} \cdot (a\boldsymbol{\nabla} \delta u)]\, \psi\, dx &= \int_{\partial\Omega} a\, \psi \frac{\partial \delta u}{\partial n}\, ds - \int_\Omega a\boldsymbol{\nabla}\psi \cdot \boldsymbol{\nabla}\delta u\, dx \\
&= \int_{\partial\Omega} a\, \psi \frac{\partial \delta u}{\partial n}\, ds - \int_{\partial\Omega} a \frac{\partial \psi}{\partial n}\, \delta u\, ds + \int_\Omega \boldsymbol{\nabla} \cdot (a\boldsymbol{\nabla}\psi)\, \delta u\, dx
\end{aligned}
\tag{11.13}
$$

to get rid of the divergence ($\boldsymbol{\nabla} \cdot \ldots$) in both. Here, we use the notation $\frac{\partial}{\partial n}$ to define the normal derivative at boundary $\partial\Omega$.

Finally, the second boundary term $a\frac{\partial \psi}{\partial n}\delta u\big|_{\partial\Omega}$ in (11.13) is zero due to the perturbation system (more specifically, its homogeneous boundary condition)

$$
\begin{aligned}
\boldsymbol{\nabla} \cdot [a\,\boldsymbol{\nabla}\delta u + \delta a\,\boldsymbol{\nabla} u] &= 0, && x \in \Omega \\
\delta u &= 0, && x \in \partial\Omega
\end{aligned}
\tag{11.14}
$$

obtained by perturbing the forward problem (11.5). And now we have (11.9) in the fully "Riesz-consistent" form

$$
\begin{aligned}
\delta\mathcal{L}(u; a, \psi) = &\int_\Omega (u - \tilde{u})\,\delta u\,dx + \int_{\partial\Omega} \psi \frac{\partial u}{\partial n}\delta a\,ds \\
&- \int_\Omega \boldsymbol{\nabla} u \cdot \boldsymbol{\nabla}\psi\,\delta a\,dx + \int_{\partial\Omega} a\,\psi \frac{\partial \delta u}{\partial n}\,ds + \int_\Omega \boldsymbol{\nabla} \cdot (a\boldsymbol{\nabla}\psi)\,\delta u\,dx
\end{aligned}
\tag{11.15}
$$

that is confirmed after grouping all terms in (11.15) by factoring δa and δu:

$$
\begin{aligned}
\delta\mathcal{L}(u; a, \psi) = &- \int_\Omega \boldsymbol{\nabla} u \cdot \boldsymbol{\nabla}\psi\,\delta a\,dx + \int_\Omega [\boldsymbol{\nabla} \cdot (a\boldsymbol{\nabla}\psi) + (u - \tilde{u})]\,\delta u\,dx \\
&+ \int_{\partial\Omega} \psi \left[\frac{\partial u}{\partial n}\delta a + a\,\frac{\partial \delta u}{\partial n}\right]\,ds.
\end{aligned}
\tag{11.16}
$$

11.2.3 Optimization Algorithm

As before, the final form of $\delta\mathcal{L}(u; a, \psi)$ obtained in (11.16) provides an analytical formula for the adjoint-based gradient derived in the L_2 functional space

$$
\boldsymbol{\nabla}_a \mathcal{L} = -\boldsymbol{\nabla} u(x) \cdot \boldsymbol{\nabla}\psi(x)
\tag{11.17}
$$

based on the solution of the adjoint PDE problem

$$
\begin{aligned}
\boldsymbol{\nabla} \cdot [a(x)\boldsymbol{\nabla}\psi(x)] &= \tilde{u} - u(x), && x \in \Omega \\
\psi(x) &= 0, && x \in \partial\Omega
\end{aligned}
\tag{11.18}
$$

to be solved to find adjoint state $\psi(x)$ with homogeneous boundary conditions.

While playing with the adjoints in Chapter 9, right at this point, we discussed the discretization phase in applications to t-domain, which is usually a 1D continuous interval. Discretization was also applied to the states, controls, and adjoints, which were all 1D functions of time. For the current problem, first, we have to discretize the space variable x subject to the structure of its domain Ω. Even in 2D, it might not be an easy task to perform. Moving to 3D and complicating shapes for Ω make this procedure rather challenging. We also note that applying FD schemes to solving PDEs is not trivial due to numerical stability and convergence issues.

> **?** *Therefore, how to proceed with the discretization phase applied for space, state, adjoint, and control variables, also for the objective, and both PDEs representing the forward and adjoint problems?*

We will discuss possible solutions with some examples in the rest of this chapter. To add more to this discussion, we now summarize all steps in the practical implementation of the "optimize–then–discretize" approach to perform numerical computations for solving the optimization problem of Example 11.2 in Algorithm 11.1 below.

Algorithm 11.1 ("Optimize–then–Discretize" for Example 11.2)

1. *Discretize spatial domain Ω for space variable x and initialize discretized state u, control a, and adjoint ψ*

2. *Obtain and store measurement (analytic/synthetic) data $\tilde{u} = \tilde{u}(\tilde{x})$*

3. *Set $k = 0$ and choose initial guess $a^0 = a_{ini}$ for control*

4. *Solve numerically (discretized) forward problem (11.5) to find state u^k*

5. *Evaluate objective $\mathcal{J}(u^k; a^k)$ subject to the chosen discretization*

6. *For $k = 1, 2, \ldots$ check optimality of a^k; if optimal $\Rightarrow$* **TERMINATE**

7. *Solve numerically (discretized) adjoint problem (11.18) to find adjoint state ψ^k*

8. *Evaluate gradient by*

$$\nabla_a \mathcal{L}(u^k; a^k, \psi^k) = -\nabla u^k \circ \nabla \psi^k, \qquad (11.19)$$

   where $\circ$ denotes Hadamard (element-wise, entrywise, or Schur) product

9. *Improve solution by finding optimal step size α^k*

$$a^{k+1} = a^k - \alpha^k \nabla_a \mathcal{L}(u^k; a^k, \psi^k) \qquad (11.20)$$

10. *Set $k \leftarrow k + 1$ and go to Step 4*

Finally, we repeat our comments made previously to reinforce them in applications to Algorithm 11.1. First, Step 9 could be practically implemented by using any gradient-based strategies and methods for optimal step size α discussed in Chapters 5–7. Second, the reader should not be confused by using gradient $\nabla_a \mathcal{L}$ obtained for Lagrangian $\mathcal{L}$ while evaluating the original objective $\mathcal{J}$. The adjoint analysis used for deriving the gradients here sets $\nabla_a \mathcal{L}$ to zero and, as such, makes the Lagrangian in (11.8) equivalent to the original objective in (11.4) by assuming that the numerical solution of the forward problem (11.5) is fairly accurate. The final comment is about our notations used for all functions (including gradients) discretized spatially over domain Ω at iteration k, e.g., a^k (with no bold font) instead of $\mathbf{a}^k$. Here and in the next chapter, we get rid of the bold font as spatial discretization will be according to the description of the finite element spaces used for particular functions. We discuss the necessity for such discretization and the techniques for its practical implementation in the following sections.

11.3   Solving PDEs in Higher Dimensions by FreeFEM

Now we are back to the question posed on p. 320. We advise the reader to review the discussion aspects of Section 3.2 (Generalized Optimization Framework) and, in particular, efficient data processing and communication functionality of the f- (in our case $\mathcal{J}$-) and d-evaluators discussed in Section 3.2.2 (Choice of Proper Software) with requirements to

- evaluate objectives $\mathcal{J}(u; a)$ and find search directions d,

- solve (systems of) (non)linear equation(s), including ODE(s) and PDE(s), and

- communicate effectively with each other.

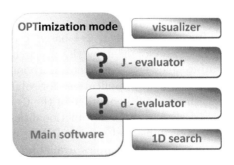

FIGURE 11.1
Generalized optimization framework: main (core) software module (updated from Figure 3.3 to focus on the parts that require further discussion.

After reviewing the structure of our practice example created in Section 11.2 (Example 11.2) and all computational steps in the solution algorithm (Algorithm 11.1), we arrive at the following **general requirements** for the searched software. In brief, it should be able to

(1) discretize 2D and 3D domains of various complexity,

(2) solve both forward and adjoint PDEs in those (discretized) domains, and

(3) accurately evaluate analytically given objectives and gradients using the same discretization.

In addition, we might also consider additional benefits of using the selected software to make the integration into the existing framework painless and as simple as possible, e.g.,

(a) a straightforward description of complex domains,

(b) automated discretization applicable to domains of different complexity,

(c) techniques to interpolate solutions between different discretizations,

(d) ability to discretize and solve PDEs of various complexity,

(e) availability of integrated solvers and libraries for linear algebra, optimization, etc., and

(f) visualization facilities and/or the ability to convert solutions into commonly used graphical formats.

The readers may already have some ideas about a particular type of software based on their personal experience, computing preferences, and the targeted problem. All of these choices may benefit at different levels. As a practical example to discuss the integration procedure, we choose FreeFEM, an open-source finite element (FE) integrated development environment; see [16] for details.

The open-source software FreeFEM, a C++-based programming environment, has a long history of development by the research group of Professor Frederic Hecht (Laboratoire Jacques-Louis Lions, Sorbonne University, Paris) and a significant record of successful applications for solving problems of various complexity. Even with minimal knowledge of the C++ syntax, the reader may start using the pre-designed FreeFEM functionalities (e.g., defining domains, creating meshes, describing and solving PDEs, etc.) after reviewing and running some of the provided examples available on the main website https://freefem.org/. To confirm the consistency of the current FreeFEM's capabilities with our requirements mentioned in the previous section, we quote the following introduction statements taken from its manual [17]:

- "FreeFEM is a partial differential equation solver for nonlinear multiphysics systems in 1D, 2D, 3D, and 3D border domains (surface and curve)."

- "Problems involving partial differential equations ... require interpolations of data on several meshes and their manipulation within one program. FreeFEM includes a fast interpolation algorithm and a language for the manipulation of data on multiple meshes."

- "FreeFEM is written in C++ and its language is a C++ idiom."

We also find convincing the facts that it currently interfaces to more than 20 various libraries and runs on Linux®, macOS®, and Windows® machines. The reader may find the latest version of the FreeFEM manual [17] added to the files with examples related to this chapter or download it directly from https://doc.freefem.org/documentation/index.html. Another good source of learning about FreeFEM and its applications is an archive of the annual FreeFEM Days Conferences; information since 2012 is available here: https://github.com/FreeFem/FreeFem-days/blob/master/README.md.

To start practicing with FreeFEM by solving simple PDE problems (in 2D or 3D), we recommend the reader to download and install the latest version of FreeFEM at https://doc.freefem.org/introduction/installation.html. We will discuss some examples and provide some advice on the FreeFEM technicalities after a brief introduction to the underlying Finite Element Method (FEM) in the next section.

11.4 Brief Introduction to Finite Elements Method

Generating finite elements and associated meshes in higher dimensions (2D and 3D) by hand is impractical and hardly achievable due to computational loads increasing enormously with the number of such elements. However, we could find some advantages in learning the basics of the FEM to understand the principles of creating meshes and discretizing PDEs over the sets of finite elements. For that reason, we use a simplified version of the forward problem (11.5) in 1D ($n = 1$) with $a(x) = -1$ as a learning example in the form of the following BVP:

$$-u''(x) = f(x), \quad 0 < x < 1,$$
$$u(0) = 0, \quad u(1) = 0. \tag{11.21}$$

To solve (11.21) numerically by *Galerkin finite element method*, we construct the *variational* (or *weak*) *formulation* for that equation by multiplying its both sides by *test function* $v(x)$, satisfying boundary conditions $v(0) = 0$ and $v(1) = 0$ (**mandatory requirement** imposed on all test functions), i.e.,

$$-u''v = fv. \tag{11.22}$$

Then, we integrate (11.22) over the domain $(0, 1)$. Integration by parts applied on the left-hand side gives

$$-\int_0^1 u''v\,dx = -[u'v]_0^1 + \int_0^1 u'v'\,dx = \int_0^1 u'v'\,dx. \tag{11.23}$$

Thus, the *weak form* of (11.21) is obtained as follows:

$$\int_0^1 u'v'\,dx = \int_0^1 fv\,dx. \tag{11.24}$$

In the next step, we generate mesh (e.g., uniform Cartesian)

$$x_i = x_0 + ih, \quad i = 1, 2, \ldots, n, \quad h = \frac{1}{n}, \quad x_0 = 0, \tag{11.25}$$

and define intervals (x_{i-1}, x_i) which we will call *finite elements*. Based on mesh (11.25), we construct a set of *basis functions* (e.g., piecewise linear (hat) functions; refer to Figure 11.2) for $i = 1, 2, \ldots, n-1$, i.e.,

$$\phi_i(x) = \begin{cases} \dfrac{x - x_{i-1}}{h}, & x_{i-1} \le x < x_i \\[2mm] \dfrac{x_{i+1} - x}{h}, & x_i \le x < x_{i+1} \\[2mm] 0, & \text{otherwise} \end{cases} \tag{11.26}$$

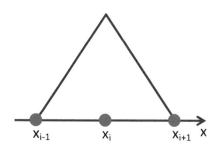

FIGURE 11.2
Piecewise linear (hat) function $\phi_i(x)$.

The approximate (or FE) solution $u_h(x)$ of the problem (11.21) now could be represented by the following linear combination of basis functions

$$u_h(x) = \sum_{j=1}^{n-1} c_j \phi_j(x), \tag{11.27}$$

where coefficients c_j are unknowns to be determined. We derive a linear system of equations for coefficients c_j by substituting exact solution $u(x)$ in the weak form (11.24) by its approximation (11.27), i.e.,

$$\int_0^1 u_h' v' \, dx = \int_0^1 \sum_{j=1}^{n-1} c_j \phi_j' v' \, dx = \sum_{j=1}^{n-1} c_j \int_0^1 \phi_j' v' \, dx = \int_0^1 f v \, dx. \quad (11.28)$$

In (11.28), we choose test function $v(x)$ as $\phi_1, \phi_2, \ldots, \phi_{n-1}$ successively to obtain the explicit form of the linear system:

$$c_1 \int_0^1 \phi_1' \phi_1' \, dx + \ldots + c_{n-1} \int_0^1 \phi_1' \phi_{n-1}' \, dx = \int_0^1 f\phi_1 dx,$$

$$c_1 \int_0^1 \phi_2' \phi_1' \, dx + \ldots + c_{n-1} \int_0^1 \phi_2' \phi_{n-1}' \, dx = \int_0^1 f\phi_2 dx,$$

$$\ldots$$

$$c_1 \int_0^1 \phi_{n-1}' \phi_1' \, dx + \ldots + c_{n-1} \int_0^1 \phi_{n-1}' \phi_{n-1}' \, dx = \int_0^1 f\phi_{n-1} dx.$$

$$(11.29)$$

To make (11.29) compact, we may rewrite this system in the matrix-vector form

$$\begin{bmatrix} (\phi_1, \phi_1) & (\phi_1, \phi_2) & \cdots & (\phi_1, \phi_{n-1}) \\ (\phi_2, \phi_1) & (\phi_2, \phi_2) & \cdots & (\phi_2, \phi_{n-1}) \\ \cdots & \cdots & \cdots & \cdots \\ (\phi_{n-1}, \phi_1) & (\phi_{n-1}, \phi_2) & \cdots & (\phi_{n-1}, \phi_{n-1}) \end{bmatrix} \begin{bmatrix} c_1 \\ c_2 \\ \cdots \\ c_{n-1} \end{bmatrix} = \begin{bmatrix} (f, \phi_1) \\ (f, \phi_2) \\ \cdots \\ (f, \phi_{n-1}) \end{bmatrix}$$

$$(11.30)$$

by defining the *bilinear* (ϕ_i, ϕ_j) and *linear* (f, ϕ_i) *forms* in the following manner:

$$(\phi_i, \phi_j) = \int_0^1 \phi_i' \phi_j' \, dx, \qquad (f, \phi_i) = \int_0^1 f\phi_i \, dx. \quad (11.31)$$

Finally, by solving linear system (11.30)–(11.31) for coefficients c_j, we obtain the finite element (approximate) solution in the form (11.27) and, if necessary, carry out a priori or a posteriori error analysis.

11.5 Solving PDEs by FreeFEM

11.5.1 Poisson's Equation in 2D

In Chapter 12, we will use FreeFEM as a PDE solver to upgrade our computational framework for solving the optimization problem of Example 11.2. Before proceeding to this step, we have to familiarize ourselves with this solver: discuss the technicalities of the coding process and analyze the solutions obtained for single PDEs. This exploration requires choosing a model that we suggest to be close enough to the forward problem (11.5). We consider the following one.

Example 11.3 Solving Elliptic PDE

Solve the given 2D Poisson's equation inside the unit circle centered at the origin

$$\nabla \cdot [a(x,y)\nabla u(x,y)] = f(x,y), \qquad (x,y) \in \Omega$$
$$u(x,y) = \phi(x,y), \qquad (x,y) \in \partial\Omega \tag{11.32}$$

for given functions

$$a(x,y) = 2 - x^2 + y^2,$$
$$f(x,y) = xy, \tag{11.33}$$
$$\phi(x,y) = 0.$$

To move this example into `FreeFEM`, we define the domain Ω as the unit circle $x^2 + y^2 \leq 1$ and parameterize its boundary $\partial\Omega$, e.g.,

$$\partial\Omega = \{(x,y) : x = \cos(t), \ y = \sin(t), \ 0 \leq t \leq 2\pi\} . \tag{11.34}$$

Here, we note that getting all boundaries parameterized is the simplest way of describing the PDE domain to `FreeFEM`. As this solver uses the FEM concept for solving PDEs, all equations should take their weak formulation discussed in Section 11.4 before being entered into the `FreeFEM` coding. This formulation for the problem (11.32) in Example 11.3 takes the following form:

$$\int_\Omega a\nabla u\nabla v \, dxdy + \int_\Omega fv \, dxdy = 0, \tag{11.35}$$

or using the partial derivative notation

$$\int_\Omega a\left(\frac{\partial u}{\partial x}\frac{\partial v}{\partial x} + \frac{\partial u}{\partial y}\frac{\partial v}{\partial y}\right) dxdy + \int_\Omega fv \, dxdy = 0. \tag{11.36}$$

In (11.35)–(11.36), both u and v functions belong to the same FE-space V_h and also

$$v(x,y) = 0, \ \forall(x,y) \in \partial\Omega$$

to satisfy the requirements imposed on the test functions. We leave the proof to the reader as homework for a better understanding of the FEM theory.

11.5.2 Coding with FreeFEM

Here, we create an example of the FreeFEM coding to solve the PDE problem of Example 11.3, including everything from meshing the domain to solving the PDE and plotting the solutions. The reader may run the provided EDP-file (FreeFEM_PDE_Example.edp), e.g., it is an executable file in Windows: double click will automatically start compiling and running the script. For editing, we could use any text editor or specialized software, e.g., Microsoft® Visual Studio® or Visual Studio Code® (free downloads available here: https://visualstudio.microsoft.com/downloads/).

FreeFEM: FreeFEM_PDE_Example.edp

```
// defining boundary of 2D domain                                    1
border C(t=0,2*pi){x=cos(t); y=sin(t); label=1;}                     2
                                                                     3
// creating & visualizing mesh: triangulated domain Th               4
mesh Th = buildmesh(C(50));                                          5
plot(Th, wait=true);                                                 6
                                                                     7
// creating FE-space Vh defined over Th                              8
fespace Vh(Th,P1);                                                   9
Vh u, v; // defining u and v as piecewise P1-continuous functions   10
                                                                    11
// defining analytical functions                                    12
func a = 2 - x^2 + y^2;                                             13
func f= x*y;                                                        14
func phi = 0;                                                       15
                                                                    16
// defining and solving PDE                                         17
solve Poisson(u,v,solver=UMFPACK)                                   18
  = int2d(Th)(a*(dx(u)*dx(v) + dy(u)*dy(v)))  // bilinear part      19
  + int2d(Th)(f*v)                            // linear part (RHS)  20
  + on(C, u = phi);                           // Dirichlet BC       21
                                                                    22
plot(u,fill=1); // plotting solution                               23
```

We encourage the reader to refer to the FreeFEM manual [17] for more details relevant to all solution steps shown in the provided code. We limit our discussion by briefly commenting on these steps in the list below.

line 2 *Defining boundary of 2D domain* (keyword border)

We describe boundary $\partial\Omega$ analytically by the parametric equations for x and y, e.g., as provided by (11.34). In case the boundary consists of several parts, i.e., when $\partial\Omega = \bigcup_{i=1}^{n} \Gamma_i$, then each curve Γ_i must be specified, and intersections of Γ_i are **not allowed** except at their endpoints. Boundaries can be referred to either by names or label numbers (in our case, label=1).

lines 5-6 *Creating and plotting mesh* (keywords `mesh`, `buildmesh`, and `plot`)
Triangulation T_h of domain Ω is generated **automatically** by using `buildmesh` and a specified number of (partial) intervals on each curve Γ_i (in our case, 50). It is also assumed that the solution domain is **on the left side** of the boundary; it is implicitly oriented by the given parameterization.

line 9 *Creating FE-space V_h defined over triangulated domain T_h* (keyword `fespace`)
FE-space is usually a space of polynomial functions on elements (in 2D, triangles) with certain matching properties at edges, vertices, etc. Some examples of `FreeFEM` elements used in 2D are

- P0 piecewise constant,
- P1 continuous piecewise linear (used in our example),
- P2 continuous piecewise quadratic;
- see the complete list in the `FreeFEM` manual [17].

line 10 *Defining u and v functions in FE-space V_h*
This line declares that u and v are approximated as follows

$$u(x,y) \approx u_h(x,y) = \sum_{j=1}^{n-1} u_j \phi_j(x,y),$$

refer to (11.25)–(11.27) for more details.

lines 13-15 *Defining analytical functions* (keyword `func`)
Here, we use the keyword `func`. However, these functions could also be defined as the FE-space functions (e.g., by adding them to line 10).

lines 18-20 *Solving PDE* (keywords `solve` and `int2d`)
We define bilinear (u,v) and linear (f,v) forms of the equation (11.32) and its weak formulation

$$(u,v) - (f,v) = 0,$$

where bilinear and linear terms **should not be under the same integral!** Refer to (11.28)–(11.31) for more details.

line 21 *Defining boundary conditions* (keyword `on`)

- For Dirichlet-type conditions, we use keyword `on` (as in our current example),
- for other conditions (Neumann and Robin), they should be applied while creating the weak form of the PDE.

line 23 *Visualizing solutions* (keyword `plot`)
There are multiple options available for detailed visualization; refer to Section 11.5.4 or the `FreeFEM` manual [17] for the complete description.

11.5.3 Solution Analysis

We could run the FreeFEM script FreeFEM_PDE_Example.edp again and experiment with various parameters to analyze the changes in the solution images. For example, we advise changing the FE-space from being represented by P1 elements to P2 (fespace Vh(Th,P2); in line 9) and mesh from being triangulated with $n = 50$ boundary partial intervals to 100 intervals (mesh Th = buildmesh(C(100)); in line 5).

Figure 11.3 provides the images for mesh and solutions $u(x, y)$ obtained with different settings. We will discuss in detail the issues related to solution accuracy, its evaluation, and its impact on the process of solving optimization problems in Chapter 12. However, even these rough experiments reveal the obvious sensitivity of the solution quality to either the smoothness of functions representing the FE-spaces or the number (also size) of the elements in the triangulated domains.

To explore more functionalities of FreeFEM and the suitability of this PDE solver for various problems in different fields, we advise the reader to experiment with the settings provided in FreeFEM_PDE_Example.edp. For instance, we see a particular benefit in practicing to solve the same PDE problem of Example 11.3 but on different domains. Figure 11.4 presents three other possible shapes obtained in the P2 FE-space with a variable number of triangle edges n placed on the boundary.

> ! *We provide the modified scripts for all three examples in the new file* FreeFEM_PDE_Example_Domain.edp.

11.5.4 Technicalities for FreeFEM Coding

At this point, we assume that the reader is comfortable enough with the FreeFEM functionality for solving various PDEs on different domains. We remind you that although it is deemed helpful, we do not target becoming fully proficient with all FreeFEM utilities. Instead, we limit the range of useful ones to start employing them for solving PDEs efficiently by investing **minimal effort** and keeping in mind our principal concept of constructing our optimization framework in the "most optimal way." Before closing this chapter and proceeding to the optimization stage, in this section, we discuss some additional technicalities that might be useful, while visualizing obtained results and saving solutions and associated meshes. However, the readers may freely expand their knowledge of other FreeFEM functions whenever solutions to their own problems require new capabilities.

We start by briefly listing some parameters for the command plot used to visualize objects specified as its first argument (e.g., meshes and any scalar functions and vector fields representing solutions):

- wait = true to wait for the keyboard (ENTER key) event,

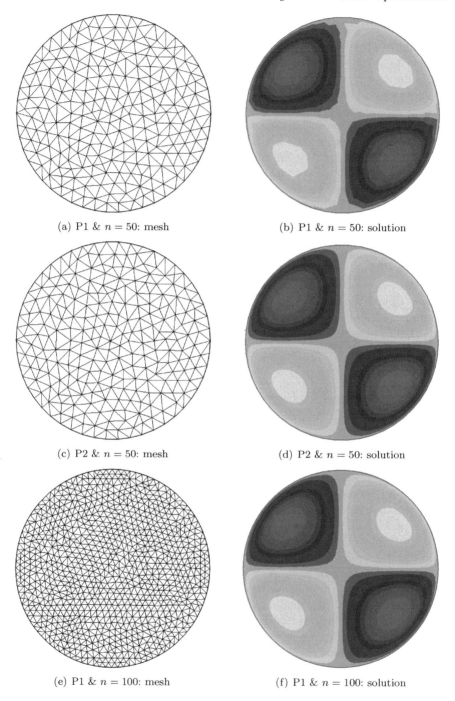

(a) P1 & $n = 50$: mesh   (b) P1 & $n = 50$: solution

(c) P2 & $n = 50$: mesh   (d) P2 & $n = 50$: solution

(e) P1 & $n = 100$: mesh   (f) P1 & $n = 100$: solution

FIGURE 11.3
FreeFEM's solutions for the PDE problem of Example 11.3 obtained with various settings for the FE-space and mesh.

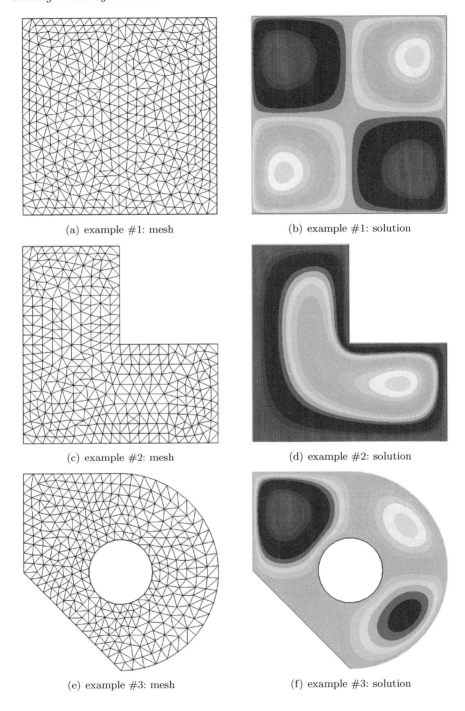

(a) example #1: mesh

(b) example #1: solution

(c) example #2: mesh

(d) example #2: solution

(e) example #3: mesh

(f) example #3: solution

FIGURE 11.4
FreeFEM's solutions for the PDE problem of Example 11.3 obtained on domains of different shapes.

- **ps** = **"file_name"** to save images in the postscript format,

- **fill** = **true** to fill the space between iso-values with color,

- **cmm** = **"string_expression"** to add a legend into the graphical window,

- **value** = **true** to plot the value of isolines by adding the color bar,

- **nbiso** = 50 to set the number of iso-values (20 by default),

- **bw** = **true** to get images in black and white, and

- **grey** = **true** to get images in grey.

Look for the complete list of **plot** parameters in the **FreeFEM** manual [17]. Figure 11.5 provides the printout of the output after executing the **plot** command in the example below; examine the file **FreeFEM_PDE_Example_Enhanced.edp** for checking the functionalities discussed in this section.

FreeFEM: plot command with added parameters

```
// plotting solution                                          1
plot(u, wait = true, ps = "solution_u.eps", fill = true,     2
   cmm = "solution u(x,y)", value = true, nbiso = 35);        3
```

A critical point for consideration is how to save created meshes. We might see it as a minor issue when running **FreeFEM** scripts for solving single PDEs. During iterative optimization, however, our computational framework will run these scripts multiple times to solve forward and adjoint problems. The time needed to create a mesh from scratch is significantly larger than the time for uploading the precomputed mesh from a file. Therefore, we could increase the computational efficacy by creating the mesh once we start solving the optimization problem and saving it in a MSH-file for future calls. Two keywords (**savemesh** and **readmesh**) save mesh and read it from the specified file as exemplified below and in the file **FreeFEM_PDE_Example_Enhanced.edp**.

FreeFEM: saving and reading mesh

```
// saving mesh in file                                        1
savemesh(Th, "mesh_pde.msh");                                 2
                                                              3
// reading mesh from file                                     4
mesh Thh = readmesh("mesh_pde.msh");                          5
```

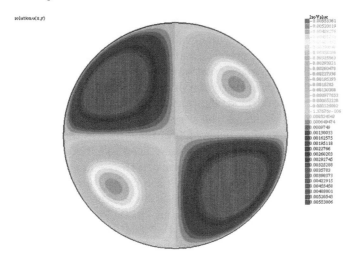

FIGURE 11.5
Example of the `plot` printout with added parameters.

To get a deeper insight into the process of creating and storing meshes, a curious reader may look inside the created mesh file `mesh_pde.msh` by opening it with any available text editor; refer to a partial printout below. The first line provides three numbers: namely, the number of vertices (942), triangles (1782), and edges ($n = 100$) on the boundary. All other lines describe uniquely the current *triangulation*. For example, line 2 provides the coordinates (0.904827052466, 0.425779291565) of the first vertex and labels it as a vertex belonging to the boundary (1; otherwise 0); refer to the manual for the complete explanation. Here, we note that the numbers provided in the examples of this section may differ depending on the used FreeFEM version.

FreeFEM: `mesh_pde.msh` **(partial printout)**

```
942 1782 100                                                    1
0.904827052466 0.425779291565 1                                 2
0.998026728428 0.0627905195293 1                                3
0.770513242776 0.637423989749 1                                 4
0.876306680044 0.481753674102 1                                 5
0.951056516295 0.309016994375 1                                 6
```

Another crucial point of discussion is the ability of FreeFEM to store the obtained solutions in the external files and manipulate them using different FE-spaces. For instance, we will benefit from saving the forward problem outcomes obtained, while evaluating the objective to be used later when we solve the adjoint problem and construct the gradient. The keyword `ofstream` saves the FE-space solution $u(x, y)$ in the file with the specified name (`solution_pde.dat`) using the prescribed precision (12 significant digits). Reading files is a fairly straightforward procedure utilizing the keyword `ifstream`. However, we should be very cautious here! We have to define a new variable (in the example below, `uNew`) that receives the solution from the file before running `ifstream`. Moreover, we need to check that this definition is associated with the FE-space consistent with the one used to create the solution stored in the file. Then, if this solution requires any interpolation between different FE-spaces (already described in the code), we could easily move $u(x, y)$ from one space (e.g., `uNew` in `Vh(Th,P2)`) to another (`uInterp` in `VhNew(Th,P1)`) by using the simple "=" command. We illustrate saving, reading, and interpolating solutions by the examples shown below and in the file `FreeFEM_PDE_Example_Enhanced.edp`.

FreeFEM: storing and interpolating solutions

```
// saving solution in file                                              1
{                                                                       2
  ofstream file("solution_pde.dat");                                    3
  int nold=file.precision(12);                                          4
  file << u[];                                                          5
}                                                                       6
                                                                        7
// reading solution from file                                           8
{                                                                       9
  ifstream file("solution_pde.dat");                                   10
  file >> uNew[];                                                      11
}                                                                      12
                                                                       13
// interpolating solution from Vh(Th,P2) to new space VhNew(Th,P1)    14
fespace VhNew(Th,P1); // new FE-space with P1 elements                15
VhNew uInterp;        // solution u(x,y) in the new FE-space          16
uInterp = uNew;       // interpolation between spaces                 17
```

Here, we stop our short trip around the `FreeFEM` functionalities in the belief that we collected **minimally but everything** to build our framework up to its ability to run the PDE-based optimization in the next chapter.

11.6   Homework Problems

1. Derive the analytical formula for adjoint-based gradients and design an iterative computational algorithm for finding optimal source term $f(x)$ in the elliptic PDE by solving the following optimization problem

$$\min_{f(x)} \quad \mathcal{J}(u(x); f(x)) = \frac{1}{2} \int_{\Omega} (u(x) - \tilde{u})^2 \, dx$$

$$\text{s.t.} \qquad \nabla \cdot [a(x)\nabla u(x)] = f(x), \qquad x \in \Omega \qquad (11.37)$$

$$u(x) = \phi(x), \qquad x \in \partial\Omega$$

   by fitting available (continuous) data $\tilde{u}(x)$ and assuming $\Omega \subseteq \mathbb{R}^n$, $n = 2, 3$, and $u, a, f : \Omega \to \mathbb{R}$.

2. Adjust Algorithm 11.1 created for Example 11.2 to enable the reconstruction of two controls, $a(x)$ and $f(x)$, simultaneously. Use the solution obtained for problem (11.37).

3. Derive analytical formulas for adjoint-based gradients and design iterative computational algorithms to identify parameters $a(x, t)$, $b(x, t)$, $c(x, t)$, and $f(x, t)$ in the parabolic PDE of the generalized problem of Example 11.1 subject to the appropriate boundary and initial conditions.

4. Create a weak formulation for the 2D Poisson's equation of Example 11.3.

5. Modify FreeFEM code FreeFEM_PDE_Example.edp to solve the PDE problem of Example 11.3 in 2D domains shaped as shown in Figure 11.4. Obtain solutions in different FE-spaces (e.g., P1, P2, etc.) and use different boundary discretization. Conclude on these solutions and make any assumptions on their possible suitability while solving an optimization problem.

6. Obtain a weak formulation of the forward problem (11.5) in Example 11.2 for boundary conditions of different types (Dirichlet, Neumann, and Robin). Modify the FreeFEM code FreeFEM_PDE_Example.edp to solve the new PDEs in 2D domains shaped as shown in Figure 11.4 or using your own domains. Conclude on the complexity of adjusting the FreeFEM script to new problems by changing

   - the domain shapes,
   - spatial discretization (triangulation),
   - type of the PDE, and
   - its boundary conditions.

READ | Where to Read More

FreeFEM Manual, [17]

FreeFEM, https://freefem.org/

Gockenbach (2011), [14]
  Chapter 5 (Boundary Value Problems in Statics), Chapter 6 (Heat Flow
  and Diffusion), Chapter 11 (Problems in Multiple Spatial Dimensions),
  Chapter 13 (More About Finite Element Methods),

Gockenbach (2006), [13]
  Chapter 1 (Some Model PDEs), Chapter 2 (The Weak Form of a BVP),
  Chapter 3 (The Galerkin Method), Chapter 4 (Piecewise Polynomials and
  the Finite Element Method), Chapter 5 (Convergence of the Finite Element Method)

Tarantola (2005), [28]
  Chapter 5 (Functional Inverse Problems)

Vogel (2002), [29]
  Chapter 2 (Analytical Tools), Chapter 3 (Numerical Optimization Tools),
  Chapter 6 (Parameter Identification)

RUN | FreeFEM Codes (with Manual) for Chapter 11

- FreeFEM_PDE_Example.edp

- FreeFEM_PDE_Example_Domain.edp

- FreeFEM_PDE_Example_Enhanced.edp

- FreeFEM-documentation.pdf

- Runing_FreeFEM_under_Linux_MacOS.txt

12

Sharing Multiple Software Environments

In the last chapter, we complete a fully functioning computational framework redesigned from running ODE-based optimization to solving PDE-based problems in 2D. We continue the transition process started in the previous chapter by introducing an external PDE solver using `FreeFEM` and discuss all stages of incorporating changes to support effective communication between the framework modules and different software environments. In addition to producing both analytic and synthetic measurement data, the created benchmark model allows checking the accuracy of the solutions produced by the PDE solver and the optimization framework as a whole. As before, the main discussion focuses on the framework transformations applied within various components and with minimal changes to the existing structure. This last transition completes and polishes the comprehension of the process of creating computational frameworks capable of solving diverse optimization problems as we use methods and strategies discussed throughout the entire study. Another focus is on the results obtained with different gradient-based methods, initial guesses, and spatial discretizations. We address the "controllability" of the observed quality and the nature of possible "defects" in the numerical solutions. Finally, we analyze the robustness of the created framework and conclude on its reliability for solving PDE-based optimization problems.

12.1 Practice Example and Benchmark Model

After a long way through various discussions and multiple practical topics, we have eventually converged on the final point in this course – completing and running our computational framework for solving PDE-based optimization problems. As mentioned in the previous chapter, we are back to Example 11.2 to practice applying various techniques discussed at different times in this book. We advise the reader to revisit Section 11.2 to review

- the formulation of the optimization problem for this example (Identifying Parameters in Elliptic PDEs; p. 317),

- its forward problem (11.5),

- derivation of the gradient formula (11.17) based on the adjoint PDE (11.18), and

- the Algorithm 11.1 for the complete gradient-based optimization utilizing the concept "optimize–then–discretize."

To evaluate the performance of our framework, we have to design a *benchmark model* for our forward problem to compare the optimization outcomes with a priori known results; refer to Section 3.5 to refresh the ideas on testing and debugging procedures. Briefly, we could formulate the *forward problem* for Example 11.2 in the following way: for given functions $a(x,y)$, $f(x,y)$, and $\phi(x,y)$, find solution function $u(x,y)$ satisfying elliptic PDE (Poisson's equation) in 2D, i.e.,

$$\begin{aligned}
\nabla \cdot [a(x,y)\nabla u(x,y)] &= f(x,y), & (x,y) \in \Omega, \\
u(x,y) &= \phi(x,y), & (x,y) \in \partial\Omega.
\end{aligned} \tag{12.1}$$

We notice that deriving such benchmark models for (12.1) is not a straightforward problem, especially for cases when functions $a(x,y)$, $f(x,y)$, and $\phi(x,y)$ are complicated analytical expressions. For having a fair trial for the optimization results, we are also not interested in taking these functions as simple expressions or constants. Instead, we use a procedure that will not require solving (12.1) in the form of a PDE and help check the accuracy of solutions obtained by the numerical PDE solver with the given spatial discretization. We complete this procedure in the following five steps.

(1) Define the control (exact) function a_{ex} to be found during optimization; we choose it to be somewhat simple, e.g., a linear function

$$a_{ex}(x,y) = x + 2y + 1, \tag{12.2}$$

also keeping in mind the control space (11.6) requirement $a(x,y) > 0$, $\forall(x,y) \in \Omega$, to ensure the PDE (12.1) is well-posed within its domain Ω.

(2) Define the domain Ω; we set it as a square box with side b (this parameter will be adjustable in the code), i.e.,

$$\Omega = \{(x,y) : 0 \le x \le b, \quad 0 \le y \le b, \quad b > 0\}. \tag{12.3}$$

(3) We could also design the analytical (exact) solution u_{ex} arbitrarily, e.g.,

$$u_{ex}(x,y) = xy(b-x)(b-y). \tag{12.4}$$

Such a solution accounts for the simplest form of the boundary conditions discussed next; however, the readers may freely experiment with other possible structures by updating formulas in steps 3, 4, and 5.

(4) Boundary conditions mentioned in step 3:

$$\phi(x,y) = 0. \tag{12.5}$$

(5) Finally, we compute the right-hand side (source) function $f(x, y)$

$$f(x, y) = (4x + 4y + 2 - b)(y^2 - by) + (2x + 8y + 2 - 2b)(x^2 - bx) \quad (12.6)$$

using (12.1), analytical solution $u_{ex}(x, y)$ and control $a_{ex}(x, y)$ defined in steps 3 and 1, respectively.

We refer to (12.2)–(12.6) as our benchmark model used in the rest of this chapter. Any modification of this model (either domain structure or functions) or new models will require another run through this simple 5-step algorithm. We also refer back to (11.35)–(11.36) for reviewing the weak formulation created for Example 11.3, which is also valid for the PDE in Example 11.2 and used for entering it into the FreeFEM coding.

To complete this benchmarking stage, we would check some objects created for our model, e.g., the domain Ω, control $a(x, y)$, and solution $u(x, y)$, visualized by the modified computational framework (with Chapter_12_opt_pde.m as main m-file), now paired with multiple FreeFEM scripts. Refer to Section 12.2 for its description and all associated technicalities. First, we explore the structure of the discretized (triangulated) domains Ω given by (12.3). We use $b = 1$ and will be interested in experimenting with rather coarse ($n = 10$) and moderately fine ($n = 50$) domain discretizations; Figure 12.1 provides images for both cases (look inside the FreeFEM file mesh.edp to see how to visualize the mesh). Their meshes consist of 240 and 5,898 elements (triangles), respectively. We assume that these numbers are distant enough to distinguish the difference in the associated solutions.

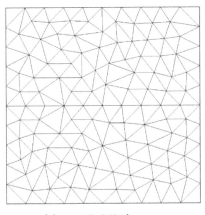
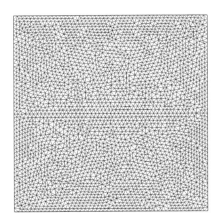

(a) $n = 10$: 240 elements   (b) $n = 50$: 5,898 elements

FIGURE 12.1
Geometry of domain Ω and mesh used for solving forward and adjoint problems for Example 11.2.

To start the iterative optimization, we have to provide the initial guess for the control. Per our multiple discussions, the best idea would be to choose this guess close enough to the optimal solution to make the fastest progress. It is possible in situations where we have assumptions about the possible solution structure. Let us start almost from scratch by assuming that we know only the average value $\bar{a}$ of the control $a(x, y)$ estimated for the current domain Ω, i.e., $a_{ini} = \bar{a}|_{\Omega} = 2.5$. Now we could compare the exact solution u_{ex} for the state variable u versus the solution obtained by the PDE solver (numerically) with $a(x, y) = a_{ini}$; refer to Figures 12.2(c,d) for these images. Figures 12.2(a,b) also visualize a_{ex} and a_{ini} confirming their different structure (linear vs. constant functions). The reader may notice that the difference between the corresponding states is less apparent. We explain it by the nature of PDEs to have **highly nonlinear structures** – **big** changes in the input parameters (in our case, control parameter a) may lead to **small** variations in their output (solution function u) and vice versa. It makes an optimization problem based on PDEs one of the hardest cases to solve. We close this section by noting that we could freely use u_{ex} as the (continuous) measurements, i.e., $\tilde{u} = u_{ex}$ in (11.4), if we choose to use data given *analytically*. Otherwise, we could run the forward problem with a_{ex} and obtain *synthetic data* as discussed in Section 9.8.2.

12.2   Updating and Tuning Optimization Framework

In this section, we discuss necessary modifications applied to the existing computational framework we left in Chapter 10, while solving the ODE-based optimization problem for the LV model of Example 9.2. This time, the transformation process requires changes for the model itself (optimization problem of Example 11.2), incorporating a new PDE solver by the external software (FreeFEM), and a careful revision of the software communication within the new framework to ensure the quality of obtained solutions. As before, we will try to optimize these transformations with minimal "invasions" to alter only the "affected" parts of the existing scripts instead of creating the framework from the ground up and giving it to the reader as the ready-to-go black-box code. We hope that playing with such assignments multiple times throughout this study helped the readers gain valuable practical experience if exercised in full. For our last problem, the process involves more changes applied within various components. However, we are more than confident that this challenge helps complete and polish the comprehension of the process of creating computational frameworks capable of solving diverse optimization problems.

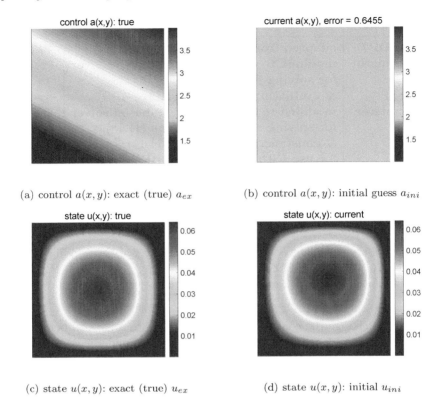

(a) control $a(x,y)$: exact (true) a_{ex} (b) control $a(x,y)$: initial guess a_{ini}

(c) state $u(x,y)$: exact (true) u_{ex} (d) state $u(x,y)$: initial u_{ini}

FIGURE 12.2
(a,b) Controls $a(x,y)$ and (c,d) states $u(x,y)$ obtained from (a,c) analytical expressions and (b,d) as (or with) initial guess $a_{ini} = 2.5$.

12.2.1 Overview of Changes

First, we start with a discussion to reflect on the large-scale changes to create a big picture of the modified framework (`Chapter_12_opt_pde.m`). As before, we advise the reader to compare its anticipated structure with the previous version used in Chapter 10; refer to Table 12.1 for a summary of all modifications.

To further help with this stage, below we note the **main aspects** to consider, while performing the transition of our framework to its new state.

(1) We might want to keep the same lists of the gradient-based techniques (namely, SD, CG, and BFGS) and algorithms for the optimal step size search (CONST, GS, and BB) from the previous version as it requires minimal changes in the entire code. However, we encourage to keep all other methods added to the LV model optimization by the readers individually following our advice as the homework in previous chapters.

element	implementation	software
main **OPT**	adapted from previous LV-model	MATLAB
$\mathcal{J}$-evaluator	modified to employ FreeFEM-based solver for solving forward problems (11.5) & evaluating objectives (11.4)	FreeFEM
d-evaluator	modified to employ FreeFEM-based solver for solving adjoint problems (11.18) & constructing gradients (11.17) for use with **SD, CG,** and **BFGS**	FreeFEM
1D α search	m-code for CONST, GS, and BB only	MATLAB
visualizer	m-code with updated subplot structure	MATLAB

TABLE 12.1
Computational elements of `Chapter_12_opt_pde.m` modified from `Chapter_10_opt_lotka_volterra_enhanced.m` to solve the PDE-based optimization problem of Example 11.2.

(2) We will change the control vector from **u** back to **a** as used before incorporating the LV model. Keeping vector **a** in this model makes the transition to the PDE model easier. However, we see changing variables (**u** to **a**) throughout the entire code as an excellent opportunity to refresh the knowledge of all components and connections established between them. `MATLAB` will "notice" a missing place with the **"Undefined function or variable"** error message giving a chance to revise this part of the framework for other necessary updates.

(3) In previous versions, we used variable `data` while calling the objective and gradient evaluators (namely, functions `fn_eval_obj.m` and `fn_eval_grad.m`). The measurement information now requires external storage (in a separate file) due to its size and use only by `FreeFEM` scripts. Therefore, we delete this variable from the input parameters of both functions and will discuss handling measurements in Section 12.2.3.

(4) We also advise removing all parts related to integrated regularization methods (gradient preconditioning, simple projection, and Tikhonov-type regularization added in Chapter 10). You have to revise files `params.m`, `direction_search.m`, and `mode_OPT.m` for the removal or maintenance necessary to support their correct functioning within the new model.

(5) Functions for evaluating objectives $\mathcal{J}(u; a)$ and computing gradients $\nabla_a \mathcal{J}$ undergo significant changes as now they include calls for the associated `FreeFEM` scripts to solve forward (11.5) and adjoint (11.18) PDE problems. We discuss the new structure of `fn_eval_obj.m` and `fn_eval_grad.m` supporting these calls in Sections 12.2.4 and 12.2.5, respectively.

(6) To complete cleaning up the framework, we must discard all parts of the LV model incapable of "recycling" to the needs of the new PDE model. For example,

  - description of parameters in `params.m`,
  - initialization of the LV model in `initialize.m`,
  - functions `fn_Lotka_Volterra_fwd.m`, `fn_Lotka_Volterra_adj.m`, `fn_precond_bvp4c.m`, and `fn_precond_fd.m` in folder `functions`, and
  - description of benchmark models for control m-file `models_ctrl.m` in folder `algorithms`.

(7) To continue, we need to add parameters to describe the new PDE model and associated optimization process. For instance, the `params.m` file now has the size of the square domain Ω (e.g., `b = 1.0`), the number of partial intervals over one boundary segment (e.g., `nx = 10`), and the specification for the chosen initial guess (e.g., `iniGuess = 0`; see Section 12.2.3 for more details) for control $a(x, y)$.

(8) Finally, the visualization part (m-code `visualize.m`) has been upgraded from one window (with six plots) to the two-window (totaling ten figures) interface. As before, the readers may easily customize the graphical output to pursue convenience, while observing the dynamics of the optimization process or after adding new functionalities.

We also discuss other conceptual changes implemented to our framework in subsequent Sections 12.2.2 through 12.2.6. Among other minor issues addressed in those sections, we note just two more ideas introduced in the new version.

- Assuming longer runtimes for each optimization iteration due to solving multiple PDEs, it might be convenient to track the optimization progress by a formal printout with the number of the last iteration completed and the value of the objective (added in the `mode_OPT.m` file). The dynamics of this printout will ensure the code is still running and not "frozen" by being caught in a loop at some point.

- We also think about collecting information on the history of the control a development in terms of its quality, e.g., proximity to the known a_{ex} evaluated at kth iteration by the L_2-norm error

$$e_{L_2}^k = \|a^k - a_{ex}\|_{L_2} = \left(\int_\Omega (a^k - a_{ex})^2 \, dx \right)^{1/2} \tag{12.7}$$

and collected in vector `aNorm`. For example, as seen in Figure 12.2(b), the constant initial guess $a_{ini} = 2.5$ has error $e_{L_2}^0 = 0.6455$ following the definition introduced in (12.7).

Finally, we preserve the core structure of the `params.m` file with the main settings propagated from those used in our computations for Chapters 9 and 10. Here, we consider the excerpt with the selected parameters closely related to the implemented changes, the description of the new PDE model, or those we modified to better fit the PDE-based optimization of Example 11.2.

MATLAB: <code>params.m</code>

Settings for PDE-based optimization of Example 11.2:

- *domain* Ω: `b = 1.0`; $\Omega = \{(x, y) : 0 \leq x \leq 1,\ 0 \leq y \leq 1\}$

- *spatial discretization:* `nx = 10`; $n = 10$, also $n = 50$

- *initial guess:* `iniGuess = 0`; 0 for constant $a_{ini} = 2.5$; refer to Section 12.2.3 for more options

- *termination #1:* `epsilonJ = 1e-4`; for ϵ_1, refer to (9.57)

- *termination #2:* `epsilonA = 1e-4`; for ϵ_2, refer to (9.58)

- *termination #3:* `kMax = 100`;

- *BB-search:* `AXini = 0`; `BXini = 1e0`; `MAXITER = 20`; `GLIMIT = 100.0`; `TOL = 1e-9`; `ITMAX = 2`;

12.2.2 Software Communication

As discussed in Chapter 11, we choose to employ `FreeFEM` as the PDE solver and now require assigning some responsibilities to its scripts, e.g.,

- generating and saving mesh,

- collecting measurement data,

- solving forward problems and evaluating objectives,

- solving adjoint problems and constructing gradients, and

- assisting with the gradient testing using κ-tests.

We recommend the motivated readers to experiment with creating their own collections of the `FreeFEM` codes for accomplishing all or some of these tasks. We also created a separate folder `freefem` where we placed the required scripts. Here, we provide a short description of all those files encouraging you to look inside the files for more comments on their structures:

- `getARGV.idp`: service file to pass command line arguments (say, numbers a, b, and c) to the EDP-files in the following format '-p1 a -p2 b -p3'; refer to the **FreeFEM** manual [17] for details and examples,

- `mesh.edp`: executable (EDP) file to describe the geometry and generate mesh for the square domain $\Omega : [0, b] \times [0, b]$ (called from **MATLAB**'s m-code `initialize.m`),

- `data.edp`: executable file to create synthetic measurements (data) $\tilde{u}$, define and store analytical functions (a_{ex}, u_{ex}, ϕ, and f), construct initial guess a_{ini} for control $a(x, y)$, and perform a simple check for the quality of numerical solutions for $u(x, y)$ (called from **MATLAB**'s m-code `measurements.m`),

- `eval_obj.edp`: executable file to solve forward PDE problem (11.5), evaluate objectives by (11.4), save current control a^k and solution (state variable) u^k in the external files for visualization in **MATLAB**, and compute L_2-norm error (12.7) for a^k (called from **MATLAB**'s function `fn_eval_obj.m`),

- `solve_pde.slv`: description for the solver (serial implementation) for the 2-order elliptic PDE in 2D (forward problem, included in the **FreeFEM** scripts `eval_obj.edp` and `eval_grad.edp`),

- `eval_grad.edp`: executable file to solve forward (11.5) and adjoint (11.18) PDE problems, construct gradients (11.17), save discretized gradients in the external files for optimization in **MATLAB**, save the current adjoint solution ψ^k and gradients in the external files for visualization in **MATLAB** (called from **MATLAB**'s function `fn_eval_grad.m`),

- `solve_adj.slv`: description for the solver (serial implementation) for the 2-order elliptic PDE in 2D (adjoint problem, included in the **FreeFEM** script `eval_grad.edp`),

- `kappa_grad.edp`: executable file to assist in computing the denominator part in the κ-test (called from **MATLAB**'s m-code `kappa_test.m`).

The schedule of **FreeFEM** files above provides initial information on the software communication – how these **FreeFEM** scripts are attached to and called from **MATLAB**'s codes and functions. Once they start running, these scripts will operate with large datasets as their input or output (e.g., current control and solution functions, gradients, etc.). Using command line arguments for exchanging these bulky datasets between **FreeFEM** and **MATLAB** is possible but impractical (e.g., due to some issues in the consistency of input–output (IO) formats of both, the necessity to allocate enough operating memory or to keep access to this data after completing optimization). A simple and deemed feasible solution would be to create a folder for temporarily storing the data exchanged between **FreeFEM** and **MATLAB**. We designed earlier such a folder in our framework called `temp`, and now we are ready to use it. One of many possible "protocols" to arrange cleaning this folder automatically is to delete its old instance and create a new one periodically, e.g., once we start optimization. MATLAB's system commands rmdir and mkdir remove and make a new directory, respectively, i.e.,

- [status, cmdout] = rmdir('temp','s');

- [status, cmdout] = mkdir('temp');

where

status (integer) variable contains an **error code** (usually 0 if executed without errors), and

cmdout (string) variable provides **report on command execution** (empty if successfully).

Refer to the main m-file Chapter_12_opt_pde.m for details.

MATLAB: Chapter_12_opt_pde.m (**temp folder**)

```matlab
% deleting old & creating new optimization storages        1
[status, cmdout] = rmdir('temp','s');                      2
if ~status                                                 3
  disp('Cannot remove temp/ directory!'); return;          4
end                                                        5
[status, cmdout] = mkdir('temp');                          6
if ~status                                                 7
  disp('Cannot create temp/ directory!'); return;          8
else                                                       9
  fid = fopen('temp/opt_log.dat', 'w'); fclose(fid);      10
end                                                       11
```

Generally speaking, the purpose of the folder temp is manyfold. First, it keeps all files to support communication between different software environments (in our case, MATLAB and FreeFEM) with possibly inconsistent IO formats. Second, its content allows recreating images showing the results of completed optimization anytime afterward without a need to rerun it, which now becomes very useful due to long-running processes. We may also consider other reasons: e.g., resuming optimization after unexpected interruptions, designing parallelized versions of the PDE solvers or the entire framework, etc. Below, we create a brief description of the temp folder content (grouped according to its purpose; refer to Section 11.5.3 and FreeFEM manual [17] for more details), which might not be comprehensive once the readers decide to expand it with any additions:

(a) **mesh files:** e.g., `mesh.msh` used for solving both forward and adjoint problems,

(b) **functions in FE-space format:** `opt_a_ex.dat`, `opt_f.dat`, `opt_meas.dat`, and `opt_phi.dat` for functions a_{ex}, f, $\tilde{u}$, and ϕ, respectively (to be used by `FreeFEM` only),

(c) **files in vector format:** `opt_a.dat`, `opt_a_ini.dat`, and `opt_grad.dat` for a^k, $a^0 = a_{ini}$, and $\nabla_a \mathcal{J}$, respectively (to be used by both `FreeFEM` and `MATLAB`),

(d) **objective function** computations are stored in `opt_obj.dat`: it contains the current objective $\mathcal{J}(u^k; a^k)$ value and the L_2-norm error (12.7) for current solution a^k; see Section 12.2.4 for explanations,

(e) **visualization** in `MATLAB` format: `opt_vis_a_curr.dat`, `opt_vis_a_true.dat`, `opt_vis_grad_curr.dat`, `opt_vis_psi_curr.dat`, `opt_vis_u_curr.dat`, and `opt_vis_u_true.dat` for a^k, a_{ex}, $\nabla_a \mathcal{J}$, ψ^k, u^k, and u_{ex}, respectively,

(f) **optimization log:** `opt_log.dat`, and

(g) **debugging log** with `FreeFEM` reports: `opt_cmdout.dat`.

While files mentioned in (a) through (d) are self-explanatory by their names, the rest requires brief comments on their content.

We start with the visualization of objects created in `FreeFEM`. We still consider plotting the images of all these objects in `MATLAB` rather convenient, e.g., being able to play with the image settings, display them simultaneously in one window with the prescribed order, etc. However, by default, `FreeFEM` saves these objects in the FE-space-related format, which we cannot "decrypt" in `MATLAB` without a description of this space and the knowledge of the associated mesh. The "MATLAB-related" structure for easy plotting represents 2D objects (functions) as matrices with dimensions consistent with the shape of rectangular domains. The entries in these matrices give the numerical values (pixels) associated with the discretized 2D functions. We set the number of pixels in both x- and y-directions in the `params.m` file (e.g., `visDiscr = 200`) and ask `FreeFEM` to evaluate FE-space functions at every pixel location as shown in the example below. While the higher resolution provides images of better quality, we should warn the reader about the increased computational time required for making such "snapshots."

FreeFEM: `data.edp` (printout of visualization part for a_{ex})

```
// Saving a(x,y) true in Matlab format (visualization)                    1
```

```
real discrStep = b/visDiscr;                              2
{ofstream file("temp/opt_vis_a_true.dat");                3
  for(int j = 0; j < visDiscr; j++)                       4
  {                                                       5
    for(int i = 0; i < visDiscr; i++)                     6
    {                                                     7
      real xx = (0.5 + i) * discrStep;                    8
      real yy = (0.5 + j) * discrStep;                    9
      file << a(xx,yy) << "    ";                          10
    }                                                     11
    file << endl;                                         12
  }                                                       13
}                                                         14
```

In the current implementation of our computational framework, we limit the text output describing the dynamics of the optimization process to the number of the last iteration completed and the value of the objective provided in the MATLAB's command window. However, it might be handy (especially during the debugging phase) to collect more information from the optimization stages. For instance, it may help us narrow the search to locate possible places where the code breaks. We do it in a separate file called opt_log.dat, containing the information appended mainly by FreeFEM. We may also access this file from MATLAB in our framework whenever we need to document anything in this log: below, we provide an example of its first entries.

opt_log.dat (log with first iteration entries)

```
Creating mesh:                                            1
 Boundary: 40 segments ordered.                           2
Mesh is created successfully!                             3
==========================================================  4
Measurements' phase started . . .                        5
Measurements are made successfully!                      6
L2-norm difference (solution) = 3.27886e-006             7
==========================================================  8
Direct Solver started . . .                              9
 Objective: 8.52159282439e-006                           10
==========================================================  11
Adjoint Solver started . . .                             12
 Gradient has been computed!                             13
==========================================================  14
```

Although the optimization log opt_log.dat contains valuable information, it does not provide any clues for the errors made in the FreeFEM scripts as they use the C++-based environment. But FreeFEM shares some data (e.g., compilation logs, exit statuses, error messages, etc.) with MATLAB, and we could collect it for later analysis. We use a separate log file opt_cmdout.dat to collect all FreeFEM messages (ffDebug = true in params.m) or just its last report (ffDebug = false). Reading these reports requires extra experience with FreeFEM; refer to its manual [17]. However, to easily locate the error, we have to search for the word "error" in this log. In many cases, it has a reference to a specific line in the FreeFEM script that might cause the error. To avoid creating large logs that may induce memory issues and slow down the optimization, we recommend using the *debugging* mode (ffDebug = true) only when tuning the framework and turning it off for long optimization runs.

We could elaborate more on sharing the data between FreeFEM and MATLAB by describing how the latter could obtain the exit reports from the former and create the log opt cmdout.dat, as we consider it a part of the invaluable software communication process. But let us first exemplify the process of running the external executables, such as EDP-files, from the MATLAB environment by showing how we run FreeFEM script mesh.edp executed from m-code initialize.m by using command system.

MATLAB: initialize.m **(running FreeFEM script** mesh.edp**)**

```
% creating & saving mesh by calling FreeFEM code mesh.edp      1
paramsLine = ['-p1 ' num2str(b) ' -p2 ' num2str(nx)];          2
[status,cmdout]=system(['FreeFem++ freefem/mesh.edp ' paramsLine]); 3
if fn_stat(status, 0, cmdout, ffDebug)                         4
  return;                                                       5
end                                                             6
```

Here, we use the string variable paramsLine to pass two command line arguments (b and nx) into the C++ code, i.e., MATLAB formally executes the following command

```
>> FreeFem++ freefem/mesh.edp -p1 1.0 -p2 10
```

and then MATLAB's command system will wait until it is finished to provide status and cmdout (exit) reports. If the FreeFEM code is finished with an error (variable status returns any number except 0), function fn_stat.m will terminate MATLAB to allow debugging. As we mentioned earlier, we may look at the end of the file opt_cmdout.dat to find an error message generated by FreeFEM running in the background.

Finally, we advise the reader to explore the communication functionality of the function fn_stat.m working with the messages (status and cmdout) obtained as exit reports of all FreeFEM scripts.

MATLAB: fn_stat.m

```
function exitInd = fn_stat(exitStatus,correctStatus,cmdout,ffDebug)   2

  exitInd = false; append = 'w';                                      3
                                                                      4
  % setting debug mode                                               5
  if (ffDebug)                                                       6
    append = 'a';                                                    7
  end                                                                8
                                                                      9
  % adding cmdout info to log-file                                   10
  fid = fopen( 'temp/opt_cmdout.dat', append);                      11
  fprintf(fid, [cmdout '\n']); fclose(fid);                         12
                                                                      13
  % checking the status                                              14
  if exitStatus ~= correctStatus                                    15
    disp(['Incorrect exit status = ' num2str(exitStatus) ...        16
          ' (' num2str(correctStatus) ' expected)!']);             17
    exitInd = true;                                                 18
  end                                                                19
return                                                               20
```

12.2.3   Another Round on Measurements

It is now the right time to comment on the procedure for creating the measurement data performed by the m-code `measurements.m` paired with the FreeFEM script `data.edp` (similar to executing the `mesh.edp` file). This script receives three parameters as the command line arguments: the size of the domain Ω (b), the number of pixels for visualization (`visDiscr`), and the type of the initial guess (`iniGuess`), e.g., after running

```
>> FreeFem++ freefem/data.edp -p1 1.0 -p2 200 -p3 0
```

After pulling up the mesh from the file `mesh.msh`, `data.edp` first creates and stores (in the format of FE-space functions) a_{ex}, u_{ex}, ϕ, and f analytically defined in (12.2) and (12.4)–(12.6):

- aEx = x + 2*y + 1;

- uEx = (x^2 - b*x)*(y^2 - b*y);

- phi = 0;

- f = (4*x+4*y+2-b)*(y^2 - b*y) + (2*x+8*y+2-2*b)*(x^2 - b*x);

Then, it generates the initial guess for control $a(x, y)$ following the instruction provided by the variable `iniGuess`; refer to this part of the code below.

FreeFEM: data.edp (printout of creating the initial guess)

```
if (iniGuess==1)                                    1
    aIni  = 3*y + 1;                                2
else if (iniGuess==2)                               3
    aIni  = x + 2*y + 0.75;                         4
else if (iniGuess==3)                               5
    aIni  = x + 2*y + 0.9;                          6
else                                                7
    aIni = 2.5;                                     8
```

Therefore, the reader may easily add other guesses to the list of existing ones

$$
\begin{aligned}
a_{ini,0} &= 2.5, \\
a_{ini,1} &= 3y + 1, \\
a_{ini,2} &= x + 2y + 0.75, \\
a_{ini,3} &= x + 2y + 0.9
\end{aligned}
\tag{12.8}
$$

to experiment with the framework robustness by expanding the options of the `if/else` command. We define a_{ini} in the space of P0-elements and save it (after vectorizing) in the external file to be opened later in the main framework; note that we moved the procedure for control vector initialization from `initialize.m` to the main m-file `Chapter_12_opt_pde.m`.

After running the solver for the forward PDE problem (`PDEforward`), we have two functions representing the solution (state variable) $u(x, y)$: namely, exact (analytical or true) solution u_{ex} and numerical solution u_n. We will use the latter as the measurement (synthetic) data for all numerical experiments in this chapter, i.e., $\tilde{u} = u_n$.

Our last comment here is about the benefits of having both solutions u_{ex} and u_n – we could evaluate the difference between analytical and numerical solutions, e.g., using the L_2-norm error notation

- `L2diff = sqrt(int2d(Xh)((u-uEx)^2));`

and use it for debugging purposes. For example, found values of `L2diff` printed into optimization log `opt_log.dat` confirm the accuracy of our numerical solver used for the forward and, similarly, the adjoint problems after comparing the errors obtained for different discretizations (triangulations) of the domain Ω:

- $n = 10$: `L2diff = 3.27886e-006`

- $n = 50$: L2diff = 2.53634e-008

We cannot set this error to zero. However, we refer to the observed effect as another example of demonstrating the expected *controllability* (finer discretization implies smaller errors) of our numerical PDE solver in terms of obtaining the solution of the "controllable quality" as discussed before in Section 9.8.4.

12.2.4   Evaluating Objectives

A significantly modified structure of the objective function $\mathcal{J}(u; a)$ evaluator fn_eval_obj.m serves as one more good example of the two-way software communication pattern established between MATLAB and FreeFEM. After previous discussions about storing the exchanged data in the external files and calling the FreeFEM scripts from inside MATLAB's m-code, it might be now easy to interpret the structure of the latter: it includes saving updated control a^k in the file opt_a.dat, executing the EDP-file eval_obj.edp, and finally reading its "message" (file opt_obj.dat) containing numerical values of $\mathcal{J}(u^k; a^k)$ and $e_{L_2}^k$ found for current control a^k.

MATLAB: fn_eval_obj.m

```
function [obj aNorm] = fn_eval_obj(step, a, d, params)      1
                                                             2
  % updating control                                        3
  a = a + step*d;                                            4
                                                             5
  % saving updated control in file                          6
  fid = fopen('temp/opt_a.dat', 'w');                       7
  fprintf(fid, '%d\r\n', length(a));                        8
  for i=1:length(a)                                          9
    fprintf(fid, '%1.20e\r\n', a(i));                       10
  end                                                        11
  fclose(fid);                                               12
                                                             13
  % objective evaluation by calling FreeFEM code eval_obj.edp  14
  paramsLine = ['-p1 ' num2str(params(1)) ' -p2 ' ...       15
      num2str(params(2))];                                  16
  [status, cmdout] = system(['FreeFem++ freefem/eval_obj.edp ' ... 17
      paramsLine]);                                         18
  fn_stat(status, 0, cmdout, params(3));                    19
                                                             20
  objFunc = load('temp/opt_obj.dat');                       21
  obj = objFunc(1);                                         22
  aNorm = objFunc(2);                                       23
                                                             24
return                                                       25
```

We invite the reader to explore the structure of the associated FreeFEM script `eval_obj.edp`. Although it may look familiar after our discussions on various parts, we would still focus on a few aspects. The main issue is to decide on the proper FE-space for all functions involved in computations. On the one hand, we are interested in accurate solutions obtained, e.g., by using P2 (piecewise quadratic) elements. However, such solutions will require up to 6 coefficients or degrees of freedom (DOF) per one element (triangle) to describe them in the space of P2 polynomials (quadratic functions). Although the total number of these coefficients is lower than $6n_t$ (where n_t is the number of triangles) as triangles share their sides, this number is still huge. For example, our mesh with $n = 10$ has 240 triangles and 521 DOF to describe all P2 elements. For $n = 50$, these numbers are 5,898 (triangles or P0 DOF) and 11,997 (P2 DOF). Making all coefficients collected from all P2 elements as entries of the control vector is not rational due to its enormous size and also repeated values. On the other hand, we are interested in the size of the control as small as possible to speed up the optimization convergence and decrease the number of local optimums. Therefore, the feasible solution would be to define our controls and their gradients in P0-element FE-space while using P2 elements for all other functions. This concept is also convenient for storing controls and gradients as vectors as all their entries are associated with particular triangles. As you remember, these entries and triangles are mapped to each other by the unique description of the mesh used unchanged during the entire optimization.

After accurate solution u^k is obtained, FreeFEM code `eval_obj.edp` computes the objective function $\mathcal{J}(u^k; a^k)$ by integrating over triangulated domain Ω

- `ObjFunc = 0.5 * int2d(Xh)((u - uMeas)^2);`

and evaluates the L_2-norm error (12.7) for updating the records in the history of the control $a(x, y)$ development.

12.2.5 Evaluating Gradients

Similar to evaluating objectives, we modified the structure of the gradient $\nabla_a \mathcal{J}(u; a)$ evaluator `fn_eval_grad.m` following the same concepts of maintaining the optimality of two-way software communication established between MATLAB and FreeFEM. We invite the reader to explore this structure: it similarly includes saving updated control a^k in the file `opt_a.dat`, executing the EDP-file `eval_grad.edp`, and finally downloading the computed gradient $\nabla_a \mathcal{J}(u^k; a^k)$ from `opt_grad.dat` in a vector format, i.e., P0-element format rewritten as a vector.

MATLAB: fn_eval_grad.m

```
function grad = fn_eval_grad(a,params)                          1
                                                                2
  % saving updated control in file                             3
  fid = fopen('temp/opt_a.dat', 'w');                          4
  fprintf(fid, '%d\r\n', length(a));                           5
  for i=1:length(a)                                            6
    fprintf(fid, '%1.20e\r\n', a(i));                          7
  end                                                          8
  fclose(fid);                                                 9
                                                               10
  % computing gradient by calling FreeFEM code eval_grad.edp   11
  paramsLine = ['-p1 ' num2str(params(1)) ' -p2 ' ...          12
      num2str(params(2))];                                     13
  [status, cmdout] = system(['FreeFem++ freefem/eval_grad.edp ' ...  14
      paramsLine]);                                            15
  fn_stat(status, 0, cmdout, params(3));                       16
                                                               17
  % getting grad as a vector                                  18
  grad = load('temp/opt_grad.dat'); grad = grad(2:end);       19
                                                               20
return                                                         21
```

The reader may also notice the similarity in the structure of both FreeFEM scripts eval_grad.edp and eval_obj.edp. In addition, we make a few more comments on the former anticipating possible questions on the used strategies. Again, we define $a(x, y)$ in the P0-element FE-space to reduce the dimensionality of the control space, and we do the same for gradient $\nabla_a \mathcal{J}$ to match the dimensionality of the control vector a^k. However, to obtain accurate solutions for the forward and adjoint PDE problems, we use P2 elements for all other functions, i.e., $u(x, y)$, $\psi(x, y)$, $\phi(x, y)$, and $f(x, y)$. We then use u^k and ψ^k to compute the gradient as a FE-space function

- grad=-dx(u)*dx(psi)-dy(u)*dy(psi);

and then interpolate it (P2 → P0) to provide $\nabla_a \mathcal{J}(u^k; a^k)$ in the vectorized form. Finally, before sending this gradient to MATLAB, we have to adjust its continuous form to be valid in the discretized space by normalizing each component by the area of the associated element (triangle) as discussed in Sections 9.3.2 and 9.7.3; refer to the example below extracted from eval_grad.edp.

FreeFEM: eval_grad.edp (printout of saving adjusted gradients)

```
// Saving gradient vector in file                          1
real[int] gradV(Xh.nt);                                    2
gradV = grad[];                                            3
{                                                          4
  ofstream file("temp/opt_grad.dat");                      5
  file << gradV.n << endl;                                 6
                                                           7
  // adjusting continuous form to discretized space        8
  for (int i = 0; i < gradV.n; i++)                        9
    file << gradV[i]*Xh[i].area << endl;                   10
}                                                          11
```

12.2.6 Checking Quality of Discretized Gradients

As usual, our final step before proceeding to the optimization problem of Example 11.2 is to check the quality of the discretized gradients by testing the correctness of gradient computations and identifying factors for making them more accurate. The reader could review the procedure of applying the *kappa-test*) we used before described in Section 3.6 and exemplified by practical applications in Sections 8.6 and 9.8.4. Here, we consider the same test used in our current n_t-dimensional case assuming that the chosen P0-element FE-space is constructed on the mesh containing n_t elements (triangles, $n_t \sim 10^2 \div 10^5$), i.e., $a^k \in \mathbb{R}^{n_t}$ and for $k = 0$

$$\kappa(\epsilon) = \frac{\mathcal{J}(a^0 + \epsilon \, \delta a) - \mathcal{J}(a^0)}{\epsilon \, \langle \boldsymbol{\nabla}_a \mathcal{J}(a^0), \delta a \rangle} = \frac{\mathcal{J}(a^0 + \epsilon \, \delta a) - \mathcal{J}(a^0)}{\epsilon \int_\Omega \boldsymbol{\nabla}_a \mathcal{J}(a^0) \, \delta a \, dx}, \quad \epsilon \to 0, \qquad (12.9)$$

where $\delta a = [\Delta a_0 \ \Delta a_1 \ \ldots \ \Delta a_{n_t}]^T$ is the *perturbation* applied to current control a^0. As before, we note that in (12.9), $\boldsymbol{\nabla}_a \mathcal{J}(a^0)$ is the gradient of objective $\mathcal{J}(a^0) = \mathcal{J}(u^0; a^0)$ defined analytically with respect to control a and evaluated for the initial values of state u^0 and control a^0 using the formula (11.19), i.e., $\boldsymbol{\nabla}_a \mathcal{J}(a^0) = \boldsymbol{\nabla}_a \mathcal{L}(u^0; a^0, \psi^0)$. We also use the notation $\langle \cdot, \cdot \rangle$ for inner products in the L_2 space.

We would reiterate here that the purpose of applying the kappa-test is at least twofold. First, we need to check the consistency of the gradients obtained with analytical formulas and FD approximations and conclude if the associated computations are correct and accurate. Second, we want to verify that this accuracy is *"controllable"* by using different meshes, e.g., created with $n = 10$ (240 triangles) and $n = 50$ (5,898 triangles); refer to Figure 12.1 for their images. Figure 12.3 depicts the gradient structures obtained on both meshes with $a^0 = a_{ini}$ and corresponding adjoint fields ψ^0 used to construct these gradients. As seen in plots (c) and (d), the difference in the adjoint field structure is not trackable as we obtain both solutions (as well as solutions for u^k) in the space of smooth (quadratic) functions using P2-elements. In the previous section, we discuss the necessity of projecting gradients obtained initially in the same FE-space onto the P0 space of piecewise constant functions where the difference in the quality is noticeable; refer to plots (a) and (b).

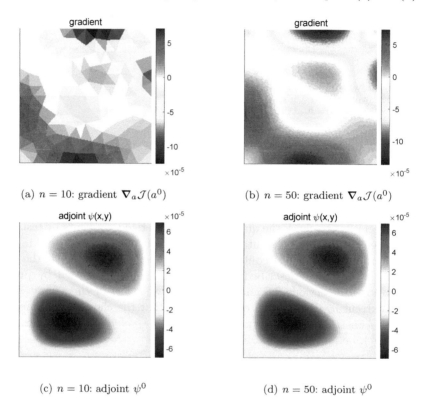

(a) $n = 10$: gradient $\nabla_a \mathcal{J}(a^0)$ (b) $n = 50$: gradient $\nabla_a \mathcal{J}(a^0)$

(c) $n = 10$: adjoint ψ^0 (d) $n = 50$: adjoint ψ^0

FIGURE 12.3
(a,b) Gradients $\nabla_a \mathcal{J}(a^0)$ and (c,d) adjoints ψ^0 obtained on meshes with (a,c) $n = 10$ and (b,d) $n = 50$.

We will check gradients obtained on both meshes by applying the kappa-test of two types.

(a) *Cheap test*. It requires **two** $\mathcal{J}$-evaluations: for fixed δa, e.g., $\delta a = a^0$, we compute $\kappa(\epsilon)$ for a range of ϵ, e.g., $\epsilon = 10^{-15} \div 10^0$.

(b) *Expensive test*. It requires $n_t + 1$ $\mathcal{J}$-evaluations in n_t steps: for fixed ϵ, e.g., $\epsilon = 10^{-6}$, we perform κ-test to check sensitivity for every component of a^0 by changing δa, i.e.,
step #1: $\delta a = [a_1^0\ 0\ 0\ \ldots\ 0]^T$,
step #2: $\delta a = [0\ a_2^0\ 0\ \ldots\ 0]^T$,
$\ldots$
step #n_t: $\delta a = [0\ 0\ 0\ \ldots\ a_{n_t}^0]^T$.

Our previous statements about the meaningfulness of choosing $\delta a = a^k$ for various k are still valid here as the magnitudes of components in a^k may vary significantly, sometimes by order. As such, this strategy helps keep uniform the applications of the κ-test to all those components.

To complete the review of the kappa-test application, we advise the reader to look into the updated m-code `kappa_test.m` and note its modification due to changes in the structure of the problem and the use of external software `FreeFEM`. This modification relates to computing the L_2 inner products in the denominator of (12.9) that requires integration over the entire domain Ω. A simple approach is to employ `FreeFEM` (by calling the separate EDP-file `kappa_grad.edp`) with its existing functionality for numerical integration in the FE-spaces.

MATLAB: `kappa_test.m` (computing $\int_\Omega \boldsymbol{\nabla}_a \mathcal{J}(a^k)\, \delta a\, dx$)

```
% L2 inner product (in denominator) by gradient                       1
paramsLine = ['-p1 ' num2str(numCtrls)];                              2
[status, cmdout] = system(['FreeFem++ freefem/kappa_grad.edp '...     3
    paramsLine]);                                                     4
fn_stat(status, 0, cmdout, ffDebug);                                  5
G = load('temp/opt_kappa.dat');                                       6
```

We run κ-test by setting `mode = 'TEST'` in `params.m` and executing the main m-file `Chapter_12_opt_pde.m` to operate the optimization framework in the TEST mode. Figure 12.4 provides the results of running a cheap version of this test on both meshes. We conclude that the gradients are computed **correctly**, as the plateau of $\kappa(\epsilon)$ in both cases is close to 1, and its width is large enough (parameter ϵ spans 8-9 orders of magnitude). We also notice that refining mesh ($n = 50$) benefits the gradient quality as the plateau moves closer to 1, as seen in Figure 12.4(b). Moreover, Figures 12.4(c) and (d) add more to this analysis. Quantity $\log_{10}|\kappa(\epsilon) - 1|$ shows that the gradient evaluations can capture two and three significant digits of accuracy for $n = 10$ and $n = 50$ cases, respectively, proving the substantial quality improvement due to the mesh refinement.

The results of the expensive version of the kappa-test performed with fixed $\epsilon = 10^{-6}$ are shown in Figures 12.5(a,b). As seen before, a few outliers are present in both cases; again, it is a common issue for problems of increased complexity. However, to conclude on the overall accuracy based on the sensitivity provided by single control components, we have to zoom out these images to have a closer view of the region around $\kappa = 1$; refer to Figures 12.5(c,d). The distribution of κ-values has obviously a higher density in the proximity to $\kappa = 1$ in the case of the finer mesh ($n = 50$), confirming its better suitability supported by the improved quality of the discretized gradients. We advise

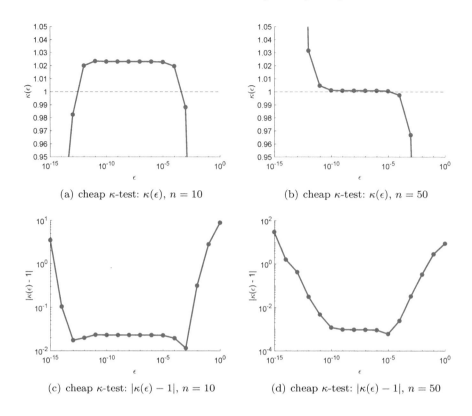

(a) cheap κ-test: $\kappa(\epsilon)$, $n = 10$ (b) cheap κ-test: $\kappa(\epsilon)$, $n = 50$

(c) cheap κ-test: $|\kappa(\epsilon) - 1|$, $n = 10$ (d) cheap κ-test: $|\kappa(\epsilon) - 1|$, $n = 50$

FIGURE 12.4
Results of the cheap κ-test applied to gradients $\nabla_a \mathcal{J}(a^0)$ computed on meshes
with (a,c) $n = 10$ and (b,d) $n = 50$.

the reader to experiment a bit more by, e.g., allowing any programming "errors" in computing objectives and gradients to see the immediate reflection
of the associated miscomputations on the results of both cheap and expensive
kappa-tests. We close this analysis with another advice for repeating these
tests throughout the optimization process to control possible loss of sensitivity due to errors accumulated in the gradients, especially in the neighborhood
of the local optimums. If implemented, we should keep in mind extra computational loads put on the framework: e.g., performing one κ-test in our case
took about 600s and 18,000s for $n = 10$ and $n = 50$, respectively.

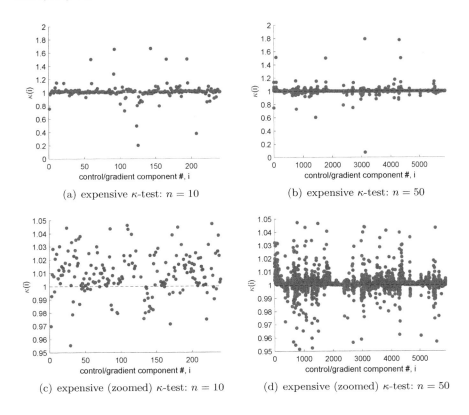

(a) expensive κ-test: $n = 10$

(b) expensive κ-test: $n = 50$

(c) expensive (zoomed) κ-test: $n = 10$

(d) expensive (zoomed) κ-test: $n = 50$

FIGURE 12.5
(a,b) Results of the expensive κ-test applied to gradients $\nabla_a \mathcal{J}(a^0)$ computed on meshes with (a) $n = 10$ and (b) $n = 50$; (c) and (d) provide the same results with a closer view of the $\kappa = 1$ region.

12.3   Analyzing Optimization Results

In Section 12.2, we complete a long discussion on the updates required for our computational framework to solve the PDE-based optimization problem of Example 11.2 based on the benchmark model created in Section 12.1. We move the framework back to its OPT mode (mode = 'OPT') and run Chapter_12_opt_pde.m with the settings discussed on p. 344. We also advise the reader to do the same and experiment by changing various parameters to see the feedback and study the behavior of the obtained solutions constrained by PDEs, one of the biggest optimization challenges. We expect you will be able to recognize and explain numerous phenomena we discussed earlier in the preceding chapters. In this section, we focus on a few main issues that will help us create a big picture and identify the directions for possible improvement.

We run it first on a "coarse" mesh (nx = 10) with the steepest descent method (method = 'SD') and limit the maximum number of optimization iterations to 100 (kMax = 100) due to the notably increased computational time compared to all problems solved previously. Figure 12.6 provides the results – selected plots from both MATLAB windows. Comparing the images created for the exact function a_{ex} in (a) and optimal control a^* in (b) reveals their similarity even with the "grainy" structure of the latter. Even for this nontrivial problem, the performance of the SD method is good: the objective drop is about four orders of magnitude, and the convergence rate is **linear** ($r = 1.0297$). The history of L_2-norm errors $e_{L_2}^k$ shows a monotonic decrease that might indicate a convergence to a "good" local minimum or, maybe, to a global one if it exists for this problem. We also note the outstanding performance of our BB algorithm. While it started the step size α search from the initial interval $[0, 10^0]$, the optimal values are within $[10^6, 10^8]$ (due to the small magnitudes, $\sim 10^{-5}$, of the gradient components). Although we have some concerns about the quality of a^* right in the center of the domain, we postpone addressing this issue until we have other results obtained with other gradient-based methods and on different meshes.

We now run a new optimization with the only change applied to the method, conjugate gradient (method = 'CG'). Figure 12.7 provides the results in the same format as for SD. The outcomes appear to be similar or very close according to various observations: e.g., the same drop in the objective, **linear** convergence ($r = 1.0374$), and monotonically decreasing L_2-norm errors. However, the readers may catch a difference in the decreasing patterns of $\mathcal{J}(a^k)$ and $e_{L_2}^k$. In the CG case, they both are smooth, while SD optimization exhibits some abrupt changes (downfalls) from time to time. We could explain it in the following manner. Conjugate gradients are "farsighted" as they use information accumulated from previous iterations; it often allows them to run optimization directly to the "good" local optimum, not being distracted by other nearby "weaker" solutions. The steepest descent is "nearsighted" as it uses only plain (unmodified) gradients and, as such, may easily change the direction by switching the optimums if "captured" by a new one. For many reasons, it is not bad; e.g., a new local optimum may be much better than the old one – in a such case, we see a corresponding downfall in the history $\mathcal{J}(a^k)$ and $e_{L_2}^k$ changes. After each downfall, BB has to adjust the search interval; compare Figures 12.6(e) and 12.7(e), illustrating this concept. It explains a better (than expected) performance of SD comparable with CG. In addition, SD is lightweight in computations, which makes it a good choice for use in PDE-based optimization.

? *Could you guess about the performance of the 2-order Newton-like BFGS method?*

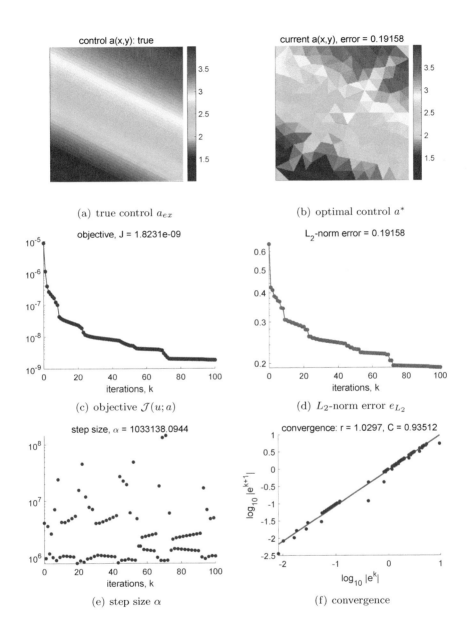

(a) true control a_{ex}

(b) optimal control a^*

(c) objective $\mathcal{J}(u; a)$

(d) L_2-norm error e_{L_2}

(e) step size α

(f) convergence

FIGURE 12.6
MATLAB window with six (selected) plots updated dynamically (finalized at $k = 100$), while solving Example 11.2 with SD and constant initial guess $a_{ini} = 2.5$ on mesh with $n = 10$.

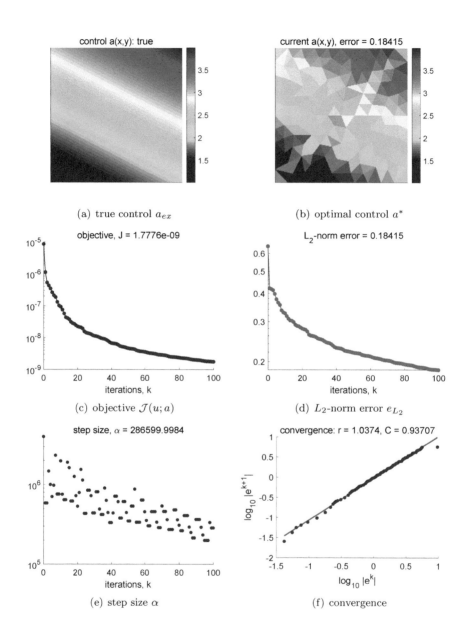

FIGURE 12.7

MATLAB window with six (selected) plots updated dynamically (finalized at $k = 100$), while solving Example 11.2 with CG and constant initial guess $a_{ini} = 2.5$ on mesh with $n = 10$.

Previously, we saw its superior performance when applied to other problems, including ODE-based optimization. We run a new optimization changing the method again (`method = 'BFGS'`), and it terminates in two iterations (there might be a few more depending on `MATLAB` and `FreeFEM` versions); see Figure 12.8 for the results. The outcome is not too bad, but it is the same as what SD and CG methods have just after their first iteration. We may explain it from different perspectives. For example, once captured by a local optimum, BFGS often has almost no chance of escape and converges to this optimal solution in a few iterations. For nonconvex problems characterized by multiple optimums, the right choice of the initial guess is critical. Another reason for such behavior is how accurately BFGS approximates its Hessians. As this accuracy often depends on the size of the control vector, problems in high dimensions may not use the benefits of quasi-Newton methods due to accumulated errors. We advise the reader to review Chapter 5 for deeper insight into other issues associated with using gradient-based strategies.

As a logical step in our current numerical phase, we now move to a finer mesh with $n = 50$ (`nx = 50`). In this round, we allow SD and CG to compete as they showed about the same performance on the coarse mesh. We refer to the results of this competition shown in Figures 12.9 and 12.10 for the steepest descent and conjugate gradient methods, respectively. The visual analysis of the obtained optimal solutions a^* in their (b) plots returns that the CG's results are of better quality and more consistent with the true solution sample a_{ex}. Although the convergence does not change significantly (still **linear**) for both cases, other criteria vote for the conjugate gradient as the "winning" method. After the same number of optimization iterations (`kMax = 100`), CG drops its objective $\mathcal{J}(a)$ down to a value of order 10^{-10} while SD manages only to 10^{-9}. In addition, the final value of the L_2-norm error e_{L_2} is also much better for CG; it is 0.18208 versus 0.24371 obtained with SD. As such, we conclude the conjugate gradient method is the better fit for this type of optimization on finer meshes.

? *Both methods applied on the finer mesh also show the "defect" in the center of the domain. Could we explain its nature and suggest any resolution methods?*

! *Before discussing this issue, we advise the reader to review Chapters 9 and 10 and particularly those parts where we address the discovered "mathematical nature" of the problem of not changing the solution for the control variable $\mathbf{u}(t)$ at the right side of its temporal interval $[0, T]$, i.e., $t = T$.*

The readers might have already got the answer to this question. Anyway, we want to bring this discussion to an end and advise first to compare the analytical formulas (9.50)–(9.51) and (11.17)–(11.18) created, respectively, for gradients to solve the LV-model and PDE-based optimization problems. In the former case, the terminal conditions $\psi_1(t) = \psi_2(t) = 0$ in the adjoint

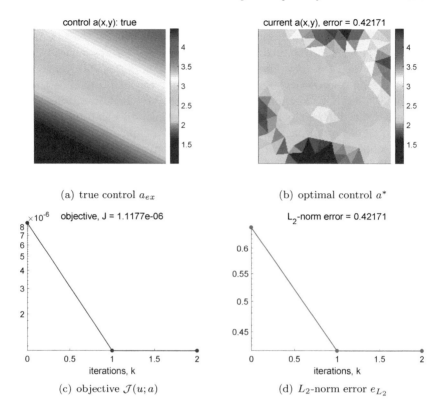

(a) true control a_{ex} (b) optimal control a^*

(c) objective $\mathcal{J}(u; a)$ (d) L_2-norm error e_{L_2}

FIGURE 12.8
MATLAB window with four (selected) plots updated dynamically, while solving
Example 11.2 with BFGS and constant initial guess $a_{ini} = 2.5$ on mesh with
$n = 10$.

system (9.51) "instruct" the gradient $\nabla_u \mathcal{L} = x_1(t)\psi_1(t) + x_2(t)\psi_2(t)$ to set
its last component to zero. It does not give any chance to the states, namely
$x_1(t)$ and $x_2(t)$, to provide their (nonzero) input into the gradient structure
when $t = T$. The situation with the current issue is similar, however, less
obvious. The gradient structure in (11.17) involves other gradients computed
over the state and adjoint fields, i.e., $\nabla_a \mathcal{L} = -\nabla u \cdot \nabla \psi$. Now we advise the
reader to turn back and explore in detail Figures 12.2(d) and 12.3(c,d), which
provide images for, respectively, $u(x, y)$ and $\psi(x, y)$ at $k = 0$. We also advise
running the optimization framework again and accurately tracing the changes
in both functions. We may notice constant changes in the adjoints as they
use information from updated control a^k and current mismatches between
measured data $\tilde{u}$ and the state u^k; refer to the PDE structure in (11.18).
However, the state $u(x, y)$ has very limited dynamics, and its image shows
a clearly identified region as a neighborhood of its maximum. Within this

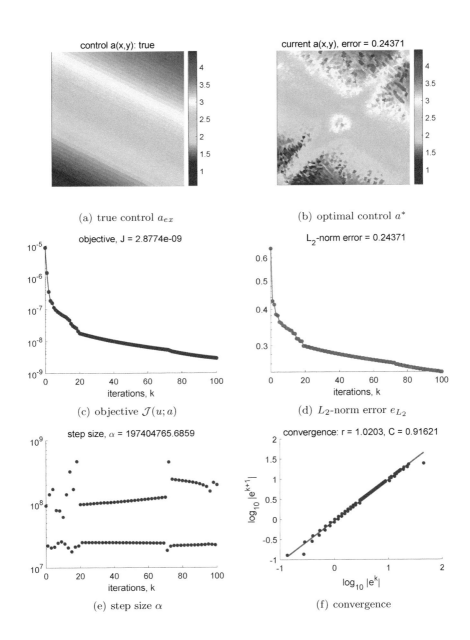

(a) true control a_{ex}

(b) optimal control a^*

(c) objective $\mathcal{J}(u; a)$

(d) L_2-norm error e_{L_2}

(e) step size α

(f) convergence

FIGURE 12.9

MATLAB window with six (selected) plots updated dynamically (finalized at $k = 100$), while solving Example 11.2 with SD and constant initial guess $a_{ini} = 2.5$ on mesh with $n = 50$.

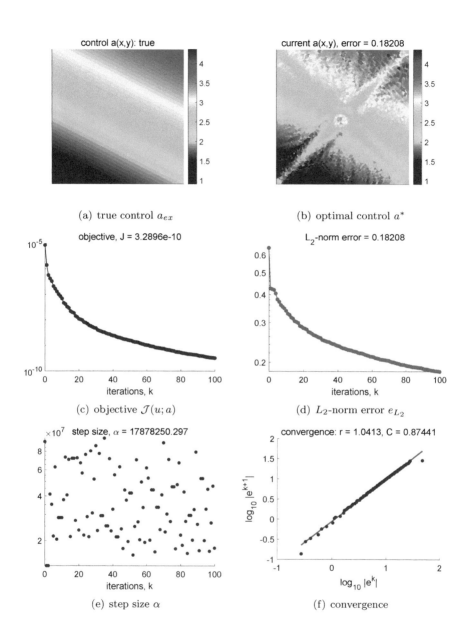

(a) true control a_{ex}

(b) optimal control a^*

(c) objective $\mathcal{J}(u; a)$

(d) L_2-norm error e_{L_2}

(e) step size α

(f) convergence

FIGURE 12.10
MATLAB window with six (selected) plots updated dynamically (finalized at $k = 100$), while solving Example 11.2 with CG and constant initial guess $a_{ini} = 2.5$ on mesh with $n = 50$.

region, ∇u is obviously very close to 0, causing the main gradient $\nabla_a \mathcal{L}$ to be almost zero in the same region. The location of this region coincides with the location of the "defect" in the image for control $a(x, y)$.

Thus, similar to the problem in the LV-model optimization, this issue is induced mathematically and behaves like a "blind spot," while we drive our optimization process. As we also clearly notice in Figures 12.9(b) and 12.10(b), the presence of this spot distorts the entire reconstruction. This distortion propagates from the center of the domain (the spot location) toward its corners; look at four "rays" in both images. We could explain it as a compensatory effect in optimization attempts to find a better fit with data, while facing an inability to improve the solution at a particular location. Now the "source" of the problem is identified, and the readers may freely experiment with any techniques discussed before that may help resolve this issue by applying, e.g., preconditioning or any types of regularization.

In the final round of our numerical experiments, we would like to check the robustness of our computational framework. We will keep the domain discretization at $n = 50$ (nx = 50) and choose the gradient method to be the CG (method = 'CG') based on the analysis of the recent results. This time, we are interested to see how the framework is sensitive to different initial guesses and if we could reconstruct the solution with at least a similar quality. As discussed in Section 12.2.3, we incorporated several options for initial guesses provided in (12.8). Now we run our optimization with $a_{ini,1}$, $a_{ini,2}$, and $a_{ini,3}$ by changing the variable iniGuess (e.g., iniGuess = 1 for $a_{ini,1}$, etc.) in the params.m file. For all three cases, we allow optimization to run for longer by setting kMax = 1000; it enables termination once the relative decrease in either the objective or solution norm reaches the bottom line of 10^{-4}. We put these new results in Figure 12.11. The added initial guesses are no longer constant functions. For example, $a_{ini,1}$ is a linear function $3y + 1$ that, in comparison with the true control a_{ex}, has a different direction for changing its values; compare Figures 12.11(a) and 12.2(a). Taking the L_2-norm error $e_{L_2}^k$ as a measure, we assume that it stands closer to a_{ex} ($e_{L_2}^0 = 0.40832$ versus 0.6455 for $a_{ini,0}$) than the constant guess used before. Figure 12.11(b) presents the results similar to those obtained before with the CG method on the same mesh after 100 iterations. Then, we repeat this experiment with $a_{ini,2}$, which is also a linear function $x + 2y + 0.75$ that moves even closer to a_{ex} ($e_{L_2}^0 = 0.25018$). The completed optimization results provided in Figure 12.11(d) reflects this fact by showing the image with significantly improved quality where the "defect" at the center starts fading. Finally, we place the initial guess $a_{ini,3} = x + 2y + 0.9$ in the proximity of a_{ex} ($e_{L_2}^0 = 0.10046$) and observe that this "defect" almost disappears; refer to Figure 12.11(f).

To combine the results obtained with different initial guesses, we repeat optimization with $a_{ini,0}$ with the same setting kMax = 1000 and create a summary report in Figures 12.12(a,b). All four cases with $a_{ini,i}$, $i = 0, 1, 2, 3$, finished with termination condition #2 when the relative decrease in the control norm $\|a^k\|_{L_2}$ reaches the determined tolerance $\epsilon_2 = 10^{-4}$ (epsilonA =

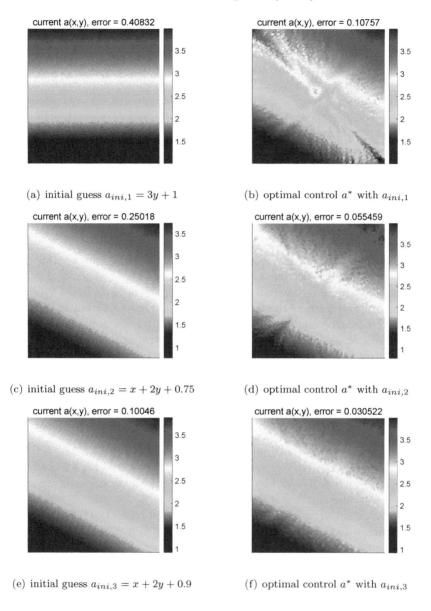

(a) initial guess $a_{ini,1} = 3y + 1$ (b) optimal control a^* with $a_{ini,1}$

(c) initial guess $a_{ini,2} = x + 2y + 0.75$ (d) optimal control a^* with $a_{ini,2}$

(e) initial guess $a_{ini,3} = x + 2y + 0.9$ (f) optimal control a^* with $a_{ini,3}$

FIGURE 12.11

The results of optimization obtained with different initial guesses.

1e-4); refer to the list of parameters on p. 344. As expected, closer proximity of the initial guess to the true control a_{ex} implies fewer optimization iterations to terminate. Also, all cases show the same level of success in fitting data,

$\sim 10^{-10}$; see Figure 12.12(a). It suggests the conclusion of convergence to the same local (or global) optimum or different local optimums (in the neighborhood of the global one) not too distant from each other. Figure 12.12(b) adds more evidence for the latter. Finally, we explore the convergence shown in all four cases. Figures 12.12(c,d) presents the results just for $a_{ini,1}$ and $a_{ini,3}$. As expected (review the Capture Theorem 5.5), once "thoroughly captured" by the local (global) optimum optimization process supplied with the conjugate gradient method exhibits its "nominal" **superlinear** convergence; refer to Figure 12.12(d) showing the convergence rate $r = 1.1446$. We advise the readers to add more experiments by playing with various types of initial guesses to complete the robustness analysis.

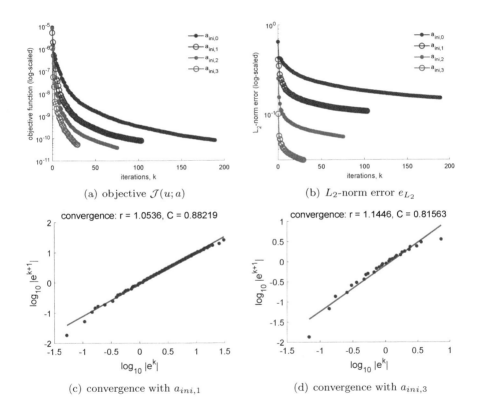

(a) objective $\mathcal{J}(u; a)$

(b) L_2-norm error e_{L_2}

(c) convergence with $a_{ini,1}$

(d) convergence with $a_{ini,3}$

FIGURE 12.12
(a,b) Summary report for the history of the objective $\mathcal{J}(u; a)$ and L_2-norm error e_{L_2} changes for optimization with $a_{ini,i}$, $i = 0, 1, 2, 3$. Plots in (c) and (d) show the results of the convergence analysis for cases $a_{ini,1}$ and $a_{ini,3}$, respectively.

Here, we conclude that the optimization framework we use to solve the PDE-based optimization problem of Example 11.2 is *robust*; it allows convergence to the exact control a_{ex} or its close neighborhood when started from different initial guesses. On the one hand, the problem itself is simple: a nice elliptic equation represents the PDE-based constraint, and the optimization uses a full range of data without any noise. It moves us toward a conclusion that this problem is well-defined. On the other hand, we use analytically defined gradients – a powerful tool for solving various optimization problems. We discretize these gradients. However, they remain consistent with the analytical formula and reasonably accurate after this discretization. We confirm it by the kappa-tests and our "controllability" experiments. Therefore, **two vital components are matched**: the problem is solvable, and the chosen method (our optimization framework) shows its ability to find the solution, proving its reliability, while dealing with complicated tasks such as PDE-based optimization.

12.4   Homework Problems

1. Modify `MATLAB` code `Chapter_12_opt_pde.m` to add new functionalities for gradient preconditioning in 2D and various regularization techniques discussed in Chapter 10 to deal with the "defect" that appeared in the center of the domain. Experiment with parameters and conclude on these methods' applicability and performance.

2. Modify `MATLAB` code `Chapter_12_opt_pde.m` to add new functionality for the control space reparameterization (upscaling), e.g., by using PCA (POD) discussed in Section 8.2.3. Discuss the modifications required for the entire framework, choice of the method-specific parameters, and conclude on the results.

3. Modify `MATLAB` code `Chapter_12_opt_pde.m` to allow dynamical switching between BFGS and SD methods to combine fast convergence of the former and possible benefits of "nearsightedness" of the latter. Discuss the modifications required for the entire framework and switching conditions. Conclude on the results.

4. Modify `MATLAB` code `Chapter_12_opt_pde.m` to solve the following optimization problem

$$\min_{f(x)} \mathcal{J}(u(x); f(x)) = \frac{1}{2} \int_{\Omega} (u(x) - \tilde{u})^2 \, dx \qquad (12.10)$$

to find optimal source function $f(x)$ in 2D elliptic PDE

$$\nabla \cdot [a(x)\nabla u(x)] = f(x), \qquad x \in \Omega$$
$$u(x) = \phi_1(x), \qquad x \in \partial\Omega_{in} \qquad (12.11)$$
$$\frac{\partial u(x)}{\partial n} = \phi_2(x), \qquad x \in \partial\Omega_{out}$$

assuming continuous availability of data $\tilde{u}(x)$ over the entire domain $\Omega \subset \mathbb{R}^2$ shown in Figure 12.13. Create your own benchmark model and solve the problem using your optimization framework with SD, CG, BFGS, different initial guesses and domain Ω discretizations. Conclude on the optimization performance.

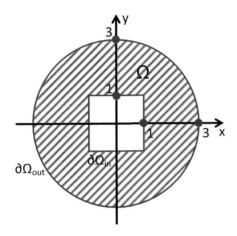

FIGURE 12.13
2D domain for homework problem 4.

5. Modify benchmark model created in Section 12.1 for Example 11.2 with the true control $a_{ex} = x + 2y$. Modify MATLAB code Chapter_12_opt_pde.m to solve the optimization problem and observe numerical oscillations that appeared as a result of violation $a > 0$ condition. Apply the simple projection or slack variable methods discussed in Chapter 10 to eliminate the negative impacts of such violations.

6. Modify MATLAB code Chapter_12_opt_pde.m to add noise to data. Observe the optimization results and discuss any techniques for possible improvement. Add these techniques to the existing framework. Conclude on their applicability and performance.

7. Create a utility for using data stored in folder temp to redraw all images "off-line" (after optimization termination). Modify MATLAB code

`Chapter_12_opt_pde.m` to use these data for re-starting optimization, e.g., after unexpected code-breaking. Discuss the potential benefits of both modifications.

[READ]   **Where to Read More**

Tarantola (2005), [28]
Chapter 5 (Functional Inverse Problems)

Vogel (2002), [29]
Chapter 2 (Analytical Tools), Chapter 3 (Numerical Optimization Tools), Chapter 6 (Parameter Identification)

[RUN]   **MATLAB and FreeFEM Codes for Chapter 12**

- **root** folder:

    - `Chapter_12_opt_pde.m`

    - `params.m`

- folder `algorithms`:

    - `direction_search.m`

    - `file_output.m`

    - `golden_section_search.m`

    - `initialize.m`

    - `kappa_test.m`

    - `measurements.m`

    - `mode_OPT.m`

    - `mode_TEST.m`

    - `stepsize_search.m`

    - `visualize.m`

- folder `freefem`:

    - `data.edp`

    - `eval_grad.edp,`

    - `eval_obj.edp,`

    - `getARCV.idp,`

    - `kappa_grad.edp,`

    - `mesh.edp,`

    - `solve_adj.slv,`

- solve_pde.slv

- folder functions:

  - fn_brent.m
  - fn_convergence_sol_norm.m
  - fn_eval_grad.m
  - fn_eval_obj.m
  - fn_min_brack.m
  - fn_stat.m
  - FUNC.m

- folder output

- folder temp

12.5 Lab Assignment #5: Review Chapters 11–12

For all problems, consider the optimization problem (12.10) of finding optimal source function $f(\mathbf{x})$ in 2D elliptic PDE (12.11), assuming continuous availability of data $\tilde{u}(x)$ over the entire domain $\Omega \subset \mathbb{R}^2$ shown in Figure 12.13, where $\partial\Omega_{in}$ is the inner square-shaped boundary, and $\partial\Omega_{out}$ is the outer circle-shaped boundary.

Problem 1: Derive the analytical expression for gradient $\nabla_f \mathcal{J}$ to solve the optimization problem (12.10)–(12.11) using a gradient-based iterative approach. Provide a full proof and a complete optimization algorithm to solve this optimization problem employing obtained adjoint-based gradients.

Problem 2: Create a benchmark model by setting $a(x,y) = x^2 + y^2$, $\phi_1(x,y) = 0$, and appropriately choosing $u(x,y)$, $\phi_2(x,y)$, and $f(x,y)$ to satisfy (12.11) and match the shape of domain Ω. Modify MATLAB code Chapter_12_opt_pde.m to perform optimization for this model over discretized (triangulated) domain Ω using

- $n_{out} = 50$ boundary elements for the outer boundary $\partial\Omega_{out}$,

- $n_{in} = 20$ boundary elements in total for all four edges of the inner boundary $\partial\Omega_{in}$, and

- constant value as initial guess f_{ini} for $f(x,y)$ obtained by averaging f_{ex}.

Present the results by the following.

1. Show domain Ω with created triangulation in FreeFEM.

2. Check your model and confirm the accuracy of the forward numerical solver by computing the L_2-norm error.

3. Show the results of running your optimization framework in the TEST mode by providing three graphs for both "cheap" and "expensive" kappa-tests and conclude on the gradient accuracy.

4. Double the number of boundary elements for all boundaries and show new graphs for the "cheap" κ-test. Compare the results with previous ones to conclude on the "controllability" of the discretized gradient quality.

Conclude on your expectations for the overall performance of the optimization process.

Problem 3: Using the optimization framework and benchmark model prepared for Problem 2, solve the problem (12.10)–(12.11) using (a) SD, CG, BFGS, $n_{out} = 50$, and $n_{in} = 20$ for domain discretization, (b) SD with doubled number of boundary elements for all boundaries. Present the results of all four optimization runs with graphs showing

- the decrease of the objective function $\mathcal{J}$,

- a comparison of exact f_{ex} and optimal f^* solutions for control,

- a comparison of exact u_{ex} and optimal u^* solutions for state,

- the convergence by the solution norm $\|f - f_{ex}\|_{L_2}$, and

- the analysis for the computational convergence

to elaborate on the optimization performance for all four cases. Finally, conclude on how these results match your expectations created in Problem 2.

For all computations, use the following:

- termination parameters for optimization: $\epsilon = 10^{-4}$ and $k_{max} = 100$,

- restarts every 5 iterations whenever BFGS is used, and

- BB method for finding optimal steps with the following parameters: initial interval $[0, 10^0]$; bracketing: `MAXITER = 20`, `GLIMIT = 100.0`; Brent: `TOL = 1e-9`, `ITMAX = 2`.

Bonus Problem: Implement the gradient preconditioning procedure by solving corresponding elliptic BVP in 2D using `FreeFEM`. Show the results of applying this technique to the current problem for multiple values of the smoothing parameter ℓ, e.g., $\ell = 0.1, 0.5, 1.0, 5.0$. Use SD, $n_{out} = 50$, and $n_{in} = 20$ for domain discretization.

Appendix: Review of Math with MATLAB

In this Appendix, we give a brief overview of some topics in calculus and linear algebra selected to help the reader refresh the knowledge of mathematical notations, definitions, and formulas we mention in this book. We also provide, whenever possible, examples of the associated computations involving existing MATLAB functionalities. For a comprehensive review, we advise consulting any textbooks in numerical linear algebra or handbooks with broader references to various math-related topics, e.g., Korn & Korn's Mathematical Handbook for Scientists and Engineers [21].

Polynomials

In general, a *polynomial* is a mathematical expression consisting of unknown variables and coefficients constructed with only operations of addition, subtraction, multiplication, and nonnegative integer exponentiation of variables. For example, the nth degree polynomial of a single unknown $x \in \mathbb{R}$ is

$$P_n(x) = a_1 + a_2 x + a_3 x^2 + \ldots + a_{n+1} x^n.$$

MATLAB example for finding the coefficients for $P_2(x)$ that is the best fit (in a least-squares sense) for the given data (note the returned coefficients in the adverse (!) order, i.e., $a_{n+1}, a_n, \ldots, a_2, a_1$)

```
>> x = [2 4 7]
>> y = [1 9 6]
>> polyfit(x,y,2)
```

Vectors

Vectors in n-dimensional vector space $\mathbb{R}^n$ (denoted by either $\mathbf{x} \in \mathbb{R}^n$ or $x \in \mathbb{R}^n$) are arrays of n scalars $x_1, x_2, \ldots, x_n$ ($x_i \in \mathbb{R}$, $i = 1, 2, \ldots n$) arranged in a column or row, i.e.,

$$\mathbf{x} = \begin{bmatrix} x_1 \\ x_2 \\ \ldots \\ x_n \end{bmatrix}, \qquad \mathbf{x}^T = \begin{bmatrix} x_1 & x_2 & \ldots & x_n \end{bmatrix},$$

where the symbol "T" represents an operation of vector transposition, e.g.,

$$\mathbf{x} = \begin{bmatrix} -1 \\ 5 \\ 10 \end{bmatrix} \in \mathbb{R}^3, \qquad \mathbf{y} = \begin{bmatrix} 2 & 0 & -1 & 4 \end{bmatrix} \in \mathbb{R}^4.$$

In this book, all vectors have a default column-wise structure unless stated otherwise.

MATLAB example:
```
>> x = [-1; 5; 10]
>> y = [2 0 -1 4]
```

Matrices and Matrix Equations

Matrices of dimensionality $n \times m$ (denoted by $A \in \mathbb{R}^{n \times m}$) are rectangular arrays (tables) of scalars containing n rows and m columns, i.e.,

$$A = \begin{bmatrix} a_{11} & a_{12} & \cdots & a_{1m} \\ a_{21} & a_{22} & \cdots & a_{2m} \\ \cdots & \cdots & \cdots & \cdots \\ a_{n1} & a_{n2} & \cdots & a_{nm} \end{bmatrix}, \qquad a_{ij} \in \mathbb{R}, \qquad \begin{array}{l} i = 1, 2, \ldots, n, \\ j = 1, 2, \ldots, m. \end{array}$$

Correspondingly, any system of linear equations

$$\begin{cases} a_{11}x_1 + a_{12}x_2 + \ldots + a_{1m}x_m = b_1, \\ a_{21}x_1 + a_{22}x_2 + \ldots + a_{2m}x_m = b_2, \\ \qquad \cdots \\ a_{n1}x_1 + a_{n2}x_2 + \ldots + a_{nm}x_m = b_n \end{cases}$$

may be written in the matrix form

$$A\mathbf{x} = \mathbf{b},$$

where $\mathbf{x}$ and $\mathbf{b}$ are column vectors of unknowns and the right-hand side constants, respectively; i.e.,

$$\mathbf{x} = \begin{bmatrix} x_1 \\ x_2 \\ \cdots \\ x_m \end{bmatrix}, \qquad \mathbf{b} = \begin{bmatrix} b_1 \\ b_2 \\ \cdots \\ b_n \end{bmatrix}.$$

Such systems (*matrix equations*) may be solved, e.g., by using the inverse matrix A^{-1}

$$\mathbf{x} = A^{-1}\mathbf{b}.$$

MATLAB example for solving linear system $A\mathbf{a} = \mathbf{y}$ in (1.4):
```
>> A = [1 2 4; 1 4 16; 1 7 49]
>> y = [1; 9; 6]
>> a = A^-1 * y
```

Eigenvalues and Eigenvalue Decomposition

Given matrix $A \in \mathbb{R}^{n \times n}$, if a scalar λ and a nonzero vector $\mathbf{v}$ satisfy

$$A\lambda = \lambda \mathbf{v},$$

then λ and $\mathbf{v}$ are called an *eigenvalue* and *eigenvector* of matrix A. All eigenvalues could be found by solving the equation

$$|A - \lambda \mathcal{I}| = 0,$$

where the notation $|\cdot|$ denotes the determinant, and $\mathcal{I}$ is the identity matrix. The set of all associated eigenvectors then could be found by solving the following matrix equations

$$(A - \lambda_i \mathcal{I})\mathbf{v}_i = \mathbf{0}, \quad i = 1, 2, \ldots, n.$$

The *eigenvalue (spectral) decomposition* of the matrix A is given by

$$A = V \Lambda V^T,$$

where V is a matrix constructed from eigenvectors as its columns, i.e., $V = \begin{bmatrix} \mathbf{v}_1 & \mathbf{v}_2 & \cdots & \mathbf{v}_n \end{bmatrix}$, and Λ is a diagonal matrix with the main diagonal represented by the vector of associated eigenvalues $\begin{bmatrix} \lambda_1 & \lambda_2 & \cdots & \lambda_n \end{bmatrix}$.

MATLAB example:
```
>> A = [1 2 5; 4 -1 0; 3 -1 -3]
>> [V Lambda] = eig(A)
```

Vector and Matrix Norms

The *N-norm of the vector* $\mathbf{x} \in \mathbb{R}^n$ is a number computed by

$$\|\mathbf{x}\|_N = \left[\sum_{i=1}^{n} |x_i|^N \right]^{1/N}, \quad N \geq 1.$$

Therefore, the 1-norm is

$$\|\mathbf{x}\|_1 = \sum_{i=1}^{n} |x_i|,$$

and the commonly used 2-norm (also called the *Euclidian norm*) is

$$\|\mathbf{x}\|_2 = \left[\sum_{i=1}^{n} x_i^2 \right]^{1/2} = \sqrt{x_1^2 + x_2^2 + \ldots + x_n^2}.$$

We may also define the *infinity norm* as

$$\|\mathbf{x}\|_\infty = \max_{1 \leq i \leq n} |x_i|.$$

Based on the definitions above, the *matrix norms* of matrix A (with entries a_{ij}) corresponding to the associated vector norms are

$$\|A\|_1 = \max_{1 \leq j \leq n} \sum_{i=1}^{n} |a_{ij}|,$$

$$\|A\|_2 = \sqrt{\lambda_{\max}(A^T A)},$$

$$\|A\|_\infty = \max_{1 \leq i \leq n} \sum_{j=1}^{n} |a_{ij}|,$$

where $\lambda_{\max}(A^T A)$ is the largest eigenvalue of matrix $A^T A$.

`MATLAB` example:
```
>> x = [1 2 5]
>> A = [1 2 5; 4 -1 0; 3 -1 -3]
>> norm(x,1)
>> norm(x,2)
>> norm(x,inf)
>> norm(A,1)
>> norm(A,2)
>> norm(A,inf)
```

Condition Number

Using any matrix norm defined above, the *condition number* of a nonsingular ($|A| \neq 0$) matrix $A \in \mathbb{R}^{n \times n}$ is defined as

$$cond(A) = \|A\| \cdot \|A^{-1}\| \geq 1.$$

If A is also symmetric, then, using the 2-norm, we could define the condition number as the ratio

$$cond(A) = \frac{\lambda_{\max}}{\lambda_{\min}},$$

where $\lambda_{\max}$ and $\lambda_{\min}$ are the largest and smallest eigenvalues of A, respectively.

`MATLAB` example:
```
>> A = [1 2 0; 2 3 -1; 0 -1 1]
>> cond(A)
```

Positive Definite Matrices

We define a symmetric ($A = A^T$) matrix $A \in \mathbb{R}^{n \times n}$ to be *positive definite* (also denoted as $A \succ 0$) if it satisfies

$$\mathbf{x}^T A \mathbf{x} > 0$$

for all nonzero vectors $\mathbf{x}$. Alternatively, a positive definite matrix has all eigenvalues as positive numbers, i.e.,

$$\forall \lambda_i > 0, \quad i = 1, 2, \ldots, n.$$

Similarly, we define matrix A to be

- *positive semidefinite* $(A \succeq 0)$ if $\mathbf{x}^T A \mathbf{x} \geq 0$ $(\forall \lambda_i \geq 0)$,
- *negative definite* $(A \prec 0)$ if $\mathbf{x}^T A \mathbf{x} < 0$ $(\forall \lambda_i < 0)$, and
- *negative semidefinite* $(A \preceq 0)$ if $\mathbf{x}^T A \mathbf{x} \leq 0$ $(\forall \lambda_i \leq 0)$.

`MATLAB` example for printing all eigenvalues λ_i:
```
>> A = [1 0 0; 0 1 0; 0 0 1]
>> eig(A)
```

Gradient

For a real-valued (scalar) function $f(\mathbf{x}) = f(x_1, x_2, \ldots, x_n)$ of n variables, the *gradient* is defined as a vector of the partial derivatives, i.e.,

$$\nabla f(\mathbf{x}) = \begin{bmatrix} \frac{\partial f}{\partial x_1} & \frac{\partial f}{\partial x_2} & \cdots & \frac{\partial f}{\partial x_n} \end{bmatrix}^T.$$

In this book, we use a subscript notation once we need to define a subset of variables with respect to which the gradient is computed. For example, for a real-valued function $g(\mathbf{x}, \mathbf{u}) = g(x_1, x_2, \ldots, x_n, u_1, u_2, \ldots, u_m)$ of $n + m$ variables, where $\mathbf{x} = [x_1 \, x_2 \, \ldots \, x_n]^T \in \mathbb{R}^n$ and $\mathbf{u} = [u_1 \, u_2 \, \ldots \, u_m]^T \in \mathbb{R}^m$, the gradient with respect to (all components of) vector $\mathbf{u}$ is computed as

$$\nabla_{\mathbf{u}} g(\mathbf{x}, \mathbf{u}) = \begin{bmatrix} \frac{\partial g}{\partial u_1} & \frac{\partial g}{\partial u_2} & \cdots & \frac{\partial g}{\partial u_m} \end{bmatrix}^T.$$

`MATLAB` example (symbolic computations):
```
>> syms x1 x2 x3
>> f(x1,x2,x3) = 3*x1*x2*cos(x3) - 2*x1*x3*sin(x2)
>> gradient(f,[x1,x2,x3])
```

Hessian

For a real-valued (scalar) function $f(\mathbf{x}) = f(x_1, x_2, \ldots, x_n)$ of n variables, the *Hessian matrix* is defined as a matrix of mixed partial derivatives $\frac{\partial^2 f}{\partial x_i \, \partial x_j}$, i.e.,

$$H(\mathbf{x}) = \nabla^2 f(\mathbf{x}) = \begin{bmatrix} \frac{\partial^2 f}{\partial x_1^2} & \cdots & \frac{\partial^2 f}{\partial x_1 \, \partial x_j} & \cdots & \frac{\partial^2 f}{\partial x_1 \, \partial x_n} \\ \cdots & \cdots & \cdots & \cdots & \cdots \\ \frac{\partial^2 f}{\partial x_i \, \partial x_1} & \cdots & \frac{\partial^2 f}{\partial x_i \, \partial x_j} & \cdots & \frac{\partial^2 f}{\partial x_i \, \partial x_n} \\ \cdots & \cdots & \cdots & \cdots & \cdots \\ \frac{\partial^2 f}{\partial x_n \, \partial x_1} & \cdots & \frac{\partial^2 f}{\partial x_n \, \partial x_j} & \cdots & \frac{\partial^2 f}{\partial x_n^2} \end{bmatrix}.$$

`MATLAB` example (symbolic computations):
```
>> syms x1 x2 x3
>> f(x1,x2,x3) = 3*x1*x2*cos(x3) - 2*x1*x3*sin(x2)
>> hessian(f,[x1,x2,x3])
```

Jacobian

For a vector-valued (vector) function

$$\mathbf{f}(\mathbf{x}) = \mathbf{f}(x_1, x_2, \ldots, x_n) = \begin{bmatrix} f_1(x_1, x_2, \ldots, x_n) \\ f_2(x_1, x_2, \ldots, x_n) \\ \ldots \\ f_m(x_1, x_2, \ldots, x_n) \end{bmatrix},$$

where all functions $f_i(x_1, x_2, \ldots, x_n)$, $i = 1, 2, \ldots, m$, are scalar functions of n variables, the *Jacobian matrix* is defined as a matrix of the partial derivatives $\frac{\partial f_j}{\partial x_i}$, i.e.,

$$\mathcal{J}(\mathbf{x}) = [\boldsymbol{\nabla}\mathbf{f}(\mathbf{x})]^T = \begin{bmatrix} \frac{\partial f_1}{\partial x_1} & \cdots & \frac{\partial f_j}{\partial x_1} & \cdots & \frac{\partial f_n}{\partial x_1} \\ \cdots & \cdots & \cdots & \cdots & \cdots \\ \frac{\partial f_1}{\partial x_i} & \cdots & \frac{\partial f_j}{\partial x_i} & \cdots & \frac{\partial f_n}{\partial x_i} \\ \cdots & \cdots & \cdots & \cdots & \cdots \\ \frac{\partial f_1}{\partial x_n} & \cdots & \frac{\partial f_j}{\partial x_n} & \cdots & \frac{\partial f_n}{\partial x_n} \end{bmatrix}.$$

`MATLAB` example (symbolic computations):
```
>> syms x1 x2 x3
>> f1 = 2 * x1 * x2 + x3
>> f2 = x1 + 3 * x2 - x3
>> f3 = x1 * x2 * x3
>> jacobian([f1,f2,f3],[x1,x2,x3])
```

Bibliography

[1] Ugur G. Abdulla, Vladislav Bukshtynov, and Saleheh Seif. Cancer detection through electrical impedance tomography and optimal control theory: Theoretical and computational analysis. *Mathematical Biosciences and Engineering*, 18(4):4834–4859, 2021.

[2] Paul R. Arbic II and Vladislav Bukshtynov. On reconstruction of binary images by efficient sample-based parameterization in applications for electrical impedance tomography. *International Journal of Computer Mathematics*, 99(11):2272–2289, 2022

[3] K. Aziz and T. Settari. *Petroleum Reservoir Simulation*. Applied Science Publishers, 1979.

[4] D. Bertsekas. *Nonlinear Programming*. Athena Scientific, 3rd edition, 2016.

[5] D. Bertsimas and J. Tsitsiklis. *Introduction to Linear Optimization*. Athena Scientific, 1997.

[6] Stephen Boyd and Lieven Vandenberghe. *Convex Optimization*. Cambridge University Press, 1st edition, 2004.

[7] Richard P. Brent. *Algorithms for Minimization Without Derivatives (reprinted from original book published in 1973)*. Dover Publications, 2002.

[8] V. Bukshtynov and B. Protas. Optimal reconstruction of material properties in complex multiphysics phenomena. *Journal of Computational Physics*, 242:889–914, 2013.

[9] V. Bukshtynov, O. Volkov, L.J. Durlofsky, and K. Aziz. Comprehensive framework for gradient-based optimization in closed-loop reservoir management. *Computational Geosciences*, 19(4):877–897, 2015.

[10] V. Bukshtynov, O. Volkov, and B. Protas. On optimal reconstruction of constitutive relations. *Physica D: Nonlinear Phenomena*, 240(16):1228–1244, 2011.

[11] R.L. Burden and J.D Faires. *Numerical Analysis: Numerical Solutions of Nonlinear Systems of Equations*. Thomson Brooks/Cole, 2005.

[12] Jennifer B. Erway and Philip E. Gill. A subspace minimization method for the trust-region step. *SIAM Journal on Optimization*, 20(3):1439–1461, 2010.

[13] Mark S. Gockenbach. *Understanding and Implementing the Finite Element Method*. SIAM, 2006.

[14] Mark S. Gockenbach. *Partial Differential Equations: Analytical and Numerical Methods*. SIAM, 2nd edition, 2011.

[15] Igor Griva, Stephen G. Nash, and Ariela Sofer. *Linear and Nonlinear Optimization*. SIAM, 2nd edition, 2009.

[16] F. Hecht. New development in FreeFem++. *Journal of Numerical Mathematics*, 20(3-4):251–265, 2012.

[17] Frederic Hecht. FreeFEM Documentation. `https://doc.freefem.org/documentation/index.html`. Accessed: 2022-08-01.

[18] Desmond J. Higham and Nicholas J. Higham. *MATLAB Guide*. SIAM, 2nd edition, 2005.

[19] J. Kennedy and R. Eberhart. Particle swarm optimization. In *Proceedings of ICNN'95 - International Conference on Neural Networks*, volume 4, pages 1942–1948, 1995.

[20] Priscilla M. Koolman and Vladislav Bukshtynov. A multiscale optimization framework for reconstructing binary images using multilevel PCA-based control space reduction. *Biomedical Physics & Engineering Express*, 7(2):025005, 2021.

[21] Granino A. Korn and Theresa M. Korn. *Mathematical Handbook for Scientists and Engineers: Definitions, Theorems, and Formulas for Reference and Review*. Dover Civil and Mechanical Engineering, 2000.

[22] R.T. Marler and J.S. Arora. Survey of multi-objective optimization methods for engineering. *Structural and Multidisciplinary Optimization*, 26(6):369–395, 2004.

[23] MATLAB. *Version 9.4.0 (R2018a)*. The MathWorks Inc., Natick, Massachusetts, 2018.

[24] Jorge J. Moré and D. C. Sorensen. Computing a trust region step. *SIAM Journal on Scientific and Statistical Computing*, 4(3):553–572, 1983.

[25] J. Nocedal and S. J. Wright. *Numerical Optimization*. Springer, 2nd edition, 2006.

[26] William H. Press, Saul A. Teukolsky, William T. Vetterling, and Brian P. Flannery. *Numerical Recipes: The Art of Scientific Computing*. Cambridge University Press, 3rd edition, 2007.

[27] Trond Steihaug. The conjugate gradient method and trust regions in large scale optimization. *SIAM Journal on Numerical Analysis*, 20(3):626–637, 1983.

[28] A. Tarantola. *Inverse Problem Theory and Methods for Model Parameter Estimation*. SIAM, 2005.

[29] Curtis R. Vogel. *Computational Methods for Inverse Problems*. SIAM, 2002.

[30] O. Volkov, V. Bukshtynov, L.J. Durlofsky, and K. Aziz. Gradient-based Pareto optimal history matching for noisy data of multiple types. *Computational Geosciences*, 22(6):1465–1485, 2018.

Index